COLLEGE
GEOMETRY

COLLEGE GEOMETRY

David C. Kay

The University of Oklahoma

HOLT, RINEHART and WINSTON, INC.

New York   Chicago   San Francisco   Atlanta

Dallas   Montreal   Toronto   London   Sydney

TO J. E. GUDEKUNST

a vigorous teacher of geometry
who was dedicated to his students

Preface

Ever since the printing of Saccheri's book *Euclides ab omne naevo vindicatus* in 1733, which marked the beginning of an unprecedented effort on the part of mathematicians to prove the parallel postulate, there has been the need to state the axioms, theorems, and proofs of geometry more carefully. Indeed, the source of much difficulty which students have with geometry is the very lack of rigor, contrary to almost any course or text in abstract algebra, for example, where pictures cannot adequately describe what is being proved. A trend has grown, therefore, to present the subject of geometry in the manner of a course in abstract algebra.

But a price must be paid. The infatuation with reformulating geometric principles in terms of set theory can have a deadening effect on students whose backgrounds do not evoke the same infatuation, and it is often done at the expense of some very exquisite ideas of geometry and the cultural values of the ages. It is most unfortunate that students are thereby given the impression—which they will one day transmit to *their* students—that geometry is a rather sterile and irrelevant subject.

This book is an outcome of an attempt to build a one-semester course in geometry suitable for mathematics majors and prospective or veteran secondary school teachers who have completed a year of calculus (continuity and limits appear occasionally) which would effectively portray the beauty and charm of geometry amidst an axiomatic structure. To accomplish this goal, a balance was established between the legalistic and the fanciful, the presentation being done in the manner of the SMSG writers but with a good measure of classical material, once the domain of the traditional course "college geometry," included. A small but reasonable amount of elementary geometry is developed from twelve judiciously chosen axioms, the primary aim being to point up the vast influence which the axioms of parallelism have. Then the development continues by the use of models where one has the means to omit the "obvious" and where learning seems to take place more effectively. Thus, the classical is not omitted nor treated as ancient, but is presented anew in a modern setting.

In the author's courses where the book has been used in the form of notes for the past five semesters, the axiomatic part usually occupies the first half of the semester. The material normally covered consists of Chapters 2, 3, 4, selected portions of 6, 7, and 8, all of 9, and a sampling of 10 and 11. The logical inter-dependence of the major topics has been included as a possible guide for the teacher in designing a course which emphasizes his own preferences (placed immediately preceding the Contents). An effective post-analytic geometry course might be built on Chapters 7–11, for example. A short course could consist of Chapters 2, 3, 4, 6, 9, and 11. Several optional sections (indicated by an asterisk) have been included which can be used for the purpose of enrichment.

A word about Chapter 1—this chapter was written primarily to be read by the student and not to be taught in the classroom. First, because the reader may not be adequately prepared for some topics which appear there but would neverthe-less be able to gain some appreciation with a casual reading. Since first impressions are important in any subject, the aim was to include at the beginning of the book a few of the more entertaining aspects of elementary geometry which would provide a more colorful setting in which the axiomatic study takes place. It is important for the student to realize that Chapter 1 is not a prerequisite for the remainder of the book, but rather an informal introduction to geometry. Second, class time is at a premium, so the thought was to let geometry speak for itself for a while through the printed word. In his own classes the author never devotes more than one or two lectures to the topics in Chapter 1.

Hints and some answers are given along with the exercises, and brief solutions to all odd-numbered problems appear following the Appendixes. Each section of exercises is graded, with the problems everybody can do at the beginning and the more difficult ones at the end. The most challenging problems are starred. With a few minor exceptions, the development does not depend on results which are given in the exercises; however, a considerable amount of theory is to be found in the exercises.

I extend my appreciation to those who made this project possible: to those who read the manuscript, Victor Klee and W. T. Fishbach who did a most thorough job, and L. M. Kelly whose criticism was most valuable; to Judy Haug for the expert job of typing the manuscript; to the University of Oklahoma Faculty Research Committee for their services; to Holt, Rinehart and Winston, Inc. for their patience; and certainly, to my wife and children who made a good many sacrifices on my behalf during the writing stages.

Finally, I express my gratitude to the following publishers who have permitted me to use certain selections to supplement the exercises (the numbers in brackets indicate the corresponding references in the bibliography): Allyn and Bacon, Inc., Boston [9]; Barnes and Noble, Inc., New York [6]; Dover Publications, Inc., New York [14]; Graylock Press, Baltimore, Md. [27]; Holt, Rinehart and Winston, Inc., New York [30]; John Wiley and Sons, Inc., New York and London [4]; McGraw-Hill Book Co., Inc., New York [31]; National Council of Teachers of Mathematics, Washington, D.C. [*Eighteenth Yearbook*, 1945]; and Random House, New York [*Hungarian Problem Book II*, compiled by Jozsef Kurschak, trans. by Elvira Rapaport; © Copyright 1963 by Yale University].

David C. Kay

Norman, Oklahoma
January 1969

LOGICAL INTERDEPENDENCE OF MAJOR TOPICS

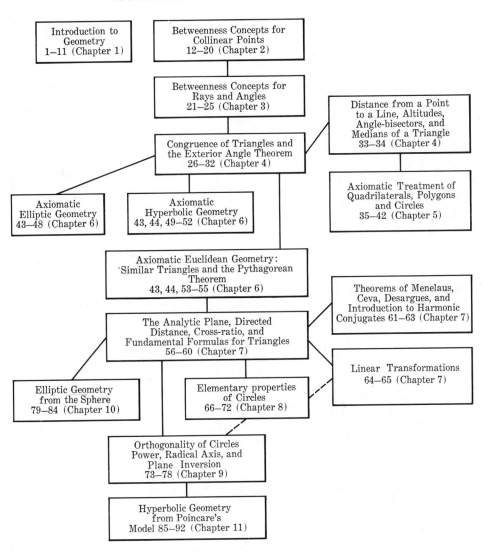

Introduction to
Geometry
1–11 (Chapter 1)

Betweenness Concepts for
Collinear Points
12–20 (Chapter 2)

Betweenness Concepts for
Rays and Angles
21–25 (Chapter 3)

Distance from a Point
to a Line, Altitudes,
Angle-bisectors, and
Medians of a Triangle
33–34 (Chapter 4)

Congruence of Triangles and
the Exterior Angle Theorem
26–32 (Chapter 4)

Axiomatic Treatment of
Quadrilaterals, Polygons
and Circles
35–42 (Chapter 5)

Axiomatic
Elliptic Geometry
43–48 (Chapter 6)

Axiomatic
Hyperbolic Geometry
43, 44, 49–52 (Chapter 6)

Axiomatic Euclidean Geometry:
Similar Triangles and the Pythagorean
Theorem
43, 44, 53–55 (Chapter 6)

Theorems of Menelaus,
Ceva, Desargues, and
Introduction to Harmonic
Conjugates 61–63 (Chapter 7)

The Analytic Plane, Directed
Distance, Cross-ratio, and
Fundamental Formulas for Triangles
56–60 (Chapter 7)

Linear Transformations
64–65 (Chapter 7)

Elliptic Geometry
from the Sphere
79–84 (Chapter 10)

Elementary properties
of Circles
66–72 (Chapter 8)

Orthogonality of Circles
Power, Radical Axis, and
Plane Inversion
73–78 (Chapter 9)

Hyperbolic Geometry
from Poincare's
Model 85–92 (Chapter 11)

Contents

PART III   ABSOLUTE GEOMETRY
AND CONCEPTS OF PARALLELISM

PART IV   DEVELOPMENT OF GEOMETRY FROM MODELS

PART I

PROLOGUE

1

Famous Theorems of Geometry

It is a commentary on the age in which we live that elegance and beauty are often sacrificed for the pragmatic and utilitarian. This is apparently no less true in the field of mathematics, where in the eyes of a great many students the elegance of geometry seems to have been replaced by the efficient methods of the calculus and the lightning speed of the electronic computer. But the mastery of many fields of mathematics (including the calculus) continues to be contingent upon one's mastery of the properties of Euclidean space. And it is widely recognized that the methods of geometry provide great insight into a variety of mathematical problems.

Thus a good introduction to geometry might consist of pointing out its significance to other branches of mathematics. It might also be desirable to present the testimony of those who have actually found the subject stimulating. But an even more effective introduction to geometry would seem to be provided by geometry itself—its exquisite theorems and methods of reasoning are extremely enticing and need no external motivation. It is then our purpose here to explore a few of the more charming ideas of geometry, those which have become well-known and are now a part of the folklore of the subject.

In order to give the disparity of topics some perspective, they have been arranged in chronological order, and some discussion about history will frequently appear. However, it should be realized that our treatment is too incomplete to be regarded as even a brief history of geometry—other sources should be consulted for this.

3

Although it was written primarily for its entertainment value, this chapter may include topics which are difficult for some readers. For them we offer the following advice: skim these pages rapidly at first, work those problems which look feasible, then return to the more difficult topics at a later, more appropriate time in your study. But above all, enjoy yourself.

1.   Theorems from Antiquity: The Pythagorean Theorem

There can be no question that the most famous theorem of all times is the *Pythagorean theorem*, attributed to a Greek school of geometers known as the *Pythagoreans*. This unique school existed around 500 B.C., but certain special cases of the theorem bearing its name were known much earlier and had been used as tools in ancient problems of engineering and construction—such as the "3-4-5 triangle," for example, which provided an accurate measurement for a right angle. It is believed that Pythagoras, or one of the Pythagoreans, first recognized and proved the theorem in its full generality. It may be stated in the following algebraic form:

If a, b, and c are the sides of a right triangle, where c is the length of the hypotenuse, then $a^2 + b^2 = c^2$.

Figure 1

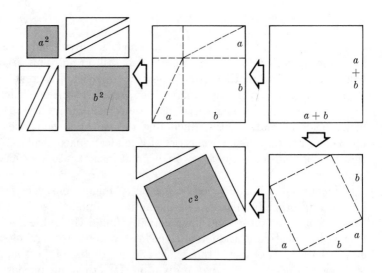

Just what proof was used at first by the Pythagoreans is not known conclusively, but it is speculated that it was a "dissection" type of proof as illustrated in Fig. 1.

In that early period, there was hardly any knowledge of algebra of the kind we know today. The symbolism "a^2" had not yet been invented, so numerical problems were often posed geometrically. A brief examination of Euclid's *Elements* (Book II) makes this eminently clear: in that volume are the details of solving a quadratic equation, entirely by geometry. And the Pythagorean theorem had to be stated: *The square on the hypotenuse of a right triangle equals the sum of the squares on the other two sides.* The reader may be familiar with this version of the theorem, illustrated for the special 3-4-5 case in Fig. 2. It suggests quite a different avenue of proof from that more commonly used today—in terms of *areas* of plane figures. It is probable that one of the Pythagoreans invented the intricate area proof that appears in Euclid's *Elements* and which used to be found in high school geometry texts. (See in this connection Exercise 1 below.)

Figure 2

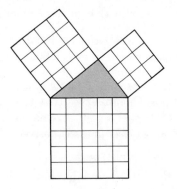

An interesting generalization of this area version of the theorem, which applies to any triangle, was proposed by Pappus (A.D. 300), another Greek mathematician. In Fig. 3, we have parallelograms I and II described externally on the respective sides AB and AC of triangle ABC. The lines containing the sides of these parallelograms opposite AB and AC will meet at some point P. Then parallelogram III

Figure 3

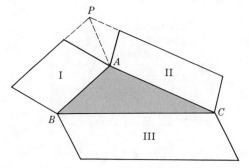

is constructed on side BC so that the side adjacent to BC is both equal and parallel to segment PA. If now I, II, and III also denote the areas of these three parallelograms,

$$I + II = III,$$

a relation which is *independent of the particular parallelograms* I *and* II *chosen on sides* AB *and* AC. This interesting theorem may be readily proved from the principle:

parallelograms having equal bases and equal altitudes have equal areas. The reader might enjoy attempting his own proof of this relation and deriving the Pythagorean Theorem from it. (See Exercise 2 below.)

2. Euclid and the Golden Rectangle

A great invention of the Greeks was a certain rectangle whose proportions have made it famous—the *golden rectangle.* Euclid (300 B.C.) suggested the problem of constructing the regular pentagon with compass and straightedge. This ultimately involves the ratio

$$\mu = \frac{1 + \sqrt{5}}{2} = 1.61803 \cdots \qquad (2.1)$$

called the *golden ratio,* which has proved to be significant many times both in mathematics and in its applications. One observes that the value μ is the positive root of the quadratic equation $x^2 - x - 1 = 0$, so the characteristic property of the number—from which much of its usefulness is derived—is the relation

$$\mu^2 = \mu + 1, \qquad (2.2)$$

or the companion formula, derived by dividing both sides of (2.2) by μ,

$$\mu = \frac{1}{\mu} + 1. \qquad (2.3)$$

The *golden rectangle* is then defined as a rectangle *whose sides are in the ratio* $a/b = 1/\mu$ (see Fig. 4). The Greeks regarded this figure as divinely inspired and held that among all rectangles the golden rectangle was the most pleasing to the eye.

Figure 4

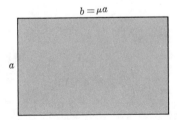

$b = \mu a$

a

One of the interesting properties of this rectangle is that if a square be cut off at one end (as shown in Fig. 5) the end remaining will be *another* golden rectangle—whose sides are $1/\mu$ times those of the original rectangle. One may easily obtain this result from the relation (2.3).

It is interesting to see how the Pythagorean theorem becomes the basis for a simple construction of the golden rectangle by compass and straightedge. Begin with a square $ABCD$ with M the midpoint of side AB (Fig. 6). Inscribe an arc with M as center and MC as radius, cutting line AB at point E. By using the Pythagorean theorem for triangle MBC it can be readily shown that $AE = \mu\,AD$. Thus the completed rectangle $AEFD$ is the desired golden rectangle. This construction is not a recent discovery—it dates all the way back to Euclid, who used it to obtain his compass, straightedge construction of the regular pentagon (see Exercise 28 below).

Figure 5

Figure 6

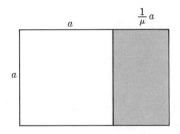
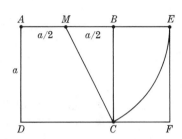

3. Heron, Ptolemy, and Brahmagupta

The familiar formula for the area of a triangle in terms of its sides,

$$K = \sqrt{s(s - a)(s - b)(s - c)}$$

(where s is one half the perimeter and a, b, and c are the lengths of the sides), dates back to the first century A.D. and is attributed to Heron of Alexandria. There is some evidence that its proof was known to Archimedes (290 B.C.), so it is probable that the formula itself was discovered long before Heron's time. The reader may be familiar with the trigonometric proof of the formula (as found in H. S. M. Coxeter, *Introduction to Geometry* [4,[1] p. 12], for example). It may also be derived synthetically as in Heron's original proof (see H. Eves, *A Survey of Geometry* [9, p. 40]).

A surprising generalization of Heron's formula is due to the Hindu mathematician Brahmagupta (A.D. 630), which gives the area of a cyclic quadrilateral in terms of its sides (s is the semiperimeter of the quadrilateral) as

$$K = \sqrt{(s - a)(s - b)(s - c)(s - d)}. \tag{3.1}$$

If $d = 0$ then (3.1) gives Heron's formula. The derivation of (3.1) is not so involved as one might expect. One method is to obtain, by law of cosines, Bretschneider's formula

$$16K^2 = 4m^2n^2 - (a^2 - b^2 + c^2 - d^2)^2,$$

where m and n represent the diagonals, and then apply the known relation for a cyclic quadrilateral (Fig. 7)

$$mn = ac + bd.$$

(See Exercises 7, 8, and 9.)

---

[1] The numbers in brackets indicate the corresponding references listed in the bibliography.

Figure 7

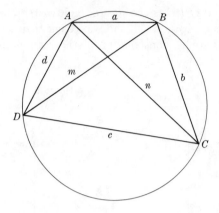

This last relation involving the sides and diagonals of a cyclic quadrilateral is another famous theorem of geometry, known as *Ptolemy's theorem* (after the Greek astronomer Claudius Ptolemy, A.D. 100). As in the case of many theorems of geometry, Ptolemy's theorem may be derived quite easily by trigonometry. Consider the quadrilateral $ABCD$ inscribed in circle O. First, for any chord x of circle O, if r is the radius and θ is the central angle subtending x, we observe from triangle OXM (Fig. 8)

$$\sin \frac{\theta}{2} = \frac{x}{2r}. \tag{3.2}$$

Figure 8

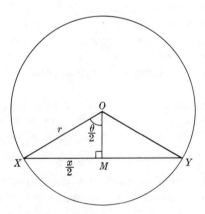

Apply this formula repeatedly to the sides of the quadrilateral (see Fig. 9 for notation):

$$\frac{a}{2r} \cdot \frac{c}{2r} + \frac{b}{2r} \cdot \frac{d}{2r} = \sin \frac{\alpha}{2} \sin \frac{\gamma}{2} + \sin \frac{\beta}{2} \sin \frac{\delta}{2}. \tag{3.3}$$

This form suggests the use of the identity

$$\sin x \sin y = \tfrac{1}{2} \cos (x - y) - \tfrac{1}{2} \cos (x + y). \tag{3.4}$$

Then (3.3) becomes

Figure 9

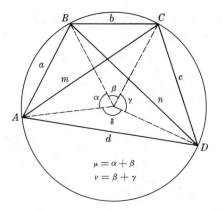

$$\frac{ac + bd}{4r^2} = \tfrac{1}{2} \cos \frac{\alpha - \gamma}{2} - \tfrac{1}{2} \cos \frac{\alpha + \gamma}{2} + \tfrac{1}{2} \cos \frac{\beta - \delta}{2} - \tfrac{1}{2} \cos \frac{\beta + \delta}{2}. \quad \textbf{(3.5)}$$

But $\delta = 2\pi - (\alpha + \beta + \gamma)$. Using the identities $\cos(\pi - x) = \cos(x - \pi) = -\cos x$, we have

$$\frac{ac + bd}{4r^2} = \tfrac{1}{2} \cos \frac{\alpha - \gamma}{2} - \tfrac{1}{2} \cos \frac{\alpha + 2\beta + \gamma}{2}$$

$$= \tfrac{1}{2} \cos \frac{\mu - \nu}{2} - \tfrac{1}{2} \cos \frac{\mu + \nu}{2}$$

$$= \sin \frac{\mu}{2} \sin \frac{\nu}{2}$$

$$= \frac{m}{2r} \cdot \frac{n}{2r}.$$

That is,

$$ac + bd = mn, \quad \textbf{(3.6)}$$

the desired formula.

4. Early Projective Geometry: Desargues' Theorem

The works of Pappus seem to be the last significant contribution to geometry in antiquity. Creativity in geometry remained dormant until the beginning of the seventeenth century when Johann Kepler (1571–1630) suggested the use of continuity in geometry. This marked the beginning of an age, which continues into the present time, called *modern geometry*. Among the most significant efforts of that early period was the invention of analytic geometry in 1637 by René Descartes. At about this time, activity began in an area of mathematics now known as *projective geometry*.

One of the early founders of this discipline was Girard Desargues (1593–1662), a Frenchman, who set out to discover the strictly nonmetrical properties of geometry. Among his discoveries was the following theorem, named after him:

If the corresponding vertices of two triangles ABC and XYZ lie on concurrent lines, the corresponding sides, if they intersect, meet in collinear points. (See Fig. 10.)

Figure 10

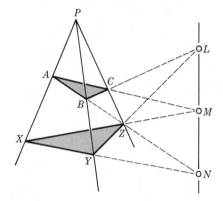

This property turns out to be crucial in any study of projective geometry.

Early efforts in projective geometry were devoted to an understanding of the problem which the artist faces in painting a three-dimensional scene on a flat piece of canvas. It was early theorized that the artist "projects" what he sees onto his canvas by essentially "intersecting" it with his line of sight. This is why the parallel rails of a railroad track, for example, are not drawn as parallel lines at all, but as *intersecting* lines (Fig. 11).

Figure 11 (a) The artist at work.

(a)

Figure 11 (b) Projecting parallel lines into intersecting lines.

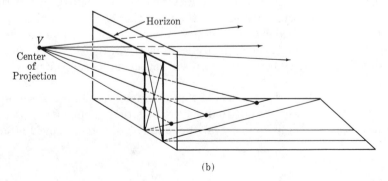

(b)

We may use this simple observation to obtain a proof of Desargues' theorem. First, the theorem will be altered slightly to give us a more familiar principle: *If lines* AX, BY, *and* CZ *meet at* P, *with* AB *parallel to* XY *and* BC *parallel to* YZ, *then* AC *is parallel to* XZ (Fig. 12). This is simple to prove: by hypothesis, $PA/PX = PB/PY$, and $PB/PY = PC/PZ$; therefore $PA/PX = PC/PZ$ and thus line AC is parallel to XZ.

Figure 12

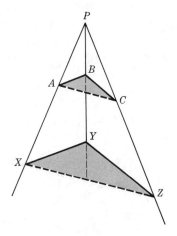

If now triangles ABC and XYZ are situated in plane π so that lines AX, BY, and CZ meet at P, and the lines AB and XY, BC and YZ, AC and XZ meet at L, M, and N, respectively, set up a plane of projection π' and a center of projection V, so that line LM becomes the "horizon" (Fig. 13). Thus, lines AB and XY project

Figure 13

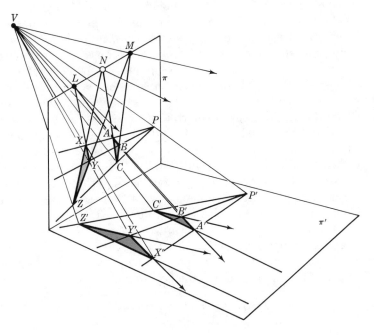

into parallel lines $A'B'$ and $X'Y'$, and similarly, $B'C'$ is parallel to $Y'Z'$. By the preceding theorem, $A'C'$ is parallel to $X'Z'$, and therefore AC and XZ must meet on the horizon. Then L, M, and N must be collinear.

5. The Triumph of Modern Geometry

For over 2000 years following Euclid's epochal works a recurring challenge to mathematicians was to find a proof for the parallel postulate within Euclid's own system, that is, to show that the parallel postulate was dependent on Euclid's other axioms. Such proofs had been attempted even in ancient times by no lesser men than Archimedes and Proclus, and serious attacks on the problem have continued even into modern times. It became characteristic of such attempts to first reveal the flaws in previous arguments (usually involving some outside assumption not proved) and then to introduce an original idea. But during this long period of history, few mathematicians even dreamed that what they were trying to accomplish was utterly impossible.

Part of the problem stemmed from Euclid's basic premise—that the "self-evident" axioms of geometry are dictated by physical reality and by one's intuitive observations of the world about him, and that every logical deduction from those axioms is merely another ramification of that reality. It thus became widely theorized that since there is only one universe, *there can be only one consistent geometry*. As a result, Euclid became firmly rooted in scholarly thinking.

The first notable attempt to establish Euclid's postulate was carried out by G. Saccheri in 1733. As others before him had done he pointed out certain errors in earlier proofs. But then he introduced a significantly new approach to the problem—the indirect method of proof, used by Euclid himself. He was thus the first to investigate the effect of assuming an actual *denial* of the parallel postulate. Saccheri failed, mainly because of his overwhelming persuasion that what he was trying to do *could* be done, a conviction which led him to "rationalize" at certain points, and to commit basically the same error as his predecessors.

Saccheri's work marked the beginning of a monumental effort in modern times by some of the ablest mathematicians in history to prove the parallel postulate. A more careful investigation than Saccheri's was undertaken by J. Lambert (1728–1777), who actually entertained some doubt as to the demonstrability of the postulate.[2] A. Legendre (1752–1833), a renowned pioneer in analysis, followed with an even more significant effort which extended throughout most of his lifetime. His accomplishments were printed in the 12 successive editions of the book entitled *Eléments de Géométrie*, which were published during the period 1794–1823. Many of Legendre's proofs are of permanent value, but as noted by him, most of them do nothing more than establish certain non-Euclidean properties from certain other non-Euclidean properties. He regarded his final effort in the last edition of his book as a solution to the problem, but later even it was found to be incorrect.

Then emerged the thinking of the great German mathematician, Karl Friedrich Gauss (1777–1855), who was the first to become completely persuaded that some-

---

[2] T. L. Heath, *The Thirteen Books of Euclid's Elements* [14, p. 212].

thing was fundamentally wrong. He had observed many "geometries" on surfaces of various shapes, which were perfectly valid, yet possessed non-Euclidean properties. He began to suspect that the parallel postulate of Euclidean geometry was actually *independent* of the remaining axioms. Upon carrying out an intensive investigation of his own, not altogether unlike Saccheri's, he became convinced that he had the answer, and that therefore nobody could ever prove the postulate of parallels.

This was no small conclusion—the very foundations of nineteenth century mathematics, science, and philosophy were at stake. Scholars were not ready to forsake those foundations, and would not be for at least another fifty years. Gauss was once persuaded to write:

> It may take very long before I make public my investigations on this issue; in fact this may not happen in my lifetime for I fear the scream of dulliards if I make my views explicit.

Thus Euclid continued to dominate the thinking of most learned men. Few paid any attention to the work of János Bolyai (1802–1860) and N. Lobachevski (1793–1856) who independently, but almost simultaneously, carried through a complete development of the consequences of a denial of the parallel postulate. Their results included a non-Euclidean type of trigonometry and even an extensive theory of surfaces, all apparently void of contradictions of any kind. Thus was born the geometry known today as *hyperbolic geometry*.

The final triumph came at last when Henri Poincaré (1854–1912) and others were able to construct models in Euclidean geometry of the elaborate system envisioned earlier by Bolyai and Lobachevski. The model of Poincaré consists of the interior points of some circle ω (Fig. 14) where the "lines" are assumed to be arcs of circles orthogonal to ω or any diameter of ω (to be explored in greater detail in Chapter 11). The very existence of the Poincaré model is itself proof that the conjecture of the ages was false, that Euclidean geometry is not the only consistent geometry. Ironically, the model also completely exonerates Euclid in his act of assuming the parallel postulate. For it shows that if Euclidean geometry is to be derived in its entirety, some form of the parallel postulate *must* be assumed.

In light of the modern axiomatic approach, it is easy to minimize this achievement with the amusing observation that it took more than 2000 years for mathe-

Figure 14

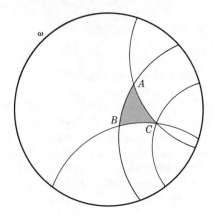

maticians to establish the independence of the parallel postulate, and only 10
to do so for the rest. The real difficulty, however, as we have pointed out, was not
so much in the complexity of the problem but with the philosophy held by those
who tackled it. Aside from the mathematical discovery itself, the great achievement of the nineteenth century was the freeing of mathematics from Euclid. The
systems of Bolyai and Poincaré became the prime examples that *consistency in
mathematics is independent of its portrayal of the physical world*, a deeply significant
thought.

6. The Isoperimetric Problem

One famous problem of modern geometry which we shall mention only briefly,
is known as the *isoperimetric problem:*

Among all simple closed curves in the plane having equal perimeters, find the one enclosing the greatest area.

From the study of such problems has emerged a branch of mathematics known
as the *calculus of variations*. The Swiss mathematician Jacob Steiner (1796–1863)
proposed several ingenious arguments to show that the solution to this problem
must be a *circle*. One of those arguments is particularly elegant.

Let K be any simple closed curve in the plane, and suppose that K is *not* a
circle. It may be assumed that K is a *convex curve*—that it has the property that
the line segment AB joining any two points A and B on the curve lies on or in the
interior of that curve. For if not, a reflection in line AB leads to a curve K' having
the same perimeter but enclosing a larger area [see Fig. 15(a)]. It may be further
assumed that if A is a point on K, and B is halfway around K from A, then line AB
equally divides the area enclosed by K. For, in Fig. 15(b), if I > II then by reflec-

Figure 15

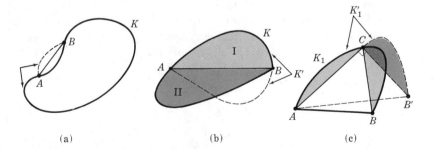

(a)   (b)   (c)

tion in AB a new curve K' exists (as indicated) having the same perimeter as before
but enclosing an area equal to 2I > I + II. Now since one of the halves K_1 of K
is not a semicircle, there is C on K_1 such that angle ACB is not a right angle. But
then the right triangle ACB' with $B'C = BC$ has area greater than that of triangle
ABC (why?) and hence Fig. 15(c) indicates a curve K_1' which is "one half" of a new
curve K' having the same perimeter as K but enclosing a greater area.

It should be pointed out that this argument, although clever, does not completely solve the isoperimetric problem, for it takes the *existence* of the solution for granted. Rather, it proves only that *if* K *is a simple closed curve in the plane which is not a circle it can be replaced by one which has the same perimeter and encloses a greater area.* The reader should examine the following reasoning, which is very similar but leads to an obviously false conclusion:

If n is any positive integer except 1 then it is clear that n^2 is a positive integer which is even larger. That is, no positive integer different from 1 can be the largest. Therefore, *1 is the largest positive integer!* (C. R. Wylie, *Foundations of Geometry* [31, p. 7].)

For a more complete discussion of the existence aspect of extremum problems in general, see R. V. Benson, *Euclidean Geometry and Convexity* [2, pp. 96–103].

7. Morley's Theorem

Many theorems of modern geometry are entertaining indeed. Frank Morley (1860–1937), the father of the novelist Christopher Morley, made the remarkable discovery that the intersections of adjacent pairs of angle trisectors in a triangle are the vertices of an equilateral triangle (Fig. 16). Impressed with the beauty of this theorem, Morley showed it to his friends, and it quickly spread to the rest of the mathematical world as an interesting item for gossip. However, the proof of the theorem did not appear in published form until 1914.

A proof of Morley's theorem may be based on the law of sines and the result:

Lemma: If a, b, x, and y are positive real numbers and

(a)
$$0 < a + b < 180,[3]$$

(b)
$$x + y = a + b,$$

(c)
$$\frac{\sin x}{\sin y} = \frac{\sin a}{\sin b},$$

then $x = a$ and $y = b$.

PROOF: Set $x + y = \theta$. Then $\sin y = \sin(\theta - x) = \sin\theta \cos x - \cos\theta \sin x$, or

$$\frac{\sin y}{\sin x} = \sin\theta \cot x - \cos\theta.$$

Similarly, since $a + b = \theta$,

$$\frac{\sin b}{\sin a} = \sin\theta \cot a - \cos\theta.$$

Condition (c) then implies that $\sin\theta \cot x = \sin\theta \cot a$, and since $\sin\theta \neq 0$, $\cot x = \cot a$. But the function $\cot x$ is strictly decreasing for x in the range $0 < x < 180$, so it follows that $x = a$, and therefore $y = b$. ∎

---

[3] In keeping with later conventions all arguments will be measured in *degrees* rather than in *radians*.

Figure 16

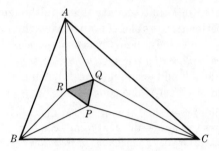

We shall also make use of the following trigonometric identity left to the reader to demonstrate (see Exercise 15 below):

$$\sin x = 4 \sin \frac{x}{3} \sin \left(60 + \frac{x}{3}\right) \sin \left(60 - \frac{x}{3}\right). \qquad (7.1)$$

Now consider any triangle ABC with the trisectors AR and AQ of angle A and BR and CQ of angles B and C as shown in Fig. 17. The law of sines in triangle AQR implies

$$\frac{\sin x}{\sin y} = \frac{AQ}{AR}. \qquad (7.2)$$

Figure 17

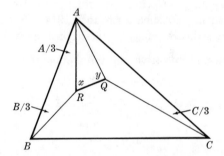

Similarly,

$$\frac{AQ}{AR} = \frac{AQ}{AC} \cdot \frac{AC}{AB} \cdot \frac{AB}{AR} = \frac{\sin (C/3) \sin B \sin (180 - A/3 - B/3)}{\sin (180 - A/3 - C/3) \sin C \sin (B/3)}, \qquad (7.3)$$

which in view of (7.2) becomes

$$\frac{\sin x}{\sin y} = \frac{\sin (C/3) \sin B \sin (A/3 + B/3)}{\sin (A/3 + C/3) \sin C \sin (B/3)}. \qquad (7.4)$$

Using $A + B + C = 180$ we get

$$\frac{\sin x}{\sin y} = \frac{\sin (C/3) \sin (60 - C/3) \sin B}{\sin (B/3) \sin (60 - B/3) \sin C}. \qquad (7.5)$$

Now apply (7.1) to $\sin B$ and $\sin C$ in (7.5):

$$\frac{\sin x}{\sin y} = \frac{\sin (60 + B/3)}{\sin (60 + C/3)}. \qquad (7.6)$$

If $a = 60 + B/3$ and $b = 60 + C/3$, then $a + b = 120 + (B + C)/3 = 180 - A/3$. Thus, $0 < a + b < 180$. Since x, y, and $A/3$ are the interior angles of triangle ARQ, $x + y = 180 - A/3 = a + b$. The conditions (a), (b), and (c) of the lemma are thereby fulfilled, so

$$x = 60 + \frac{B}{3} \quad \text{and} \quad y = 60 + \frac{C}{3}. \tag{7.7}$$

From the original figure for Morley's theorem (Fig. 16), the further relations are inferred:

$$\angle BRP = 60 + \frac{A}{3}, \quad \angle BPR = 60 + \frac{C}{3},$$

$$\angle CQP = 60 + \frac{A}{3}, \quad \angle CPQ = 60 + \frac{B}{3}. \tag{7.8}$$

We shall let the reader finish the proof from this point on, since it is now simply a matter of deducing the remaining angles in Fig. 16.

8. The Nine-Point Circle and Feuerbach's Theorem

An elementary proposition of Euclid is often taught in high school geometry courses:

The line segment joining the midpoints of two sides of a triangle is parallel to and has length one half that of the third side.

It is this proposition which is useful in obtaining the existence of a certain intriguing circle associated with a triangle.

Consider any triangle ABC, with *orthocenter H* (the point of concurrency of the altitudes), and let the following three groups of points be determined (see Fig. 18):

(a) L, M, N—the midpoints of the sides,
(b) D, E, F—the feet of the altitudes on the sides,
(c) X, Y, Z—the midpoints of HA, HB, and HC.

Figure 18

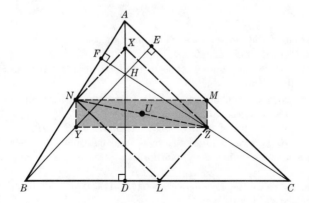

With a few simple observations one can establish the fact that these nine points *all lie on a circle*. Note that Euclid's proposition applied to triangle HBC proves that lines YZ and BC are parallel and $YZ = \frac{1}{2}BC$. For the same reason, lines NM and BC are parallel and $NM = \frac{1}{2}BC$. Hence lines NM and YZ are parallel and equal in length. Therefore, $MNYZ$ is a parallelogram, and since line NY is parallel to line AH it follows that $MNYZ$ is a rectangle; let U be the intersection of its diagonals. Then points M, N, Y, and Z lie on a circle centered at U. In the same manner one observes that $LNXZ$ is a rectangle. Since U is the midpoint of NZ, it is the intersection of the diagonals of rectangle $LNXZ$. Hence L, M, N, X, Y, and Z all lie on the same circle, with U as center. To finish the proof, observe that XL is a diameter of circle U and that angle XDL is a right angle. Thus D lies on circle U, and in exactly the same manner it follows that E and F lie on circle U.

The circle we have just obtained is part of the folklore of elementary geometry. It is known as the *nine-point circle* of triangle ABC. The supply of remarkable properties associated with it seem to be inexhaustible. Chief among these are the following, to be proved at a later time (see Fig. 19):

Figure 19

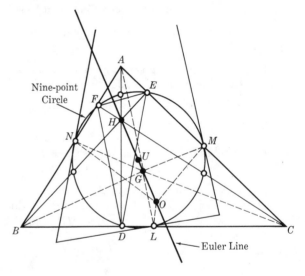

(a)   The radius of the nine-point circle is one half that of the *circumcircle*—the circle passing through A, B, and C.

(b)   The center U is the midpoint of segment HO, where O is the center of the circumcircle (called *circumcenter*).

(c)   As a corollary of (b), U lies on the *Euler line* of the triangle—the line of collinearity of the orthocenter, centroid, and circumcenter.[4]

(d)   If G is the centroid of triangle ABC, then $HU/UG = HO/OG = 3$. Thus, the segment HG is divided internally by U and externally by O in the same ratio. (HG is said to be divided *harmonically*.)

---

[4] Named after its discoverer, Leonhard Euler (1707–1783). (The *centroid* is the point of concurrency of the medians of the triangle.)

(e) The tangents to the nine-point circle at L, M, and N are parallel to the respective sides of the *orthic triangle DEF*.

Now consider the four circles which are tangent to the sides of triangle ABC, called *tritangent circles*. Another property of the nine-point circle is known as *Feuerbach's Theorem*, after the German mathematician Karl Feuerbach (1800–1834), and its intricacy surpasses all the others (illustrated in Fig. 20):

(f) *The nine-point circle is tangent to each of the four tritangent circles of triangle ABC.*

Figure 20

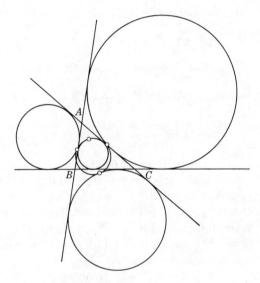

9. The Fibonacci Sequence and the Golden Ratio

Sometimes a number which was originally involved in the solution of a problem in geometry turns up in rather odd places. For example, the number π originated as the ratio of the circumference of a circle to its diameter. How unlikely it appears then, that the rational sum

$$1 - \frac{1}{3} + \frac{1}{5} - \frac{1}{7} + \frac{1}{9} - \frac{1}{11} + \cdots$$

should have as its limit the value $\frac{1}{4}\pi$. Conversely, concepts in number theory and analysis often turn out to be of geometric significance.

A case in point is the *Fibonacci sequence*, studied as early as 1202[5]

$$1, 1, 2, 3, 5, 8, 13, 21, \cdots,$$

Since the rule of formation evidently consists of adding two successive terms to

---

[5] Coxeter [4, p. 165].

obtain the next, the sequence may be defined recursively by the formulas

$$f_1 = 1, \qquad f_2 = 1, \qquad f_n = f_{n-1} + f_{n-2} \qquad (n \geq 3). \tag{9.1}$$

Note that (9.1) is a basis for defining the series "backwards" to include negative values of n. Set $-k = n - 2$ to obtain $f_{-k+2} = f_{-k+1} + f_{-k}$ and therefore $f_{-k} = f_{-k+2} - f_{-k+1}$, which may be taken as the recursive definition of f_{-k} for $k = 0, 1, 2, \cdots$. Thus (9.1) also holds for the extended sequence, a few values of which are tabulated below:

n	$\cdots$	-10	-9	-8	-7	-6	-5	-4	-3	-2	-1	0	1	2	3	4	5	6	7	8	9	10	11	12	13	$\cdots$
f_n	$\cdots$	-55	34	-21	13	-8	5	-3	2	-1	1	0	1	1	2	3	5	8	13	21	34	55	89	144	233	$\cdots$

Kepler observed a peculiar behavior of f_n: the ratios of successive terms,

$$\frac{1}{1} = 1.0, \qquad \frac{2}{1} = 2.0, \qquad \frac{3}{2} = 1.5, \qquad \frac{5}{3} = 1.6 \cdots, \qquad \frac{8}{5} = 1.60, \qquad \frac{13}{8} = 1.62 \cdots,$$

$$\frac{21}{13} = 1.615 \cdots, \qquad \frac{34}{21} = 1.614 \cdots, \qquad \frac{55}{34} = 1.617 \cdots, \qquad \frac{89}{55} = 1.618 \cdots,$$

tend to the limit $\mu = 1.61803 \cdots$ (the *golden ratio*) as n tends to $+\infty$.

Actually this result may be anticipated by dividing both members of the third equation in (9.1) by f_{n-1} and setting $x = \lim_{n \to \infty} (f_n/f_{n-1})$. The resulting equation is therefore $x = 1 + 1/x$, and it follows that $x = \mu$. Of course the existence of x was taken for granted here so as it stands the proof is inadequate. However, one can obtain the general formula for f_n with very little difficulty, and then the above limit can be evaluated directly.

Use (2.2) repeatedly to obtain

$$\mu^2 = \mu + 1$$
$$\mu^3 = \mu(\mu + 1) = \mu^2 + \mu = (\mu + 1) + \mu = 2\mu + 1$$
$$\mu^4 = \mu(2\mu + 1) = 2\mu^2 + \mu = 2(\mu + 1) + \mu = 3\mu + 2$$
$$\mu^5 = \mu(3\mu + 2) = 3\mu^2 + 2\mu = 3(\mu + 1) + 2\mu = 5\mu + 3$$
$$\cdot$$
$$\cdot$$
$$\cdot$$
$$\mu^n = f_n\mu + f_{n-1}$$
$$\mu^{n+1} = \mu(f_n\mu + f_{n-1}) = f_n(\mu + 1) + f_{n-1}\mu = f_{n+1}\mu + f_n.$$

Apply (2.3) in a similar manner to prove

$$\frac{1}{\mu^k} = f_{-k}\mu + f_{-k-1}.$$

Thus by mathematical induction the formula

$$\mu^n = f_n\mu + f_{n-1} \tag{9.2}$$

is valid for all integers n. The additional formula,

$$f_{-n} = (-1)^{n+1}f_n, \tag{9.3}$$

which is evident from the above table, may also be proved for all n. Then use (9.2) and (9.3) as follows:

$$(-\mu)^{-n} = (-1)^n \mu^{-n} = (-1)^n(f_{-n}\mu + f_{-n-1}) = -f_n\mu + f_{n+1}.$$

But

$$-f_n\mu + f_{n+1} = -f_n\mu + f_n + f_{n-1} = -f_n(\mu - 1) + f_{n-1} = -f_n \cdot \frac{1}{\mu} + f_{n-1}.$$

That is,

$$(-\mu)^{-n} = -f_n \cdot \frac{1}{\mu} + f_{n-1}. \tag{9.4}$$

Next, subtract (9.4) from (9.2):

$$\mu^n - (-\mu)^{-n} = f_n\left(\mu + \frac{1}{\mu}\right) = f_n\sqrt{5}.$$

Therefore,

$$\boxed{f_n = \frac{\mu^n - (-\mu)^{-n}}{\sqrt{5}} = \frac{1}{\sqrt{5}}\left(\frac{1 + \sqrt{5}}{2}\right)^n - \frac{1}{\sqrt{5}}\left(\frac{1 - \sqrt{5}}{2}\right)^n,} \tag{9.5}$$

—a formula originally discovered by J. P. M. Binet in 1843. Thus we have found that the golden ratio is intimately related to the Fibonacci sequence.

10.   Dissection of the Rectangle into Squares

We end our introduction to geometry by discussing a particularly appealing area known as *dissection theory*, a field that includes a great variety of problems, among which is the following simple example:

A farmer who lives on a square plot of land decides to retire. He takes out a quarter square from the southeast corner for himself, leaving the rest for his four sons (Fig. 21). How may the remainder be divided so that the sons will all receive lots which are the same size and shape (that is, *congruent*)?

Figure 21

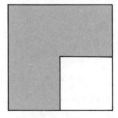

If the reader finds this entertaining he will appreciate to some extent the great fascination dissection problems hold for geometers.

One of the classic problems in dissection theory is the dissection of a rectangle into *squares*. If the dissection is carried out in such a way that no two squares are the same size, that dissection is termed *perfect*. It has been proved that any perfect dissection of a rectangle must employ *at least nine squares*, and that there are only *two* using *exactly nine.*[6] The following method can be used to obtain a dissection having *ten* squares: Begin with two adjacent squares having respective sides x and y, and arrange three more squares about them in spiral-like fashion, as shown in Fig. 22. One obtains a "notched" rectangle, with the dimensions of the notch being $(x + 2y) \times (y) \times (4x) \times (3x + y)$. Now construct another notched rectangle, this time generated by squares having respective sides u and v; the dimension of the notch will be $(u + 2v) \times (v) \times (4u) \times (3u + v)$. The two notched rectangles may be pieced together to form a rectangle provided the notches have three consecutive dimensions equal. Hence we impose the conditions

Figure 22

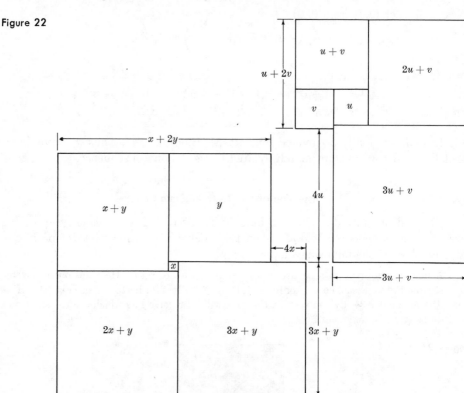

$$y = u + 2v, \qquad 4x = v, \qquad 3x + y = 4u. \tag{10.1}$$

The terms u and v may be easily eliminated to obtain the equation

$$35x = 3y. \tag{10.2}$$

It follows that for some constant t,

$$x = 3t, \qquad y = 35t. \tag{10.3}$$

Substitute these into (10.1) and we obtain

$$u = 11t, \qquad v = 12t. \tag{10.4}$$

---

[6] See Eves [9, pp. 260–272] for a more complete discussion.

It is clear that we may choose $t = 1$. The resulting values for x, y, u, and v in (10.3) and (10.4) then lead to the dissection shown in Fig. 23, *a perfect dissection of a rectangle into ten squares*. To obtain a nine-square dissection, considerably more ingenuity is required.

Figure 23

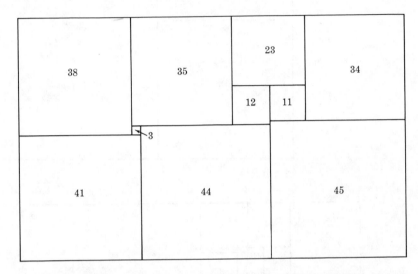

A special case of the perfect dissection of a rectangle into squares is the perfect dissection of a *square* into squares, a much more difficult problem. As late as 1900 it was not known whether this could be done. In 1939, a perfect dissection of a square into 55 squares was published, and a year later a dissection which employed only 26 squares was accomplished jointly by R. C. Brooks, C. A. B. Smith, A. H. Stone, and W. T. Tutte. In 1948, T. H. Willcocks presented a perfect dissection of the square into 24 squares (his solution may be found in Eves [9, p. 264]), and at present, the question whether 24 is the least number possible remains unanswered.

We shall indicate a method of obtaining a perfect "squared" square which makes use of the Fibonacci sequence discussed in the preceding section, a method due to P. J. Federico.[7] Begin with two adjacent squares of unit dimension and arrange further squares in sequence as indicated in Fig. 24. We notice that the dimensions of succeeding squares make up the numbers in the Fibonacci sequence, and that by stopping with the square f_n there is obtained an $(f_n) \times (f_{n+1})$ rectangle, which, but for the presence of the two unit squares, is perfectly dissected into squares. These rectangles are referred to as *Fibonacci rectangles*. Now one seeks a perfectly squared rectangle obtained by some other method, which is similar to one of the Fibonacci rectangles. If such a rectangle can be found, then the two are brought to the same size by multiplying by the appropriate factors, and chances are good that no two squares of the resulting squared rectangles will have the same

[7] P. J. Federico, "A Fibonacci Perfect Squared Square," *American Mathematical Monthly*, 71 (1964), pp. 404–406.

Figure 24

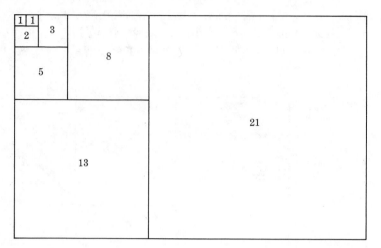

Figure 25

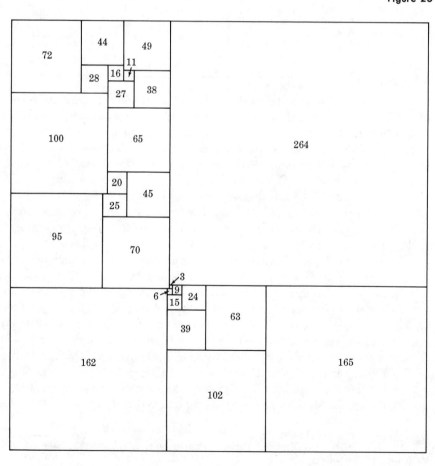

dimensions. There is an extensive list of the known perfect squared rectangles,[8] and in that list appears a rectangle with sides 165 and 267; there is an n for which

$$\frac{f_{n+1}}{f_n} = \frac{267}{165},$$

namely, $n = 10$. Since the corresponding Fibonacci rectangle is one third the size of the perfect squared rectangle, the former is enlarged by a factor of 3. One then removes the corner duplicated square and places the other rectangle into this corner (as shown in Fig. 25), thereby forming a larger square. The result is *a perfect squared square having 26 pieces.*

11. Conclusion

The reader has now had a glimpse of geometry. He has seen geometry through the eyes of the ancients, the geometers of the seventeenth, eighteenth, and nineteenth centuries, and finally through the eyes of geometers living today. He has examined a few of the better known theorems coming from each period. It is hoped that by now the reader has begun to derive a sense of appreciation for geometry, and to acquire an attitude which will best suit him for further study in this subject. The mood is aptly set in the following story told of Euclid: Someone who had begun to read geometry with Euclid, when he had learned the first theorem, asked Euclid, "But what shall I get by learning these things?" Euclid called his slave and said, "Give him threepence, since he must make gain out of what he learns."[9]

EXERCISES

As the reader progresses through the book he frequently will be asked to return to certain problems which appear here. Therefore, since some of these may be beyond his present skill, he should attempt only those which he feels he can now solve.

1. A "dynamic" five-step proof of the Pythagorean theorem which illustrates how Euclid's formal area proof works was given by H. Baravalle (*Eighteenth Yearbook* of the National Council of Teachers of Mathematics, Washington, D.C., 1945, pp. 80–81). Study the sequence of figures in that proof shown in Fig. 26 and give your own explanation of the proof. (Figure used by permission.)
2. Prove Pappus' generalization of the Pythagorean theorem. The dotted lines of Fig. 27 should be sufficient to guide you.
3. Three pipes each 6 in. in diameter are stacked pyramid style, as shown in Fig. 28. Find the height of the stack.   *Ans: $6 + 3\sqrt{3}$ in.*
4. Three pipes are stacked, the bottom two each being 6 in. and the top 4 in. in diameter (Fig. 29). Find the height of the stack.   *Ans: 9 in.*
★5.[10] A general formula can be found which will solve Exercises 3 and 4 as special cases. Find it.   *Ans: If* a, b *represent the radii of the two circles acting as base,* a $\geq$ b, *and*

---

[8] C. J. Bouwkamp, A. J. Duijvestijn, and P. Medema, *Tables Relating to Simple Squared Rectangles of Orders Nine through Fifteen* (Eindhoven, Netherlands, 1960).

[9] Heath [14, p. 3].

[10] The more difficult problems will be starred.

Figure 26

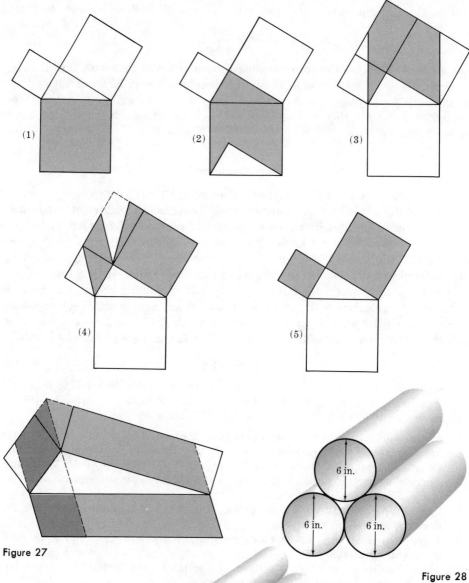

(1)　　　　　(2)　　　　　(3)

(4)　　　　　(5)

Figure 27

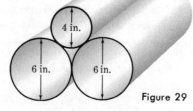

Figure 28

Figure 29

c *is that of the circle on top, then the altitude of the configuration is the maximum of* 2a *and* h, *where*

$$h = b + c + \frac{4ab\sqrt{c(a+b+c)} + (a-b)(ab-ac+bc+b^2)}{(a+b)^2}.$$

6. Four balls, each having a diameter of 6 in., are placed with one on top of the others, each ball touching the other three (Fig. 30). How high are the balls stacked?

Ans: $6 + 2\sqrt{6}$ *in.*

7. Using the law of cosines, prove the following formula relating the sides, diagonals, and the angle θ between the diagonals of any quadrilateral (see Fig. 31):

Figure 30   **Figure 31**

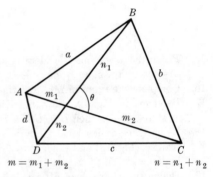

$$a^2 - b^2 + c^2 - d^2 = 2mn\cos\theta.$$

8. Starting with the area formula $K = \frac{1}{2}a'b'\sin\theta$ for a triangle in terms of two sides and the angle between them, use Fig. 31 to derive Bretschneider's formula for the area of a quadrilateral (Section 3), using the result of Exercise 7.

9. Using Ptolemy's theorem and Bretschneider's formula, prove Brahmagupta's formula for the area of a cyclic quadrilateral in terms of its sides (Section 3). *Hint:* You will find the factoring law $x^2 - y^2 = (x+y)(x-y)$ useful. Recall that $a + b + c - d = 2s - 2d$, and so on.

10. Let T be a triangle having consecutive integers $n-1$, n, and $n+1$ for sides, and whose area is also an integer. Prove that one of the altitudes divides T into two Pythagorean triangles (right triangles having integral sides) and divides the base into segments whose lengths differ by 4, the 3-4-5 right triangle being an exceptional case [Problem E 1773, *American Mathematical Monthly,* **72** (1965), p. 316].

11. **The Cevian Formula.** Derive the following formula for the length d of cevian AD drawn to side BC of triangle ABC, in standard notation:

$$d^2 = pb^2 + qc^2 - pqa^2, \qquad (11.1)$$

where $p = BD/BC$ and $q = DC/BC$. *Hint:* Write $a_1 = BD$, $a_2 = DC$, and apply the law of cosines to triangles ABD and ADC with respect to $\theta = \angle ADB$ and $\pi - \theta = \angle ADC$.

12. One circle is half the size of another. Find the locus of a point P on the smaller circle as it rolls without slipping along the inside of the larger (*Hungarian Problem Book II,* Random House: New York, 1963, p. 17; used by permission).

13. A hole 6 in. long is bored into a solid sphere with the axis of the hole passing through
the sphere's center. What is the volume of the part that is left?

14. A square house of side h has been built at the corner of a square lot of side $s > h$
(Fig. 32). What is the largest square garage which can be built on the remainder of
the lot?

Figure 32

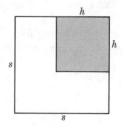

15. Prove the trigonometric identity used in the proof of Morley's theorem. *Hint:* Set
$x = 3y$, write $\sin 3y = \sin (y + 2y)$, and expand.

16. Complete the proof of Morley's theorem.

17. Referring to Fig. 19, show that if $HG = 2GO$ and U is the midpoint of segment GO
as stated above, then $HU/UG = HO/OG$.

18. Using three-dimensional analytic geometry, verify that a circle lying in the vertical
plane π (Fig. 13) and tangent to the horizon is projected from V onto a parabola lying
in plane π'.

19. Define the *skewness* of a triangle whose sides are a, b, and c, with $a \le b \le c$, to be

$$\sigma = \max \left(\frac{a}{b}, \frac{b}{c}, \frac{c}{a} \right) \cdot \min \left(\frac{a}{b}, \frac{b}{c}, \frac{c}{a} \right).$$

(For example, a triangle whose sides are 2, 12, and 13 has $\sigma = 13/12 = 1.08333 \cdots$,
while, in comparison, the triangle having sides 3, 4, and 5 has $\sigma = 5/4 = 1.25$. One
might say that the skewness measures the extent to which the sides of a triangle differ
relative to each other.) Show that $1 < \sigma < \mu$ (where μ is the golden ratio) and that $\sigma = 1$
for an isosceles triangle. If $1 < r < \mu$ prove that $\sigma = r$ for a triangle whose sides are in
the geometric progression a, ar, ar^2. Thus, σ approaches maximum for triangles whose
sides are in geometric progression with common ratio tending to the golden ratio.
[Problem E 1705, *American Mathematical Monthly*, **71** (1964), p. 680.]

20. (a) Discover the pattern which seems to prevail in the following sequence and use the
formula for f_n (9.5) to derive a formula for the general term:

$$1, 2, 4, 7, 12, 20, 33, 54, 98, \cdots, g_n, \cdots$$

(b) Prove that if F_n represents the sum of the first n terms of the Fibonacci sequence,
then $F_n = f_{n+2} - 1$.

21. Prove $\lim_{n \to \infty} f_{n+1}/f_n = \mu$.

22. Establish the formula

$$f_{n-1}f_{n+1} = f_n^2 + (-1)^n. \tag{11.2}$$

23. Sometimes formulas derived or discovered by algebraic means can be revealed quite
easily by geometric arguments. A case in point is this formula involving the Fibonacci
numbers:

$$f_1^2 + f_2^2 + f_3^2 + \cdots + f_n^2 = f_n f_{n+1} \tag{11.3}$$

Observe Fig. 33 and verify that it proves (11.3) for $n = 6$. Generalize, then try a purely algebraic argument based on induction. Which argument do you find most appealing?

Figure 33

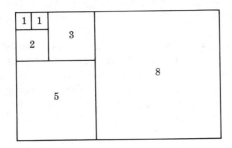

24. A well-known dissection puzzle is illustrated in Fig. 34, where an 8×8 square is dissected into four pieces, then reassembled into a 5×13 rectangle. How can one account for the extra unit square? (If the reader is already familiar with this puzzle he should go on to the next exercise.)

Figure 34

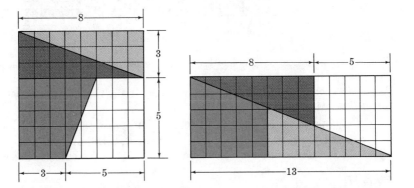

25. There exist a variety of numbers for which the puzzle mentioned in Exercise 23 works. In Fig. 35 is illustrated a generalization of the puzzle. If $x = 8$, $y = 13$, $z = 21$, we obtain a dissection of a 13×13 square whose pieces (seem to) form an 8×21 rectangle.

Figure 35

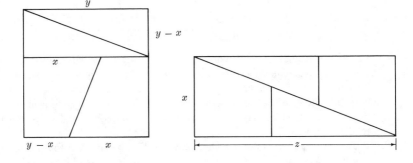

This time we *lose* a unit square! In general we seek numbers which satisfy the relations

$$\begin{cases} xz = y^2 \pm 1 \\ \quad z = x + y. \end{cases} \tag{11.4}$$

Find other integral solutions of (11.4), thus providing other realizations of the dissection puzzle. *Hint:* Use (11.2). [It can be shown that this provides the *only* integral solution set for (11.2).]

26. Explain the fallacy posed in Fig. 36, based on what is called a *Curry triangle* (see M. Gardner, *Mathematical Puzzles and Diversions* [13, pp. 144, 145]). An equilateral triangle, which we may imagine is constructed from a piece of cardboard, is dissected in the manner shown. The parts are rearranged in such a way that there apparently results a *net loss of two square units*. But if they are rearranged with half the pieces turned face down in the manner indicated, a net loss of only *one* square unit results!

Figure 36

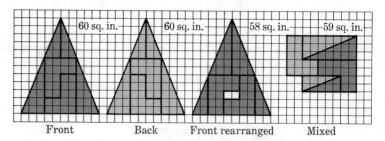

| Front | Back | Front rearranged | Mixed |

27. Solve the dissection problem involving the farmer and his four sons, as presented in Section 10.

28. *Euclid's construction of the regular pentagon.* A triangle ABC is constructed by using the following procedure:

Extend a given segment AD through D to any point B, and inscribe two arcs—one centered at B and the other centered at D, each having radius equal to AD. Let C be the intersection of the two arcs; triangle ABC is thereby determined. (See Fig. 37.)

Figure 37

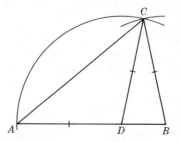

In general, triangle ABC will not be isosceles. But if B is determined so that $AB = \mu AD$ (recall in that connection the construction of Fig. 6), then it will be. (a) Prove that this is so, using the standard propositions regarding similar triangles. (b) It then follows that with this choice of B, the angles of triangle ABC are of (degree) measure 36, 72, and 72, respectively. Deduce this, and derive Euclid's compass straightedge construction of the regular pentagon.

29. *Dudeney's dissection of the equilateral triangle.* Fig. 38 shows a four-piece dissection of an equilateral triangle into a square, discovered by H. E. Dudeney (1857–1931). Segments

GA, AB, and DE are each one-half AC, and BE is the mean-proportional between AB and BF. Prove that it works. *Caution:* $FE \neq DC$.

Figure 38

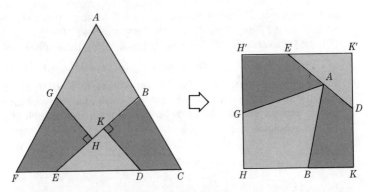

30. J. Travers devised the five-piece dissection of a regular octagon into a square shown in Fig. 39. Explain how this works. *Hint:* The center square in the octagon, with sides equal to AC, was obtained by rotation through a certain angle so that $AB = BC = DE = EF$, and so forth.

Figure 39

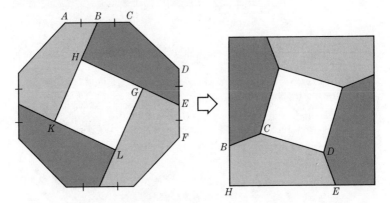

31. **The golden spiral.** In the calculus one studies various kinds of *spirals*. One of these is the *logarithmic spiral*, having polar equation $r = e^{a\theta}$, where a is constant. (Since this curve makes a constant angle $\cot^{-1} a$ with the radial vector, it is sometimes called the

Figure 40

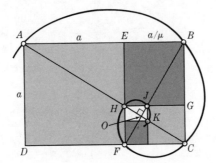

equiangular spiral.) If $ABCD$ is a golden rectangle, we cut off a square $AEFD$ and consider BF and AC, as shown in Fig. 40. We find that BF and AC are perpendicular at O (why?). Rectangle $EBCF$ is another golden rectangle, so cut off square $EBGH$ and once again HC is perpendicular to BF at some point O'. But O' must coincide with O, so it follows that A, H, and C are collinear. The process continues. Take rays OA and OF as coordinate axes, with OA as unit. By the similarity of the figures involved,

$$1 = OA = \mu OB, \qquad OB = \mu OC, \qquad OC = \mu OF, \qquad OF = \mu OH, \qquad \cdots$$

Find polar coordinates (r, θ) for the points A, B, C, F, H, J, and K and prove that the logarithmic spiral $r = e^{a\theta}$ for which $a = (-2/\pi) \log \mu$ passes through all those points, and in general will pass through three of the vertices of each of the rectangles in this infinite sequence.

PART II

FOUNDATIONS
FOR
GEOMETRY

2
Points, Lines, and Distance

The study of geometry by the deductive method necessarily begins with very primitive and very simple ideas. Accordingly, we shall start with "point" and "line," and then, bearing in mind certain intuitive notions of geometry, set down in Euclid's fashion the properties we think those objects ought to have. The axioms will be chosen in such a way, however, that it will be possible to study spherical and Euclidean geometry side by side. This will allow an axiomatic study that is more than just a formal development of familiar Euclidean principles. Indeed, at times it may appear that we are studying axiomatic *spherical* geometry, obtaining certain Euclidean ideas as byproducts.

It is of course desirable to have a set of axioms that are both *consistent* and *independent*. These questions will be considered as the development progresses, with examples presented at certain stages to illustrate both the consistency of the preceding axioms and the independence of the one to follow. Since it is not our aim to dwell merely on the foundations of geometry, some of the axioms will be made quite powerful. One might say that we are building geometry from the "walls" rather than from the "two-by-fours," prefabricating the whole structure in order to speed the construction along.

Perhaps the most obvious question at the outset is: Why axiomatic geometry? One reason is simply that geometry is easier to *understand* when it is given a more precise treatment. Another reason is the object of discussion in the two opening sections.

12. Fallacious Proofs in Geometry

Faith in Euclid's elegant system falters in the face of such apparent paradoxes as the following:

Theorem (!): Every obtuse angle is a right angle.

PROOF: Let $\angle ABC$ be an obtuse angle (Fig. 41). Then the perpendicular BE will fall within $\angle ABC$ and one may then complete the rectangle $BCDE$ as shown, with $BE = BA$. Since lines DA and DE are not parallel the perpendicular bisectors MX and NY of AD and BC will meet at some point O. Draw OA, OB, OC, and OD, and label angles as shown. Now $OD = OA$, $OC = OB$ (every point on the perpendicular bisector of a line segment is equidistant from the ends of that segment), and $AB = CD$ by construction.

Figure 41

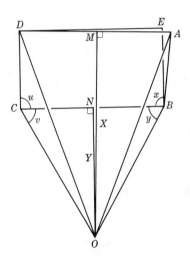

Therefore,

$$\triangle OAB \cong \triangle OCD,$$

and

$$\angle OBA = \angle OCD.$$

Since the whole of a quantity equals the sum of its parts,

$$\angle x + \angle y = \angle u + \angle v.$$

But $\triangle OBC$ is isosceles, so

$$\angle y = \angle v.$$

Therefore,

$$\angle x = \angle u. \quad \text{Q.E.D.} \quad \blacksquare$$

This example was not included merely for its shock value, nor does it stand merely as an admonishment for the reader to use more care in the construction of proofs. A far more significant lesson is to be had, for if this proof is invalid, then many of Euclid's own arguments and methods are inadequate. To illustrate, consider the following:

Theorem: The sum of the angles of a parallelogram equals the measure of two straight angles.

Figure 42

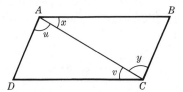

PROOF: Let $ABCD$ be a parallelogram (Fig. 42). In triangles ABC and ADC,

$$\angle x + \angle y + \angle B = 180,$$
$$\angle u + \angle v + \angle D = 180.$$

Therefore,

$$(\angle x + \angle u) + \angle B + (\angle y + \angle v) + \angle D = 360.$$

Since the whole equals the sum of its parts,

$$\angle A = \angle x + \angle u,$$
$$\angle C = \angle y + \angle v.$$

Substitution then yields

$$\angle A + \angle B + \angle C + \angle D = 360.^1 \quad \text{Q.E.D.} \; \blacksquare$$

This time, in the same "spirit of Euclid," a valid proposition has been established. What then could possibly be wrong?

13.   "The Whole Equals the Sum of Its Parts"

Observe that in the latter proof the equation

$$\angle A = \angle x + \angle u$$

played a key role. But beyond an appeal to "common sense" this equation was never justified. It seems to depend on the fact that ray AC visibly falls between rays AB and AD. Is the basis for that equation, therefore, any more concrete than that for $\angle ABO = \angle x + \angle y$ which was used in the incorrect proof?

Apparently, the additive principle "the whole of a quantity equals the sum of its parts" applies in the latter proof but not in the former. Then the problem must be to determine when to use this law and when to avoid it.

But let us examine this well-known principle more closely. Presumably it is a blanket statement which covers any situation where one has "divided" a quantity or figure into various "parts." In algebra, one might "divide" the number "5" into

---

[1] Since the fact that the quadrilateral is a parallelogram was apparently not used, the reader should be able to generalize this proof.

the parts "2" and "3," then use the principle as the basis for the equation $2 + 3 = 5$. However, this does not explain *why* $2 + 3 = 5$. One already knows this from laws and basic axioms of arithmetic. Similarly, the principle does not explain why in the previous figure $\angle A = \angle x + \angle u$.

Thus, we are at a complete loss in trying to justify those equations on the basis of previous axioms or laws, simply because Euclid provided none. A more logical development of geometry, one which relies on reason rather than mechanical drawing, is obviously desirable.

EXERCISES

1. In connection with Fig. 41 use analytic geometry to determine the coordinates (x, y) of point O if the coordinate system is chosen as shown in Fig. 43. Then prove

    (a) $\lim_{u \to 0} y = -\infty$,

    (b) $y < y'$ where (x', y') is the point of intersection of AB and ON.

    Do these results have any bearing on the "proof" given earlier that $\angle ABC = 90$?

Figure 43

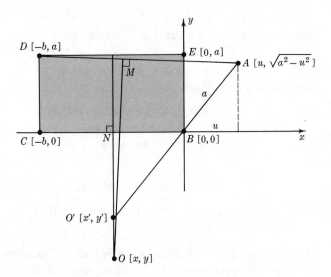

2. Discuss the fallacy in the following "proof" that every triangle is isosceles:

    In $\triangle ABC$, suppose AO bisects $\angle A$ and that OM is the perpendicular bisector of BC, with perpendiculars OD and OE (Fig. 44). Then $AO = AO$ and $\angle DAO = \angle EAO$. Since triangles DAO and EAO are right triangles, they are congruent. Therefore, $AD = AE$. Also, in right triangles BOD and COE, $OD = OE$, and $OB = OC$ (since O is equidistant from B and C). Then $\triangle BOD \cong \triangle COE$, and therefore, $DB = EC$. Thus $AD + DB = AE + EC$ or $AB = AC$.   Q.E.D.

3. Can the proof concerning the angle sum of a parallelogram be extended to quadrilaterals? Did your answer take into account a quadrilateral $ABCD$ which has $\angle A = \angle C = 30$ and $\angle B = 90$?

Figure 44

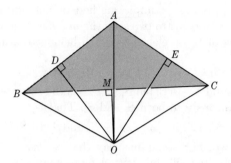

14. Set Theory

In order to facilitate the statement of the axioms and the later development of geometry, free use of the fundamental notions of set theory will be made. Mainly, this involves a knowledge of the *notation* of set theory rather than its *content*. For the sake of completeness that notation will be reviewed here.

Recall that a *set* may be regarded as a collection of *elements*. If x is an element of the set y, then one writes

$$x \in y.$$

Frequently the set being considered will be a line l, and an element belonging to it will then be some point P. This relationship would accordingly be denoted

$$P \in l.$$

If all the elements of a set S_1 are also among the elements of a set S_2, then S_1 is called a *subset* of S_2 and this is written

$$S_1 \subset S_2.$$

Note that this definition allows S_1 and S_2 to represent the *same* set of elements, in which case

$$S_1 = S_2.$$

It is clear that S_1 and S_2 represent the same set whenever it may be asserted that an element belongs to S_1 if and only if it belongs to S_2. That is, $S_1 = S_2$ *whenever both $S_1 \subset S_2$ and $S_2 \subset S_1$ hold*. Usually it is assumed that all sets being considered in a particular discussion are collections of elements from some larger set, called the *universal set*. In this case, *all sets are subsets of the universal set*.

Next, suppose that some set S is characterized by the fact that all its elements satisfy a certain property $\mathcal{P}$. This can be denoted by writing

$$S = \{X \colon\ X \text{ has property } \mathcal{P}\}.$$

In this connection, the set consisting of the elements $A_1, A_2, \ldots, A_n$ for some positive integer n is written

$$\{A_1, A_2, A_3, \ldots, A_n\}.^2$$

---

[2] This notation may not be self explanatory. If $n = 1$ it becomes $\{A_1\}$; if $n = 2$, then $\{A_1, A_2\}$ is understood, and so forth.

The above notation provides a simple method of defining the union and intersection of two sets. If S_1 and S_2 are two sets, the *union* of S_1 and S_2—written $S_1 \cup S_2$ —is the set consisting of those elements which belong to either S_1, or S_2, or both. That is,

$$S_1 \cup S_2 = \{X: \ X \in S_1 \text{ or } X \in S_2\}.$$

The *intersection* of S_1 and S_2—written $S_1 \cap S_2$—is the set consisting of those points which belong simultaneously to both S_1 and S_2. That is,

$$S_1 \cap S_2 = \{X: \ X \in S_1 \text{ and } X \in S_2\}.$$

When $A \in S_1 \cap S_2$, we shall sometimes say that S_1 *and* S_2 *intersect* or *meet in* A, and if $S_1 \cap S_2 = \{A\}$, this will be written more simply as $S_1 \cap S_2 = A$. If two sets have no points in common, they are called *disjoint*. One may also express this in terms of the *empty set* $\varnothing$: S_1 and S_2 are disjoint if and only if

$$S_1 \cap S_2 = \varnothing.$$

15.   Intuitive Spherical Geometry

Let S be a sphere centered at O and having $r = \alpha/\pi$ as radius, where α is any positive real number. Recall that a **great circle** is the intersection of S with a plane passing through O. Each great circle is a circle with center at O and radius α/π. Since its circumference would then be

$$2\pi \left(\frac{\alpha}{\pi}\right) = 2\alpha,$$

the given number α has the distinction of being *one half the circumference of any great circle*.

Now imagine a piece of string which passes through two fixed points P and Q and lying on S (Fig. 45). If the string is gradually drawn taut, it begins to assume the position of what appears to be a great circle arc PAQ. This experiment lends support to (is not a proof of) the principle that the shortest possible arc joining two points on S is the (minor) great circle arc joining them. Thus, in spherical geometry the great circles are the analogs of straight lines. The distance between two points on S is then defined as the *minimal length among the lengths of all great circle arcs having those two points as endpoints*. The number α mentioned previously is therefore the *maximum of all distances between pairs of points on S*.

Figure 45

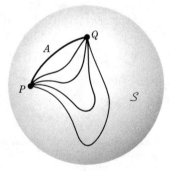

For points P and Q on S the case $PQ = \alpha$ is exceptional. This happens only when P and Q are the endpoints of some diameter (are **opposite poles**), and in this case infinitely many great circles pass through P and Q. If on the other hand $PQ < \alpha$, then P and Q are not poles and are therefore not collinear with O. In this case the plane through P, Q, and O is unique and determines a unique great circle passing through P and Q. If the term "line" is used in place of "great circle," the preceding analysis may be summarized as follows: *Each pair of points* P *and* Q *on* S *lie on at least one line, and if* PQ $< \alpha$ *that line is unique.*

16. The Undefined Terms

To attempt a definitive explanation of the meaning of "point" in geometry could lead to some frustration. Euclid himself made no elaborate attempt to define the term. His statement,

A point is that which has no part

is more a reminder than a definition. Indeed the statement is not very illuminating at all unless one has previously defined or knows precisely what is meant by "having no part." Similar troubles accompany the term "line."

Such attempts inevitably lead to a circuitous discourse where A is defined in terms of B and then B is defined in terms of A, with perhaps a few intermediate terms and definitions. There is clearly no point in pursuing such a discourse. To avoid this, we follow the modern axiomatic approach and take *point* and *line* as **undefined terms** and let the *axioms* set forth the needed properties. In so doing, however, it must be remembered that the axioms are the *only* source of information concerning these terms and all preconceived notions or conceptual images of them must be set aside. Thus, previous knowledge may actually work as a hindrance in such a study. To aid the reader as much as possible in this venture, he will be reminded of the occasional pitfalls.

We are now ready to proceed with the axioms.

17. The Axioms of Alignment and Distance

The first axiom affirms the *existence* of lines, which, in view of the second axiom, also implies the existence of points. The second axiom explicitly justifies the use of set theory.

Axiom 1: *There exist at least two lines.*

Axiom 2: *Each line is a set of points having at least two elements.*

The only conclusion which these first two axioms seem to imply is that there are at least three points and two lines. (Why?) A very simple example of a "geometry" in which these axioms are realized (called a **model**) is any three-element set $\{A, B, C\}$ where elements A, B, and C, are taken as "points" and the subsets $\{A, B\}$ and $\{A, C\}$ are taken as "lines." Fig. 46 shows the essential structure of this

Figure 46

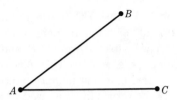

"geometry." Thus, the axioms at this point are not very interesting. However, they do indicate the fundamental structure of all models and how one must derive them. One starts with a set of points—the universal set—and forms various subsets thereof for the lines. Since the first few axioms are not very restrictive, this may be done in a variety of ways.

Presumably, we are studying "plane" geometry. The "plane" for us is merely *the set of all points being considered*. Since this terminology will be used frequently a formal definition is in order.

17.1 Definition: The **plane** is the set of all points, that is, the universal set.

In the interest of more colorful language, such terminology as "belongs to," "passes through," "lies on," and so on, will be freely used. Specifically, if a point is an element of a line, then we shall say that the point *belongs to*, or *lies on* that line, and that the line *passes through* the point. Two or more points are said to be **collinear** if they lie on the same line. It is to be noted that the axioms so far do not imply that two points are always collinear. For, in the model given above points B and C are *not* collinear.

The wealth of geometric ideas and their great usefulness is due in large part to the interplay of number and configuration. The next axiom serves to introduce the notion of distance into our logical system.

Axiom 3:   *To each pair of points* P *and* Q, *distinct or not, there corresponds a non-negative real number* PQ *which satisfies the following properties:*

(a) $PQ = 0$ *if and only if* $P = Q$, *and*
(b) $PQ = QP$.

17.2 Definition: The number corresponding to the pair (P, Q) as guaranteed by Axiom 3 is called the **distance from P to Q**, or the **distance between P and Q**.

A notable difference between Euclidean and spherical geometry is that in the latter, distances are *bounded* while in the former they are not. To take this into account we let α denote in either case the *least upper bound*[3] *of the set of all distances.* That is,

$$\alpha = \sup \{PQ: \quad P \text{ and } Q \text{ are points}\}.$$

The convention of using the symbol ∞ for the least upper bound of an *unbounded* set of non-negative reals will be followed. If the ordering $<$ as understood for the reals be extended to ∞ such that $\infty \leqq \infty$ and, for any real number a, $a \leqq \infty$, then

---

[3] In Appendix 2 will be found a brief review of this concept.

the set of distances always has a least upper bound α which formally satisfies the usual properties for least upper bounds. Hence, $\alpha = \infty$ symbolically expresses the idea that as P and Q range over all points, PQ assumes arbitrarily large values, while $\alpha < \infty$ announces the fact that PQ remains less than or equal to some fixed real number.

Now we are ready for a much more restrictive axiom, one which helps to formulate our idea of how two points "determine" a line in both Euclidean and spherical geometry.

Axiom 4:   *Each pair of distinct points* P *and* Q *lie on at least one line, and if* $PQ < \alpha$ *that line is unique.*

This axiom is clearly realized on the sphere, provided α represents the distance between poles. There, if $PQ < \alpha$, then P and Q cannot be poles (endpoints of the same diameter) and in that case there is only one great circle which passes through P and Q. But Axiom 4 is at the same time expressing the Euclidean concept of "two points determine a line," for in that case $\alpha = \infty$ and PQ is *always* less than α, which then means that the line containing P and Q is unique for *all* pairs of points P and Q. It now becomes clear, in view of the double meaning of this axiom, how the two geometries Euclidean and spherical can be handled simultaneously. One simply capitalizes on the freedom to choose α to be finite or infinite.

Two useful theorems may now be derived.

17.3   Theorem: If two distinct lines l and m both contain the points A and B then either $A = B$ or $AB = \alpha$.

   PROOF: Assume $A \neq B$. Then A and B are two distinct points, and if $AB < \alpha$, Axiom 4 implies $l = m$, a contradiction. Otherwise, $A = B$. ∎

Figure 47

Note the special interpretation of Theorem 17.3 when $\alpha = \infty$. In that case the theorem may be reworded slightly to state that *two distinct lines intersect in at most one point*. This is due to the fact that the case $AB = \alpha$ is impossible in the presence of the condition $\alpha = \infty$.

If $AB < \alpha$, Axiom 4 justifies the use of the familiar notation

$$\overleftrightarrow{AB}$$

for the *unique line determined by points* A *and* B.

A fundamental principle involving this symbolism, often taken for granted, is the following (the proof will be left as an exercise):

17.4   Theorem: If $C \in \overleftrightarrow{AB}$ and $D \in \overleftrightarrow{AB}$ with $0 < CD < \alpha$ then $\overleftrightarrow{CD} = \overleftrightarrow{AB}$.

Figure 48

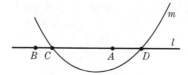

EXERCISES

1. Consider this model for Axioms 1–3: Let A, B, C, D, and E be five points in ordinary Euclidean geometry. Take these as "points" and all *triples* $\{A, B, C\}$, $\{A, B, D\}$, $\cdots$, $\{C, D, E\}$, as "lines," with ordinary Euclidean distance as "distance." (a) How many "lines" are there passing through each pair of "points"? (b) How many "lines" altogether? (c) Discuss the validity of Axioms 1–3, and show that Axiom 4 does not hold. How might one remedy the choice of "lines" to make Axiom 4 work?
2. Prove Theorem 17.4.
3. (a) Discuss the validity of Axioms 1–4 in the following model: Let the "points" be the chairs in a classroom (Fig. 49) and take as "lines" all possible *pairs* of "points." Define

$$PQ = 0 \quad \text{if} \quad P = Q$$
$$PQ = 1 \quad \text{if} \quad P \neq Q$$

as "distance." (b) If in (a) the "lines" are taken as all possible triples of points, how does this affect the validity of Axioms 1–4?

Figure 49

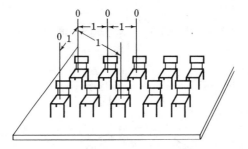

★4. Consider a sphere S of radius $2r$. A curious sort of geometry arises when one agrees to *identify* each point with its opposite pole. The "lines" are great circles as before, but for each pair of points (A, B) the "distance" from A to B is the smaller of the lengths of two minor great circle arcs, one from A to B and the other from A to the opposite pole of B. Show that $\alpha = \pi r$. Discuss the validity of Axioms 1–4 for this "geometry."

18. Coordinates for a Line

Not much can be achieved with the axioms thus far introduced. For example, suppose A and B are two points such that $AB < \alpha$. Then we know that line $\overset{\leftrightarrow}{AB}$ is uniquely determined, but our axioms do not necessarily imply that $\overset{\leftrightarrow}{AB}$ has infinitely many points as in Euclidean geometry. And more importantly, we cannot say that there exists a point C on $\overset{\leftrightarrow}{AB}$ which has the property

$$AC + CB = AB.$$

As a matter of fact, in the model of the chairs in a classroom (Exercise 3 above) we *never* have $AB + BC = AC$ for three distinct points. (Why?) In Euclidean geometry the midpoint of a line segment and its endpoints obviously satisfy this condition, so we want our system of axioms to reflect this property. Before going further let us introduce this important concept formally.

18.1 Definition: A point B is said to be **between** points A and C, and we write (ABC), whenever

(a) A, B, and C, are distinct, collinear points, and
(b) $AB + BC = AC$.

The following is an obvious implication of the definition just given:

18.2 Theorem: The relation of betweenness satisfies the property

$$(ABC) \text{ if and only if } (CBA).$$

A second fundamental property may be easily verified:

18.3 Theorem: If (ABC) then neither (BAC) nor (ACB).

PROOF: If (ABC) and (BAC), then

$$AB + BC = AC$$

and

$$BA + AC = BC.$$

Since $AB = BA$ this yields the equation

$$2AB + BC + AC = AC + BC$$

or

$$AB = 0.$$

Therefore, $A = B$ which denies one of the conditions in (ABC). A similar proof shows that (ABC) and (ACB) cannot exist simultaneously. ∎

The concept of betweenness, or *order*[4] as it is sometimes called, can be derived from a coordinate system for each line (as on a coordinate axis in analytic geometry, for example). But spherical geometry requires special attention. Using the earth's surface as a model, suppose an astronaut were to clock the mileage on his journey as he circles the earth in a perfect 100-mile-high orbit (Fig. 50). We know that the distance around the earth's surface is approximately 25,000 miles, which would be roughly the distance the astronaut would travel in one orbit. If he began his recording at Cape Kennedy, that point would be associated with the number 0 at first, then as each successive orbit is completed, with the numbers 25,000, 50,000, 75,000, and so on. Let us assume that his orbital path is eastward. When he passes over Cairo, Egypt, that point would be associated first with the number 8000 (roughly

---

[4] It is evident that betweenness is related to the order in which points occur on a line. For example, if we have the ordering A-B-C, then presumably $AB + BC = AC$, but if instead A-C-B, then $AC + CB = AB$.

the great circle distance from Cape Kennedy to Cairo) then on successive passes with the numbers 33,000, 58,000, 83,000, and so on. When he passes over Tokyo, since this is approximately 10,000 miles due *west* of Cape Kennedy, he would first assign it the number 15,000 (Why?), then on the next few passes Tokyo would be assigned the numbers 40,000, 65,000, and so on. In like manner, the astronaut could "assign" an indefinite number of "coordinates" to any point on his orbital path. This seems to suggest that the only natural coordinate system for "lines" on a sphere must be a *many-to-one* mapping of real numbers into points.

Figure 50

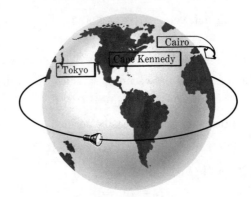

On the other hand, the coordinate system of a line in analytic geometry is a *one-to-one* mapping of real numbers into points, as it should be for any "bonafide" coordinate system. At this time we define precisely what is to be meant by the term "coordinate system."

18.4 Definition: A **coordinate system** for a line l is a one-to-one mapping f from the points of l to some set of reals, called the **coordinate set.** If the real number a belongs to the coordinate set and $f(A) = a$ (for some $A \in l$), then a is called the **coordinate** of point A and the relationship will be written simply $A[a]$. The point whose coordinate is zero, provided zero belongs to the coordinate set, is called the **origin** of the coordinate system.

Consider any line in the Euclidean plane, coordinatized as in analytic geometry (Fig. 51). We observe that all points to the "right" of the origin have positive

Figure 51

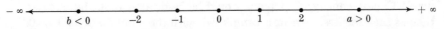

coordinates ranging from 0 to $+\infty$ (excluding $+\infty$), while those to the left have negative coordinates ranging from 0 to $-\infty$ (excluding $-\infty$). Suppose we formally substitute α for ∞ and make the given line into a circle (Fig. 52); naturally, the "point" corresponding to $\pm\alpha$ is missing. Does this suggest how one might coor-

Figure 52

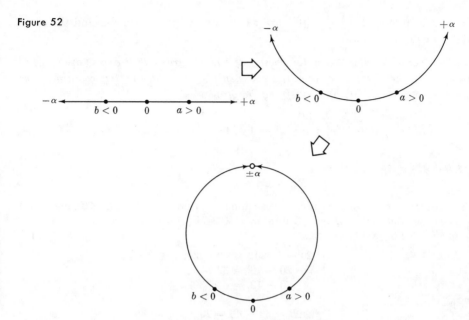

dinatize a great circle on a sphere whose maximum distance is α? In that situation (Fig. 53) the "missing" point is present, so it must be decided which of the two coordinates α or $-\alpha$ that point is to have. Suppose we take $+\alpha$. With this scheme the astronaut could therefore have coordinatized his orbital path in the manner depicted in Fig. 54, where the value of α is 12,500.

Figure 53

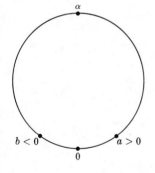

Figure 54

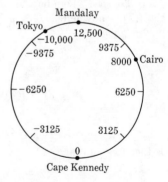

The relationship between distance and coordinates on a line in analytic geometry is clear: if $f(A) = a$ and $f(B) = b$, then $AB = |a - b|$. But on a great circle this does not work. Observe the distance from Cairo to Tokyo, for example, in Fig. 54. Since $f(\text{Cairo}) = 8000$ and $f(\text{Tokyo}) = -10,000$, the difference between coordinates is 18,000. But the great circle distance from Cairo to Tokyo is $(12,500 - 10,000) + (12,500 - 8000) = 2500 + 4500 = 7000$. Observe that in this case

$7000 = 25{,}000 - |f(\text{Tokyo}) - f(\text{Cairo})|$. It is easy to generalize this and thus formulate the desired axiom.

Axiom 5 (Coordinatization Axiom for Lines): *If* l *is any line and* P *and* Q *are any two points on* l, *there exists a coordinate system* f *for* l *having the coordinate set* $\{x: -\alpha < x \leq \alpha,\ x\ real\}$ *and satisfying the properties*

(a) $f(P) = 0$ *and* $f(Q) > 0$, *and*

(b) *for any two points* $R[r]$ *and* $S[s]$ *on* l, *if* $|r - s| \leq \alpha$, *then*

$$RS = |r - s|,$$

and if $|r - s| > \alpha$, *then*

$$RS = 2\alpha - |r - s|.$$

18.5 Example: If for some coordinate system on line l points $A(20)$, $B(-20)$, and $C(12)$ are determined find AB, BC, and AC if (a) $\alpha = 30$, and (b) $\alpha = \infty$.

Solution: For (a) observe that

$$|20 - (-20)| = 40 > 30,$$
$$|-20 - 12| = 32 > 30,$$
$$|20 - 12| = 8 < 30.$$

Hence,

$$AB = 60 - 40 = 20,$$
$$BC = 60 - 30 = 30,$$
$$AC = 8.$$

For (b) all three absolute values are less than α, so

$$AB = 40,$$
$$BC = 32,$$
$$AC = 8. \quad \blacksquare$$

18.6 Example: Let f be a coordinate system for a line l and suppose $f(X) = -a, f(Y) = \frac{1}{2}a$, and $f(Z) = a$, where $3\alpha/4 < a < \alpha$. Prove (XZY).

Solution:

$$|f(X) - f(Y)| = \frac{3a}{2} > \frac{9\alpha}{8} > \alpha. \qquad \therefore\ XY = 2\alpha - \frac{3a}{2}.$$

$$|f(Y) - f(Z)| = \frac{a}{2} < \frac{\alpha}{2} < \alpha. \qquad \therefore\ YZ = \frac{a}{2}.$$

$$|f(X) - f(Z)| = 2a > \frac{3\alpha}{2} > \alpha. \qquad \therefore\ XZ = 2\alpha - 2a.$$

It then follows that

$$XZ + ZY = (2\alpha - 2a) + \frac{a}{2} = 2\alpha - \frac{3a}{2} = XY.$$

Since X, Y, and Z are also distinct and collinear then (XZY). $\quad \blacksquare$

An important special case of Axiom 5 arises when $\alpha = \infty$:

Axiom 5′ (Coordinatization Axiom When $\alpha = \infty$): *If* l *is any line and* P *and* Q *are any two points on* l, *there exists a one-to-one correspondence between the points of* l *and the real number system such that the number corresponding to* P *is zero and the* *number corresponding to* Q *is positive, and such that for any two points* R[r] *and* S[s] *on* l,

$$RS = |r - s|.$$

EXERCISES

1. Suppose f is a coordinate system on some line l and for $A \in l, B \in l$, we have $f(A) = 10$, $f(B) = -9$. Find AB in each of the following cases: (a) $\alpha = \infty$; (b) $\alpha = 20$; (c) $\alpha = 15$; (d) $\alpha = 10$. Work this problem in two different ways: first, in the cases of finite α deduce the distances from an appropriately chosen circle; second, check your answers by reworking the problem using Axiom 5.

2. Suppose for some coordinate system on line l the points $A[-7]$, $B[20]$, $C[-10]$, $D[-19]$, $E[13]$, and $F[14]$ are determined. If $\alpha = 20$ find the distances from A to B, C, D, E, and F in the two different ways suggested in Exercise 1.

3. In analytic geometry suppose the coordinates of the points A, B, and C on the x axis are, respectively a, b, and c. How can you tell whether (ABC) holds? Does this rule work for the coordinate system on a line in spherical geometry?

4. Let A, B, and C be points on a line l having coordinates 5, -4, and 12, respectively. Determine which betweenness relation (if any) holds in each of the cases $\alpha = \infty$, $\alpha = 20$, $\alpha = 16$, and $\alpha = 15$.

5. If the coordinates of A, B, and C are -5, 5, and 9, respectively, and $\alpha = 10$, show that (ACB) holds. How does this relation change as you increase the value of α? In particular, observe the values $\alpha = 12, 14, 20$, and ∞.

6. Attempt stating and proving a theorem of your own which relates the betweenness relation to a coordinate system on some line.

7. Let f be a coordinate system on line l and $\alpha = 20$. Find the coordinate(s) of the point(s) on l whose distance from A such that $f(A) = 11$ is (a) 20, (b) 19, (c) 10, (d) 3, and (e) 1.

8. If $\alpha = \infty$ we know that on any line we can always obtain one coordinate system from another by a *translation*, a *reflection* in the origin, or combination of the two. For example, if under one coordinate system (x) the coordinate of P is 0 and that of Q is 5, then the coordinate system (x') which assigns 0 to Q and 5 to P, is related to the system (x) by the equation $x' = -x + 5$. Is a similar statement true for the case $\alpha < \infty$? Discuss it.

9. Consider "mod 2α" arithmetic, in connection with Exercise 8. We say that $x \equiv y$ (mod 2α) if and only if $x - y = 2k\alpha$, for some integer k. We say a number x *has been reduced to* y *modulo* 2α if y lies in the range $-\alpha < y \leq \alpha$ and $x \equiv y$ (mod 2α). For any two numbers a and b in the range given, define addition modulo 2α by $a \oplus b = a + b$, *reduced mod* 2α. What properties does this addition have? What bearing does this have on Exercise 8? What is $a \oplus b$ when $\alpha = \infty$?

10. In connection with the consistency of Axioms 1–4, show that with "points" as the ordinary points of a Euclidean plane, with "distance" the usual Euclidean distance, but with "lines" as the family of vertical (Euclidean) lines and the class of cubic curves of the form $y = (ax + b)^3$, one obtains a model for Axioms 1–4. How many "lines" pass through the points $(0, 0)$ and $(2, 8)$? Find the values for a and b such that $y = (ax + b)^3$ passes through $(0, 27)$ and $(-1, 1)$. How many values for a and b are possible? Does (ABC) hold for collinear points A, B, C?

11. Again in connection with Axioms 1–4, consider the distance set $\{0, 1, 2, 3, 5\}$. It is clear that one can make up an abstract set of "points" and "lines" such that Axioms 1–4

are satisfied with the given set as the distance set. What is the smallest number of "points" necessary? Can these points be placed in the plane so that the distances assigned to them will be Euclidean?

19. Betweenness Properties for Collinear Points

The properties of betweenness which are needed can be derived from the coordinatization axiom. Some of these are familiar properties known in Euclidean geometry, such as for example:

If B *is between* A *and* C, *and* C *is between* A *and* D *then* B *is between* A *and* D. (See Theorem 19.5 below.)

But many familiar properties must either be altered or discarded entirely; the reader is warned not to make hasty conclusions.

The first theorem guarantees the *existence* of a point between each given pair of points.

19.1 Theorem: If relative to some coordinate system for a line the points $A[a]$, $B[b]$, and $C[c]$ are determined such that either $a < b < c$ or $c < b < a$, and $|a - c| \leq \alpha$, then (ABC).

PROOF: It suffices to prove the assertion for the case $a < b < c$ only. Since $0 < b - a < c - a \leq \alpha$, then $|a - b| < \alpha$. By Axiom 5,

$$AB = b - a.$$

Also, $(c - b) + (b - a) = c - a \leq \alpha$, and since $c - b > 0$ and $b - a > 0$ it follows that $c - b < \alpha$. Therefore,

$$BC = c - b.$$

Finally, $0 < c - a \leq \alpha$ implies

$$AC = c - a,$$

and therefore,

$$AB + BC = AC. \quad \blacksquare$$

The next theorem is a partial converse of the previous one. The example illustrated in Fig. 55 shows that the converse does not hold without some qualifi-

Figure 55

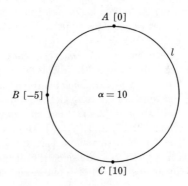

A [0]

B [−5]

$\alpha = 10$

l

C [10]

cations. For, A, B, and C are three points on a line having coordinates $a = 0$, $b = -5$, $c = 10$, where $|a - c| = 10 \leq \alpha$. Although (ABC), neither $a < b < c$ nor $c < b < a$ holds.

19.2 Theorem: If relative to some coordinate system for a line the points $A[a]$, $B[b]$, $C[c]$ are determined, with $AC < \alpha$ and with one of those points as origin, then (ABC) implies either $a < b < c$ or $c < b < a$.

PROOF: Observe that in any case, $a \neq b$, $b \neq c$, and $a \neq c$. There are then six possible orders for the real numbers a, b, and c, two of which are the desired inequalities. Let each of the other four possibilities be considered. First suppose either $b < a < c$ or $c < a < b$. If $|b - c| \leq \alpha$, then Theorem 19.1 implies (BAC), which is impossible with (ABC). Hence $|b - c| > \alpha$, and neither b nor c can vanish. Therefore $a = 0$, and it follows that

$$AB = |b|, \qquad BC = 2\alpha - |b - c|, \qquad \text{and } AC = |c|.$$

Consider the case $b < 0 < c$. Then

$$AB = -b, \qquad BC = 2\alpha - (c - b), \qquad \text{and } AC = c.$$

(ABC) implies

$$-b + 2\alpha - c + b = c,$$

or

$$c = \alpha.$$

But this is a contradiction of the hypothesis $(AC < \alpha)$. The case $c < 0 < b$ is analogous; it leads to the false conclusion $c = -\alpha$ (since no coordinate equals $-\alpha$).

Finally, suppose either $a < c < b$ or $b < c < a$. As before, $|a - b| > \alpha$ must follow and hence $c = 0$. If $a < 0 < b$, then $AB = 2\alpha - (b - a)$, $BC = b$, and $AC = -a$. Thus

$$2\alpha - b + a + b = -a,$$

or

$$a = -\alpha,$$

is an impossibility. The other case implies a similar contradiction, so the only possibilities left are $a < b < c$ or $c < b < a$. ∎

NOTE: The example illustrated in Fig. 55 shows that Theorem 19.2 is false without the restriction $AC < \alpha$.

Given three points on a line, it cannot always be asserted that one of them lies between the other two as it can in Euclidean geometry. Recall Example 18.5, where A, B, and C are distinct points on l but neither (BAC), (ABC), nor (ACB). The following theorem is the best that can be done in this respect. The proof will be left as an exercise.

19.3 Theorem: If A, B, and C are three distinct collinear points such that the sum of at least one pair of the numbers AB, BC, or AC does not exceed α, then either (BAC), (ABC), or (ACB), the cases being mutually exclusive.

19.4 *Observation:* If $\alpha = \infty$, the preceding theorem becomes the familiar Euclidean proposition: *Among three distinct collinear points, one point always lies between the other two.*

The next two theorems involve the ordering of *four* points on a line. When all four betweenness relations

$$(ABC), \qquad (ABD), \qquad (ACD), \qquad (BCD)$$

occur, it is convenient to write simply

$$(ABCD).$$

19.5 *Theorem:* If (ABC) and (ADB), then $(ADBC)$.

PROOF: Since $AB < AB + BC = AC \leqq \alpha$, the line $\overleftrightarrow{AB}$ is uniquely determined. Then $C \in \overleftrightarrow{AB}$ and $D \in \overleftrightarrow{AB}$ and one may take a coordinate system on $\overleftrightarrow{AB}$ such that a, b, c, and d are the coordinates of A, B, C, and D with $a = 0$ and $b > 0$. Since $AB < \alpha$ and (ADB), Theorem 19.2 implies $0 = a < d < b$. Also, either $c = \alpha$ (if $AC = \alpha$) or Theorem 19.2 applies to the relation (ABC). Thus in either case, $0 = a < b < c$. Therefore,

$$0 = a < d < b < c \leqq \alpha.$$

Theorem 19.1 then implies both (ADC) and (DBC).

This section is concluded with the following useful theorem:

19.6 *Theorem:* If A, B, C, and D are four distinct collinear points, and if (ABC), then either (DAB), (ADB), (BDC), or (BCD), the cases being mutually exclusive.

PROOF: Choose coordinates on $\overleftrightarrow{AB}$ such that a, b, c, and d are the coordinates of A, B, C, and D and $b = 0$, $c > 0$. Then it follows that $a < 0 < c$, for if either $0 < a < c \leqq \alpha$ or $0 < c < a \leqq \alpha$, then either (BAC) or (BCA)—impossible. Therefore, either $d < a < 0$, $a < d < 0$, $0 < d < c$, or $0 < c < d$. By Theorem 19.1 then, either (DAB), (ADB), (BDC), or (BCD). ∎

EXERCISES[5]

1. Prove that if a and b belong to a coordinate set for some line and have the same sign (either both are non-negative or both are nonpositive), then

$$|a - b| \leqq \alpha.$$

   Hence if A and B are the points whose coordinates are a and b,

$$AB = |a - b|.$$

   NOTE: This result is often useful in the application of Axiom 5. It was used in some of the proofs of the preceding section.

2. Show that Theorem 19.1 is not valid without the restriction $|a - c| \leqq \alpha$ by finding an example of your own which has $a < b < c$ but not (ABC).

---

[5] The reader may find a review of the basic rules for inequalities and absolute value helpful. The properties used here may be found in Appendix 1.

3. Prove Theorem 19.3. *Hint:* If $AB + BC \leq \alpha$, then take B as origin and use the inequality $|a - c| \leq |a| + |c|$.

4. Let "point" be a point on the sphere (Fig. 56) and "line" be any great circle. Define the "distance" between each pair of points to be the Euclidean length of the *chord* joining those points. (a) Find α and determine which of the axioms thus far introduced are satisfied for this "geometry." (b) What can be said about the betweenness concept here?

Figure 56

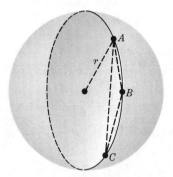

5. If A, B, and C are distinct collinear points with $AC = \alpha$, prove that (ABC). Later this will be strengthened to the assertion: *If* AC = α *and* B *is any point distinct from* A *and* C, *then* (ABC). What does one need to know in order to justify this statement?

6. Is this proposition valid: *If* (ABC) *and* (BCD), *then* (ABCD)?

7. Using the sphere as model, find whether it is reasonable to expect the result: *If* A *and* B *are any two points, then there is* C *such that* (ABC).

8. Show that if l and m are any two lines and $l \subset m$, then $l = m$.

★9. Show that if the points $A[a]$, $B[b]$, and $C[c]$ are determined on some line, with $a \leq -\alpha/2$ and $b \geq \alpha/2$, then $a < b < c$ implies (ACB).

20. Segments and Rays: The Segment-Construction Theorem

From the betweenness concept emerges the important geometric objects *segment* and *ray*.

20.1 Definition: If A and B are two distinct points and $AB < \alpha$, the **segment** joining A and B, denoted $\overline{AB}$, is the set consisting of A, B, and all points between A and B. The **ray** from A passing through B, denoted $\overrightarrow{AB}$, is the set consisting of A, B, all points between A and B, and all points X, such that B is between A and X.

Thus the segment $\overline{AB}$ and ray $\overrightarrow{AB}$ are defined by the following expressions:
$$\overline{AB} = \{X: \ X = A, X = B, \text{ or } (AXB)\}$$
$$\overrightarrow{AB} = \{X: \ X = A, X = B, (AXB), \text{ or } (ABX)\}.$$

Points A and B are called the **endpoints** of segment $\overline{AB}$ and all other points of the segment are called **interior points**. The number AB is called the **length** of segment $\overline{AB}$.

For ray $\overrightarrow{AB}$, A is called the **origin,** and any point X on $\overrightarrow{AB}$ such that

$$0 < AX < \alpha$$

is called an **interior** point.

These ideas are quite familiar in Euclidean geometry, but for the sphere they may not be so obvious. Consider the example shown in Fig. 57(a). It is clear that (AXB) holds if and only if X is some interior point on the minor arc joining A and B. Hence, the segment $\overline{AB}$ is precisely the minor arc ACB. In Fig. 57(b) consider point X such that (ABX). X may evidently assume any position on the arc shown, excluding the points of $\overline{AB}$. With some thought one becomes convinced that the ray $\overrightarrow{AB}$ consists of those points which lie on the semicircle ABA'—*including* A'(!). While this example is still fresh in mind, the reader should consider these two questions: What points make up $\overrightarrow{BA}$? Does $\overrightarrow{AB} \cup \overrightarrow{BA} = \overleftrightarrow{AB}$, as it does in Euclidean geometry?

Figure 57

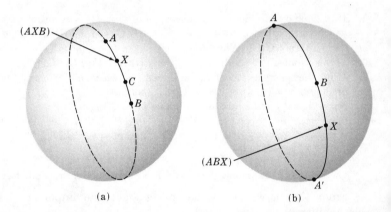

(a)   (b)

It does not require a great deal of imagination to conclude that the next theorem will be for us an indispensable tool.

20.2 Segment-Construction Theorem: If $\overline{AB}$ and $\overline{XY}$ are any two segments and $AB \neq XY$, there is a point C on $\overleftrightarrow{AB}$ such that $AC = XY$, and (ACB) if $XY < AB$, or (ABC) if $XY > AB$ (Fig. 58).

Figure 58

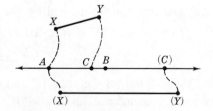

PROOF: Choose coordinates on $\overleftrightarrow{AB}$ such that A and B correspond to 0 and $b > 0$. Set $c = XY$. By hypothesis, $0 < c < \alpha$, so there exists $C \in \overleftrightarrow{AB}$ having coordinate C. Since $|0 - c| = c < \alpha$, it follows that $AC = c = XY$. If $XY < AB$, then $0 < c < b$ and by Theorem 19.1 (ACB). Similarly, if $XY > AB$ then $0 < b < c$ and (ABC). ∎

The preceding theorem leads immediately to two useful corollaries. They involve a concept which will first be defined formally.

20.3 Definition: If a point M lies on a segment $\overline{AB}$ such that (AMB) and $AM = MB$ it is called the **midpoint** of that segment, and it or any line passing through it is said to **bisect** the segment.

20.4 Midpoint-Construction Theorem: The midpoint of any line segment exists and is unique.

PROOF: If $\overline{AB}$ is the given segment, then since there obviously exists a segment $\overline{XY}$ of length $\frac{1}{2}AB$, by Theorem 20.2 there is M on $\overleftrightarrow{AB}$ such that $AM = XY = \frac{1}{2}AB$, and since $XY < AB$ implies (AMB), then $MB = AB - AM = \frac{1}{2}AB = AM$. To prove uniqueness, suppose M' is a second midpoint. Let A, M, M', and B take on the coordinates 0, m, m', and $b > 0$. Since $AB < \alpha$ and $(AM'B)$, Theorem 19.2 implies $0 < m < b$ and $0 < m' < b$. Hence

$$m' = AM' = \tfrac{1}{2}AB = AM = m$$

and $M = M'$. ∎

The next theorem is a formulation of the concept of "extending a segment its own length."

20.5 Segment-Doubling Theorem: If $\overline{AB}$ is any segment of length less than $\alpha/2$ there is an interior point C on ray $\overrightarrow{AB}$ such that B is the midpoint of $\overline{AC}$.

PROOF: Since $2AB < \alpha$ there is a segment XY whose length is $2AB$. By Theorem 20.2 there is C on $\overleftrightarrow{AB}$ such that $AC = XY$, and since $XY > AB$, (ABC). Thus,

$$BC = AC - AB = XY - AB = 2AB - AB = AB.$$

Therefore, B is the midpoint of $\overline{AC}$. Also, $C \in \overrightarrow{AB}$ and $AC < \alpha$ so C is an interior point on ray $\overrightarrow{AB}$. ∎

Recall that in analytic geometry the positive x axis is a ray with origin at zero. That is, the set of all points on the x axis having non-negative coordinates is a ray. This result may be proved here for lines in general.

20.6 Theorem: If h is any ray, there is a coordinate system on the line which contains h such that h consists of all those points having non-negative coordinates.

More specifically, if $h = \overrightarrow{AB}$ and f is the coordinate system for $\overleftrightarrow{AB}$ such that $f(A) = 0$ and $f(B) > 0$, then

$$h = \{X: \quad f(X) \geqq 0\}.$$

PROOF: This is a *set* equation which may be proved by showing (1) all points belonging to h also belong to $\{X: \quad f(X) \geqq 0\}$, and (2) all points belonging to $\{X: \quad f(X) \geqq 0\}$ also belong to h. To prove (1) let $X \in h$. Then either $X = A$, $X = B$, (AXB), or (ABX). In the first two cases, from the definition of f stated in the hypothesis, $f(X) \geqq 0$. Consider (AXB). Let $f(B) = b$ and $f(X) = X$. Then by Theorem 19.2, $0 < b < x$. That is, $f(X) > 0$. If (ABX) and $AX < \alpha$, then again by Theorem 19.2, $f(X) > 0$. Finally, $AX = \alpha$ if and only if $f(X) = \alpha > 0$. To prove (2) let X be a point on $\overleftrightarrow{AB}$ such that $f(X) \geqq 0$. Again letting $f(X) = x$ and $f(B) = b$, then either $x = 0$, $x = b$, $0 < x < b$, or $0 < b < x$, implying, in turn, either $X = A$, $X = B$ (AXB), or (ABX), the latter two by Theorem 19.1. Therefore, $X \in \overrightarrow{AB} = h$. ∎

20.7 Corollary: If C is an interior point of ray $\overrightarrow{AB}$, then $\overrightarrow{AC} = \overrightarrow{AB}$.

PROOF: Let f be a coordinate system for $\overleftrightarrow{AB}$ such that $f(A) = 0$ and $f(B) > 0$. By Theorem 20.6,

$$\overrightarrow{AB} = \{X: \quad f(X) \geqq 0\}.$$

Since $C \in \overrightarrow{AB}$, $f(C) \geqq 0$. By Theorem 17.4, $\overleftrightarrow{AC} = \overleftrightarrow{AB}$, so f is a coordinate system for $\overleftrightarrow{AC}$ such that $f(A) = 0$ and $f(C) > 0$ ($f(C) \neq 0$ or else $AC = AA = 0$). Therefore Theorem 20.6 implies

$$\overrightarrow{AC} = \{X: \quad f(X) \geqq 0\}.$$

Hence, $\overrightarrow{AB}$ and $\overrightarrow{AC}$ represent the same set and therefore, $\overrightarrow{AB} = \overrightarrow{AC}$. ∎

This section will conclude with a concept which will be needed later for the definition of straight angles. In Euclidean geometry it is clear that two distinct rays from the same origin form a line only when they extend in opposite directions, as illustrated in Fig. 59. This suggests the following definition:

20.8 Definition: Two rays h and k are said to be **opposite** or to **oppose** one another if they have the same origin and their union is a line.

Figure 59

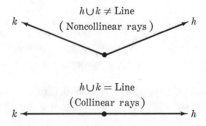

$h \cup k \neq$ Line
(Noncollinear rays)

$h \cup k =$ Line
(Collinear rays)

The question immediately arises whether there exist opposite rays in the axiomatic development. The next theorem provides the answer.

20.9 Theorem: If A, B, and C be any three points such that (ABC), then

$$\overrightarrow{BA} \cup \overrightarrow{BC} = \overleftrightarrow{AB},$$

and the rays $\overrightarrow{BA}$ and $\overrightarrow{BC}$ are accordingly opposite rays (Fig. 60).

Figure 60

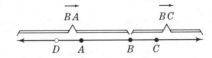

PROOF. This is immediate from Theorem 19.6 (the details are left to the reader). ∎

Theorem 20.9 guarantees that *to each ray* h *there corresponds a unique opposite ray* h′. Let $h = \overrightarrow{AB}$. Since $AB < \alpha$ there is a segment $\overline{XY}$ such that

$$AB < XY < \alpha,$$

and thus by the segment-construction theorem there is C on $\overleftrightarrow{AB}$ such that (BAC).

Figure 61

Set $h' = \overrightarrow{AC}$. Then by Theorem 20.9 $h \cup h'$ is a line and hence h' opposes h. To show uniqueness, suppose h'' is also opposite h. Then if l is the unique line through A and B,

$$l = h \cup h''.$$

Figure 62

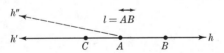

Now $C \in l$ but (BAC) implies $C \notin h$. (Why?) Therefore, $C \in h''$, and since $0 < AC < \alpha$, C is an interior point of h''. By Corollary 20.7, $h'' = \overrightarrow{AC} = h'$. ∎

EXERCISES

1. On a sphere of radius r find the least upper bound for the (shortest) distances measurable on that sphere. If A and B are two points of the sphere such that $AB = \pi r$, determine the set of all points X such that (AXB). Does this point out the desirability of the restriction $AB < \alpha$ in the definition of "segment" given above? (Why?)
2. Prove that if A and B are any two points, there exists C such that (ACB).
3. Can a line segment be trisected (divided into three equal parts)? How can this be

accomplished axiomatically? Define the concept rigorously and quote precisely the axioms and theorems you must use.

4. Prove that if $AB < \alpha$, $\overline{AB} \subset \overrightarrow{AB} \subset \overleftrightarrow{AB}$.

5. Show that $\overrightarrow{AB} \cap \overrightarrow{BA} = \overline{AB}$.

6. On a sphere we never have $\overrightarrow{AB} \cup \overrightarrow{BA} = \overleftrightarrow{AB}$—there will always be points on $\overleftrightarrow{AB}$ not on $\overrightarrow{AB} \cup \overrightarrow{BA}$. Identify those points.

7. Show that $\overrightarrow{AB} \cup \overrightarrow{BA} = \overleftrightarrow{AB}$ if $\alpha = \infty$.

8. Show that if A, B, and C are any three points on a ray, then one of the conditions (ABC), (ACB), or (BAC) holds.

9. Prove Theorem 20.9.

3

Angles, Angle Measure, and Plane Separation

The list of axioms will now be enlarged to include the familiar ideas concerning angles, as well as the apparently trivial fact that each line has two "sides."

21. Angles: A Duality Between Segments and Angles

It is interesting—and useful—that a family of rays which have a common origin is dual to a family of collinear points, and that this therefore leads to a duality between angles and segments and their respective measures. Consequently, the betweenness properties established previously all hold for rays from a common origin, and we do not have to relearn those concepts in the case of rays and angles. First, the particular concept of "angle" to be used must be agreed upon (there are several). We shall adopt the following:

21.1 Definition: An **angle** is the set of points belonging to two rays, distinct or not, having a common origin. If the rays coincide, the angle is said to be **degenerate;** if the rays are opposite, the angle is said to be **straight.** When we use the term "angle" without the qualifying adjectives "degenerate" or "straight," we shall mean a nondegenerate, nonstraight angle. Other familiar terminology associated with angles will be used: The rays forming the angle are called its **sides** while the common origin of those rays is called its **vertex.** In connection with this, if two or more lines meet in the same point they are said to be **concurrent,** and if two or more rays have the same origin, they are called **concurrent rays.**

Several acceptable notations for an angle are possible, unlike the situation for a segment where the two endpoints characterize the set. If $h = \overrightarrow{AB}$ and $k = \overrightarrow{AC}$ (Fig. 63), the angle thus formed is the set

$$h \cup k.$$

Figure 63

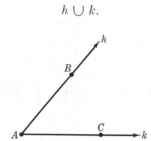

It would therefore be logical to use the notation $\overline{hk}$ (resembling segment notation). Then when angle measure is introduced hk could be used for the *measure* of the angle, consistent with the precedent already set for segments. Alternatively, the same angle can be denoted $\angle BAC$, and its measure, $\angle BAC$. If no ambiguity is involved, one can use simply $\angle A$ for the angle and $\angle A$ for its measure.

Roughly speaking, an angle is the "dual" of a segment. This was partly evident in the notation $\overline{hk}$ for angle just introduced. To be more precise, we shall say that a family of concurrent rays (called a **pencil** of rays) is **dual** to a set of collinear points if there is a one-to-one correspondence between the rays and the points. This leads to the

21.2 *Principle of Duality*: To dualize any previous statement regarding points, collinearity, and distance, the terms *point, collinear, line,* and *distance* are replaced by the respective terms *ray, concurrent, pencil,* and *angle measure.* (The resulting statement is then sometimes paraphrased to make it less awkward and more concise.)

NOTE: Since by Axiom 4 two points are *always* collinear, the dual of *two points* is a pair of concurrent rays. One may then associate an *angle*[1] with the *dual of two points*.

To illustrate this principle, let the concept of betweenness for points be dualized to yield the corresponding concept of betweenness of rays (in doing so we anticipate the concept of angle measure, but that will be introduced shortly):

21.3 *Definition*: A ray u is said to lie **between** rays h and k, and we write (huk), whenever

(a) h, u, and k are distinct, concurrent rays, and
(b) $hu + uk = hk$.

An alternate form of this definition is the following:

---

[1] The reader should note the perhaps subtle distinction between *two rays* and *angle* (the union of two rays).

21.3′ Definition: A ray $\overrightarrow{OB}$ is said to lie between rays $\overrightarrow{OA}$ and $\overrightarrow{OC}$, and we write $(\overrightarrow{OA}\ \overrightarrow{OB}\ \overrightarrow{OC})$, whenever

(a) the rays are distinct, and

(b) $\angle AOB + \angle BOC = \angle AOC$.

(See Fig. 64.)

Figure 64

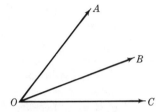

The reader should now be able to do his own "dualizing" to see what he comes up with. It will not be necessary to dualize all we have considered up to this point, but it is entertaining to experiment with the idea by dualizing a concept chosen at random. (See Fig. 65 for a few simple examples; note that if the figure to be dualized does not consist of collinear points its dual is not defined.)

Figure 65

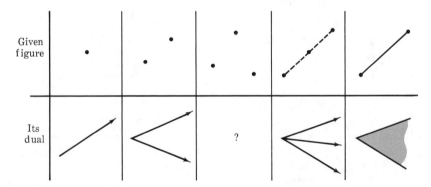

22. Betweenness Properties for Rays: Properties of Angle Measure

It is merely a formality to dualize the axioms and concepts of distance in the preceding sections to obtain the desired (Euclidean) concepts of angle measure. It naturally follows that, having dualized the necessary axioms and definitions, the duals of all *theorems* are valid, and if there be any room for doubt one has only to dualize the *proof* to obtain a proof for the dualized theorem. For completeness, we now set down in detail the various axioms, definitions, and theorems which result by dualizing previous axioms, definitions, and theorems.

Axiom 6 (Dual of Axiom 3): *To each angle* $\overline{pq}$ *degenerate or not, there corresponds a non-negative real number* pq, *satisfying the properties*

(a)   pq $= 0$ *if and only if* p $=$ q, *and*

(b)   pq $=$ qp.

(The number pq mentioned here will be called the **measure** of the angle $\overline{pq}$.)

Analogous to the discussion of distance, define β to be the *least upper bound of the set of measures of all angles*. One could agree to leave open the possibility $\beta = \infty$ (as has been done with respect to α), but this would unnecessarily complicate matters.[2] Thus we take as our next axiom:

Axiom 7: β *is the measure of any straight angle.*

There are several logical choices for the value of β. $\beta = \frac{1}{2}$ would correspond to the use of *revolutions* to measure angles, $\beta = \pi$ would correspond to the use of *radians*, while $\beta = 180$ corresponds to degree measure. We shall keep with traditional developments of geometry by choosing $\beta = 180$.

With this value of β replacing α, the coordinatization axiom for lines can then be dualized as follows:

Axiom 8 (Coordinatization Axiom for Rays—Dual of Axiom 5): *If O is the common origin of a pencil of rays and* p *and* q *are any two rays in that pencil, there exists a coordinate system* g *for pencil O whose coordinate set is the set* $\{x: -180 < x \leq 180,$ x *real*$\}$ *and satisfying the properties*

(a) $g(p) = 0$ *and* $g(q) > 0$, *and*

(b)   *for any two rays* r *and* s *in that pencil, if* $g(r) = x$ *and* $g(s) = y$, *then*

$$rs = |x - y|$$

in the case $|x - y| \leq 180$, *and*

$$rs = 360 - |x - y|$$

in the case $|x - y| > 180$.

22.1   Example: Observe Fig. 66 which shows five rays with their respective coordinates. Find the measures of $\angle BOE$ and $\angle COE$ in two different ways: (a) By appealing to Fig. 66 and using the familiar ideas of angles, and (b) by the direct application of Axiom 8.

    Solution: By carefully inspecting Fig. 66 one observes that $\angle BOE = 121$ and $\angle COE = 179$. By the coordinatization axiom on the other hand,

$$|30 - (-91)| = 121 < 180, \quad \text{so } \angle BOE = 121,$$

and

$$|90 - (-91)| = 181 > 180, \quad \text{so } \angle COE = 360 - 181 = 179. \blacksquare$$

22.2   Dual of Theorem 19.1: If h, k, and u are three concurrent rays whose coordinates are, respectively, x, y, and z, with either $x < y < z$ or $z < y < x$ and $|x - z| \leq 180$, then (hku).

---

[2] A peculiar sort of geometry results if one assumes $\beta = \infty$. The ambitious reader might enjoy investigating this geometry and discovering some of its properties.

Figure 66

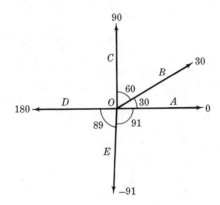

22.3   Dual of Theorem 19.2:   If h, k, and u are three concurrent rays whose coordinates are, respectively, x, y, and z, with one of x, y, and z being zero and $hu < 180$, then (hku) implies either $x < y < z$ or $z < y < x$.

The circumstance in which all four relations (hku), (hkv), (huv), and (kuv) hold for the rays h, k, u, and v is denoted

$$(hkuv).$$

22.4   Dual of Theorems 19.5 and 19.6:   The betweenness relation for rays obeys the following two fundamental properties (Fig. 67):

(a)   If (hku) and (huv), then $(hkuv)$.
(b)   If (hku) and v is any other ray concurrent with h, k, and u, then either (vhk), (hvk), (kvu), or (kuv), the cases being mutually exclusive.

Figure 67

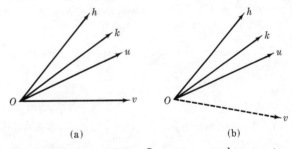

(a)   (b)

22.5   Angle-Construction Theorem—Dual of Theorem 20.2:   Let two nondegenerate, nonstraight angles $\angle ABC$ and $\angle XYZ$ be given, $\angle ABC \neq \angle XYZ$. Then there exists a ray $\overrightarrow{BD}$ such that $\angle ABD = \angle XYZ$, with $(\overrightarrow{BA}\ \overrightarrow{BD}\ \overrightarrow{BC})$ if $\angle XYZ < \angle ABC$, and $(\overrightarrow{BA}\ \overrightarrow{BC}\ \overrightarrow{BD})$ if $\angle XYZ > \angle ABC$. (See Fig. 68.)

22.6   Angle-Doubling Theorem—Dual of Theorem 20.5:   If $\angle ABC$ is any angle whose measure is less than 90, it may be *doubled on the* $\overrightarrow{BA}$ *side of* $\overrightarrow{BC}$, that is, there exists a ray $\overrightarrow{BD}$ such that $\angle ABC = \angle ABD$ and $(\overrightarrow{BC}\ \overrightarrow{BA}\ \overrightarrow{BD})$. (See Fig. 69.)

Figure 68

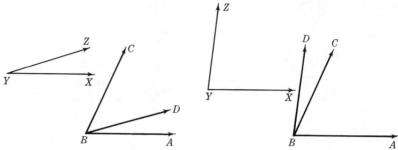

Figure 69

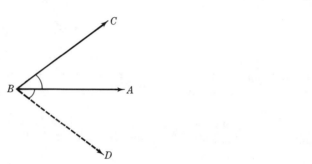

22.7 Definition: The ray $\overrightarrow{BA}$ in Theorem 22.6 is called the **bisector** of $\overline{\angle CBD}$.

22.8 Angle-Bisection Theorem—Dual of Theorem 20.4: Each angle has a unique bisector.

One last result will be mentioned which is quite useful, sometimes tacitly assumed in elementary treatments. The proof will be left as an exercise for the reader (see Exercise 10 below).

22.9 Lemma: If h, k, and h' are any three distinct rays such that h' is opposite h, then (hkh').

EXERCISES

1. Find an example of three rays having the property that no one of them is between the other two.
2. Find the measure of angles $\overline{\angle DOE}$ and $\overline{\angle AOE}$ of Fig. 66 directly from Axiom 8.
3. Find the measures of the six possible pairs of angles formed by the concurrent rays $h[-179]$, $k[-61]$, $u[5]$, and $v[120]$.
4. Suppose h, k, and u are concurrent rays with coordinates 20, -40, and -160, respectively. If v is concurrent with h, k, and u, find which of the relations (vhk), (hvk), (kvu), or (kuv) holds (as guaranteed by Theorem 22.4) if v has coordinate (a) 90, (b) 175, (c) -175, (d) -30, (e) 0, (f) -40, and (g) 21. Check your answers in (a) and (b) by calculating the measures of the angles involved.
5. Dualize the phrase "five collinear points."

6. Dualize the *point equation* $AB + BC = \alpha$ so it becomes an *angle equation*. (Assume A, B, and C are collinear.)
7. Dualize $XY + YZ > XZ + ZW$ for collinear points X, Y, Z, and W so the dualized inequality concerns angles.
8. Prove Theorem 22.5 by dualizing the previous proof of Theorem 20.2, replacing "point" by "ray" and "distance" by "angle measure," rephrasing at certain places to effect better clarity.
9. Define "angle trisection" and prove that the trisectors of a given angle always exist.
10. Prove that if $hk = 180$ and u is any ray concurrent with h and k, then (huk). (Dual of Exercise 5, Section 19; note that this proves Lemma 22.9.)

23. Two Models for Axioms 1–8: The Sphere and Pseudosphere

It has not yet been pointed out explicitly how the last three axioms are realized in spherical geometry. The purpose of this section is to complete that detail and at the same time introduce another geometry possible on the sphere, also a model for the previous axioms.

To create a measure for angles in spherical geometry, one need only rely on a precedent set in elementary calculus. In calculus the measure of the "angle" between two intersecting *curves* is ordinarily taken to be the Euclidean measure of the angle between the *tangents* at the point of intersection. If $\overline{BAC}$ is a spherical angle (Fig. 70), then one may take as its measure the Euclidean measure of the angle determined by the tangent rays to the spherical rays $\overrightarrow{AB}$ and $\overrightarrow{AC}$. (The proper orientation of those tangent rays is understood; thus in the figure, $\angle BAC$ is θ, not ϕ.) Since the tangents to all the spherical rays at A lie in a plane, it is evident that this concept of angle measure agrees with that postulated in Axioms 6–8. Thus spherical geometry is a model for all the previous axioms (with $\alpha < \infty$).

Figure 70

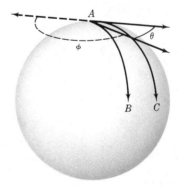

Another geometry can be constructed by using the following simple device: *Let each point on the sphere be identified with its opposite pole, and retain all other spherical concepts consistent with this identification.*[3] Thus, "lines" are still great cir-

---

[3] See in connection Exercise 4, Section 17.

cles, "distance" is still "shortest arc length," and "angle measure" is still determined by the tangents. The result is a geometry which is radically different from spherical geometry. We shall call it **pseudospherical geometry,** and the sphere with this geometry will be called the **pseudosphere.** It may be easily verified that Axioms 1, 2, 4, and 6–8 hold in pseudospherical geometry since the identification of opposite poles does not affect those axioms. (Indeed, Axiom 4 has the stronger form: *Each two points determine a unique line.*) But the concept of distance on the pseudosphere must be treated with care.

The peculiarities of this distance concept may be illustrated as follows: Take the earth's surface as a model and imagine that each individual has his own *personal ambassador* on the extreme opposite side of the earth. Suppose further, that each person is in constant communication with his "ambassador" and that no other means of communication is possible, except by direct conversation. If person A wants to talk to person B he then has only to travel the great circle distance to B, or *to B's ambassador,* whoever is closest. The maximum distance any one person must travel is therefore *one fourth the circumference of a great circle.* The distance of travel in each case would be precisely the *pseudospherical distance.* Having convinced ourselves of this it is then clear that Axiom 3 is satisfied for pseudospherical geometry, and that $\alpha = \frac{1}{2}\pi r$, where r is the radius of the sphere with which we began.

The remaining axiom is Axiom 5, the coordinatization axiom. In order for a coordinate system to be valid for a great circle in pseudospherical geometry it must be such that the endpoints of each diameter have the *same coordinate,* and that the maximal coordinate occurs at a point which is one quarter the way around the circle from the origin. But this is clearly possible (see Fig. 71), and with some reflection one becomes convinced that the relation between "distance" and "coordinates" mentioned in Axiom 5 yields the pseudospherical distance. Thus Axiom 5 is satisfied. Pseudospherical geometry is therefore *another model for Axioms 1–8.*

Figure 71

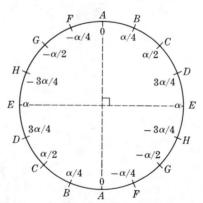

But the pseudosphere is disturbing in many respects. Consider point A and line l, $A \in l$, as shown in Fig. 72. An unhappy thought is that we are apparently unable to decide on which side of l point A lies! For, A and its opposite pole A'

occupy *both* hemispheres determined by l. However, this may be only an apparent difficulty. It is quite possible that the concept of a "side" of a line in this geometry simply does not conform to our persistently Euclidean mode of thinking (that is, it may be that a "side" of l is some other portion of the sphere besides a hemisphere). Perhaps each line has only *one* side. This would certainly seem to explain the present paradox. Unfortunately, the problem goes deeper than this. First let us remove the vagueness associated with the term "side of a line."

Figure 72

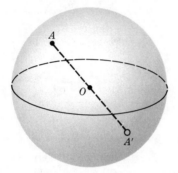

23.1   Definition: Two points A and B are said to be on the **same side** of a line l if and only if A and B do not belong to l and there is no point X on l such that (AXB). If two points not on l are not on the same side of l they are said to be on **opposite sides** of l.

If this concept is to have any meaning or usefulness to geometry it must satisfy the property: *If* A *lies on the same side of line* l *as* B, *and* B *lies on the same side of* l *as* C, *then* A *lies on the same side of* l *as* C. But no such concept is possible in pseudospherical geometry. Consider the circle shown in Fig. 73, with B and C trisecting the arc $AA''A'$. If this circle is a great circle on a pseudosphere, then $A = A'$ and $AB = BC = AC = 2\alpha/3 < \alpha$. Note that point B is *not* the midpoint of segment $\overline{AC}$ as it would be in spherical geometry. The midpoint of $\overline{AC}$ is instead the point M whose coordinate is $-\alpha/3$ as shown in the figure. This of course follows

Figure 73

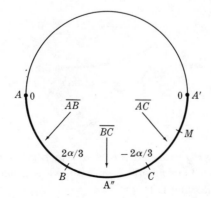

from Axiom 5. After careful consideration one discovers that the "segments" $\overline{AB}$, $\overline{BC}$, and $\overline{AC}$ are as indicated in Fig. 73. Now consider those same points on the pseudosphere, and let line l pass through M (see Fig. 74). Then by definition A and B lie on the same side of l, and B and C lie on the same side, *but* A *and* C *lie on opposite sides!*

Figure 74

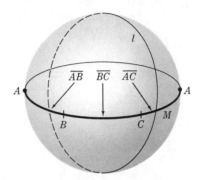

Because of this complication it would be exceedingly difficult to include pseudo-spherical geometry in our study. Since the axioms thus far introduced do not exclude this geometry we seek an additional axiom that will.

24.   Plane Separation: The Postulate of Pasch

If a jet airliner were to fly the shortest route from one point in the Northern Hemisphere to another, it is not altogether trivial that the path of flight itself remains in the Northern Hemisphere. This finds some support in the following analysis, based in part on the previous axioms: Suppose the path does cross the equator. As the flight proceeds from A to B there must be some *first* point P where the path crosses. Then at this point the path passes over into the Southern Hemisphere, but in order to arrive at B, the path must cross the equator again at some point Q, $P \neq Q$. (See Fig. 75.) But this yields two great circle arcs whose endpoints

Figure 75

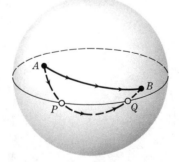

are P and Q. Therefore, $PQ = \alpha$ and $AB > \alpha$, which is impossible.

This property which hemispheres seem to have is greatly significant. It may be formulated in the following manner:

24.1   Definition: A set S in the plane is called **convex** if for each pair of points A and B belonging to S such that $AB < \alpha$, $\overline{AB} \subset S$. (See Fig. 76.)

Figure 76

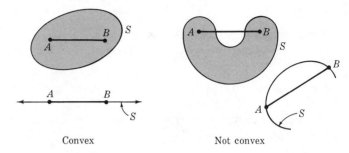

Convex   Not convex

Convexity itself is a relatively simple concept, but it leads to many interesting and important properties of the plane. The theory of convexity is now a recognized field of mathematics which is rich with elegant propositions and unanswered questions. (For a good introduction to this subject the reader might consult either R. V. Benson, *Euclidean Geometry and Convexity* [2] or I. M. Yaglom and V. G. Boltyanskiĭ, *Convex Figures* [33].) One of the fundamental theorems in the theory of convexity is the following:

24.2   Theorem: The intersection of two convex sets is convex.

PROOF: Let S and T be convex sets and suppose $A \in S \cap T$ and $B \in S \cap T$, with $AB < \alpha$. Then, by definition of set intersection, $A \in S$ and $B \in S$, and by the definition of convexity, $AB \subset S$. But in the same manner, $AB \subset T$; therefore, it follows that $AB \subset S \cap T$. ∎

The thought that each hemisphere in spherical geometry is convex and that in the Euclidean plane the points lying on each side of a line form a pair of convex sets leads us to the next axiom (the previous section shows that it cannot be established as a theorem).

Axiom 9 (Plane Separation Principle):   *There corresponds to each line* l *in the plane two regions (sets)* H_1 *and* H_2 *with the properties:*

(a)   *each point in the plane belongs to one and only one of the three sets* l, H_1, *and* H_2,
(b) H_1 *and* H_2 *are each convex sets, and*
(c)   *if* $A \in H_1$ *and* $B \in H_2$ *such that* AB < α *then* l *intersects segment* $\overline{AB}$.

24.3   Definition: The sets H_1 and H_2 mentioned in Axiom 9 are called the **half planes determined by** *l*, and are said to be **opposite** each other.

Among the first consequences of Axiom 9 is that if $A \in l$, $B \in H_1$, and $AB < \alpha$ (with the same notation as before), then all the interior points of both the *segment* $\overline{AB}$ and the *ray* $\overrightarrow{AB}$ also belong to H_1. For, if some interior point P lay in l or H_2

then l would pass through some point X on $\overline{BP}$ such that $AX < \alpha$ (proof?) and hence

$$l = \overset{\leftrightarrow}{AX} = \overset{\leftrightarrow}{AB}.$$

But this would be a contradiction since $B \notin l$.

The foregoing observation is extremely useful so it will now be stated more formally.

24.4 Theorem: If a segment has exactly one of its endpoints lying on a line l, then all other points of that segment lie in the half plane of l which contains the other endpoint. Similarly, if the origin of a ray lies on l, then all the interior points of the ray lie in the half plane of l which contains any other point of that ray.

This result will now be used to prove a rather attractive theorem that is at the same time fundamental to our development.

24.5 Theorem: If A and B are two points such that $AB = \alpha$, then every line which passes through A also passes through B.

PROOF: Let l be any line through A and let m contain A and B. If $m = l$, then $B \in l$ and we are finished; assume therefore, that $m \neq l$ (Fig. 77). For any $P \in m$ such that $P \neq A$, $P \neq B$, then $0 < AP < \alpha$. (Why?) Therefore, l cannot pass through P (or else $l = \overset{\leftrightarrow}{AP} = m$). For some coordinate system on m having A as origin, determine the points $C[\alpha/4]$ and $D[-\alpha/4]$. Then (CAD), and hence $A \in \overline{CD}$. Since $C \notin l$, $D \notin l$, then $C \in H_1$ and $D \in H_2$, where H_1 and H_2 are the two half planes determined by l (Axiom 9). Locate $E[3\alpha/4]$ and $F[-3\alpha/4]$. Since E and F are interior points on the rays $\overset{\rightarrow}{AC}$ and $\overset{\rightarrow}{AD}$, by Theorem 24.4 $E \in H_1$ and $F \in H_2$. But $EF = \alpha/2 < \alpha$. Then l must intersect $\overline{EF}$ at an interior point P, and by the opening comment in the proof that point is either A or B. But (EAF) is clearly not the case since $AE = AF = 3\alpha/4$, so $P \neq A$. Therefore, $P = B$, and l passes through B. ∎

There are a number of useful corollaries (the proofs will be left as exercises):

24.6 Corollary: If two lines intersect at A and $\alpha < \infty$, they also intersect at a point B such that $AB = \alpha$.

Figure 77

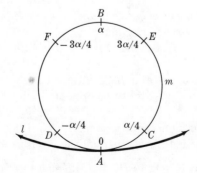

24.7 Corollary: If $\alpha < \infty$, to each point A there corresponds a *unique* point A' in the plane such that $AA' = \alpha$.

24.8 Definition: If A and A' are two points such that $AA' = \alpha$, they are called **extreme points,** and each is called the **extremal** of the other.

We have then arrived axiomatically at a characteristically spherical property —if $\alpha < \infty$—, that of extremal points, whose analog on the sphere are the pairs of opposite poles.

24.9 Corollary: Two points A and B lie on the same side of a line l if and only if they lie in the same half plane determined by l. Equivalently, two points A and B lie in the same half plane if and only if there exists no point $X \in l$ such that (AXB).[4]

A consequence of the plane separation principle (with the strengthening afforded by Corollary 24.9) is the famous postulate of Pasch, which is merely the statement of the "obvious" principle that if a line cuts one side of a triangle and does not pass through any of its vertices it must also cut another side (Fig. 78). Moritz Pasch (1843–1930) recognized its necessity and was among the first to notice that it had been tacitly assumed by Euclid. This principle will be stated without reference to triangles, however, since it is valid whether or not the points A, B, and C are collinear.

Figure 78

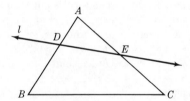

24.10 Postulate of Pasch:[5] If A, B, and C are any three points in the plane, D is any other point such that (ADB), and l is any line passing through D but not through A, B, or C, then l either passes through a point E such that (AEC) or through a point F such that (BFC), the cases being mutually exclusive.

Figure 79

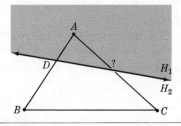

---

[4] Refer to Definition 23.1. The reason for placing this apparent fact here is that the case $AB = \alpha$ may now be handled more easily.

[5] It is interesting that this postulate and the plane separation principle are equivalent—each may be proved from the other (see Exercise 13 below).

PROOF: (See Fig. 79.) Let H_1 and H_2 be the two half planes determined by line l with $A \in H_1$ and $B \in H_2$ (Axiom 9). Then by Axiom 9, since C does not lie on l, C must either lie in H_1 or in H_2, but not both. If $C \in H_2$, there is a point $E \in l$ such that (AEC), and if $C \in H_1$, there is a point $F \in l$ such that (BFC), but not both. ∎

It is convenient to have a functional notation for the two half planes determined by a line. The *half plane determined by* l *and containing point* A *is denoted*

$$H_1(A, l),$$

while the *half plane not containing* A *is denoted*

$$H_2(A, l).$$

In view of Corollary 24.9, $H_1(A, l)$ stands for the *set of all points which lie on the same side of* l *as* A, while $H_2(A, l)$ stands for the *set of points on the opposite side of* l *as* A. We shall take advantage of this notation in defining the interior of an angle, the topic of the next section.

EXERCISES

1. Make a sketch to show a direct violation of the postulate of Pasch on the pseudosphere for a triangle whose vertices are not collinear. *Hint:* Study Fig. 74; a slight adjustment will do the trick.
2. Provide a clear explanation why Axiom 9 rules out pseudospherical geometry.
3. Show directly why Corollary 24.7 does not hold on the pseudosphere.
4. Generalize Theorem 24.2 relative to the number of convex sets given. What is the "best" result which can be proved in this direction?
5. Prove that if $AB = \alpha$ and C is any other point in the plane, then (ACB). (See Exercise 5, Section 19.)
6. Prove that a line is convex.
7. Does the argument involving the Northern and Southern Hemispheres in a previous discussion give a clue as to how one might prove the convexity of "hemispheres" from the previous Axioms 1–8?   *Ans.: No.*
8. Using the notation $H_k(P, l)$ $(k = 1, 2)$, denote the two half planes of l in terms of the points A, B, and C (six answers in all) if $A \in H_1$, $B \in H_1$, and $C \in H_2$.
9. Prove Corollaries 24.6 and 24.7.
10. Prove that extreme points lie on opposite sides of any line not passing through them.
11. Prove Corollary 24.9.
12. Prove that the union of a half plane and the line which determines it is a convex set.
★13. If the postulate of Pasch were assumed, how might one go about defining the half planes determined by a line? Prove Axiom 9 from the postulate of Pasch, using your definition of "half plane."

25. The Interior of an Angle: The Crossbar Principle

In any representative drawing of an angle it is not difficult to picture what the "inside" or "interior" of that angle should be. Any child can point to the region and correctly identify it. But *defining* the concept without recourse to figures could prove rather frustrating—were it not for the concept of half planes. A singularly elegant way to handle this problem appears to be provided by the following definition:

25.1   Definition: The **interior** of a nondegenerate, nonstraight angle $\overline{\angle ABC}$ is the set

$$H_1(A, \overleftrightarrow{BC}) \cap H_1(C, \overleftrightarrow{BA}).$$

Any point lying in the interior of an angle is called an **interior point** of that angle. If a point is not an interior point and does not lie on the sides of an angle, it is called an **exterior point**. (See Fig. 80.)

Figure 80

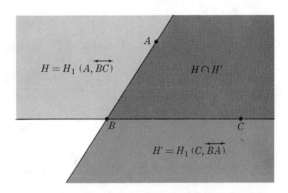

A moment's reflection convinces one that the definition just presented agrees with the intuitive idea concerning the "inside" of an angle. (Consider the sequence shown in Fig. 81.) The interior of an angle on a sphere is obtained by intersecting

Figure 81

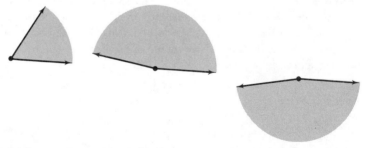

two hemispheres, which together with the sides of the angle, produces a region called a **lune**. (See Fig. 82 for the spherical analog of Fig. 81.)

One of the apparent facts about interiors of angles is that if

$$U \in \text{Interior } (\angle \overline{ABC}),$$

then

$$(\overrightarrow{BA} \; \overrightarrow{BU} \; \overrightarrow{BC})$$

Figure 82

(Fig. 83). Careful consideration of this statement shows that it ultimately involves a relationship between the order concepts for collinear points and concurrent rays. For, from Fig. 84 it becomes reasonably clear that

$$B \in H_1(A, \overleftrightarrow{OC}) \cap H_1(C, \overleftrightarrow{OA})$$

if and only if

$$(ABC).$$

Figure 83

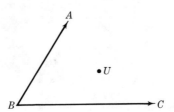

Thus the validity of the previous assertion is related to the question: Does (ABC) imply (hku)? It should be eminently clear that none of the previous axioms include or imply such a relationship. Indeed, as the development now stands, the two order concepts are *completely independent*. It is relatively easy to construct a model for Axioms 1–9 in which there occurs a denial of this relation (see Exercise 2 below). Hence, to provide the needed connection between point order and ray order, another axiom is required.

Figure 84

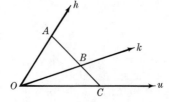

Axiom 10: *If the concurrent rays* p, q, *and* r *meet line* l *at the respective points* P, Q, *and* R, *and* l *does not pass through the origin of* p, q, *and* r, *then* (PQR) *if and only if* (pqr). (See Fig. 85.)

Figure 85

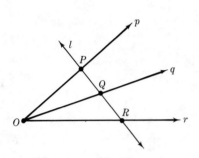

Figure 86

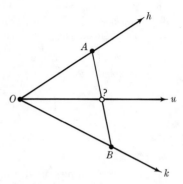

Lemma 22.9 will be used repeatedly in the proof of the following important consequence of Axiom 10.

25.2 The Crossbar Principle: If A and B are any two interior points on the respective sides h and k of any (nondegenerate, nonstraight) angle, then any ray u which lies between h and k must contain an interior point C such that (ACB). (See Fig. 86.)

PROOF: Let h' and u' be the rays opposite h and u, respectively, and choose a coordinate system for the rays at O such that h is the origin and k has coordinate $y > 0$; let the coordinates of u and u' be x and x' (Fig. 87). Now (huk) and $hk < 180$ implies $0 < x < y$ by Theorem 22.3. Also, if $x' > 0$, then since either $x' < x < 180$ or $x < x' < 180$ it would follow (by Theorem 22.2) that either $(u'uh')$ or $(uu'h')$, both of which imply $uu' < 180$, a contradiction. Therefore, $x' < 0$.

Figure 87

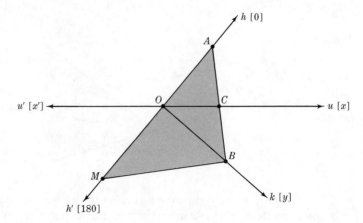

Since $OA < \alpha$, we may choose $M \in h'$ such that (AOM). Then by the postulate of Pasch the line $u \cup u'$ passes through a point C such that (ACB)—the case shown in Fig. 87—or a point D such that (BDM). Therefore, the cases with their respective implications (by Axiom 10) may be arranged in the following manner:

$D \in u$, (BDM).
Then (kuh').
But also (hkh')
so by Theorem 22.4(a),

$$(hkuh')$$

which denies (huk).

$D \in u'$, (BDM).
Then $(ku'h')$.
Since (hkh')
then

$$(hku'h').$$

Therefore, (hku') and by
Theorem 22.3, since
$hu' < 180$,

$$0 < y < x'$$

a contradiction.

$C \in u'$, (ACB).
Then $(hu'k)$.
Since $hk < 180$
then Theorem 22.3 implies

$$0 < x' < y,$$

a contradiction.

The only remaining possibility is $C \in u$ with (ACB). To finish the proof, $0 < OC < \alpha$ must hold or else $\overleftrightarrow{AC}$ would pass through O and thus would coincide with $h \cup h'$. In that case $B \in h \cup h'$, so either $k = h$ or $k = h'$, denying that $\overline{hk}$ was both nondegenerate and nonstraight. ∎

NOTE: In view of this result and the fact that Axiom 10 was used in only one direction, namely, that if (PQR) then (pqr), it follows that Axiom 10 can be weakened to state only that direction. The converse follows from the postulate of Pasch, using the above argument.

The following result settles the issue raised earlier:

25.3   Corollary: If a ray concurrent with the sides of an angle passes through an interior point of the angle, then that ray lies between the sides of the angle.

PROOF: Let $B \in$ Interior $\overline{\angle AOC} = H_1(A, \overleftrightarrow{OC}) \cap H_1(C, \overleftrightarrow{OA})$. Put $\overrightarrow{OA} = h$, $\overrightarrow{OB} = u$, and $\overrightarrow{OC} = k$ (Fig. 88). Let $OA' = h'$ be the ray opposite h, choosing A'

Figure 88

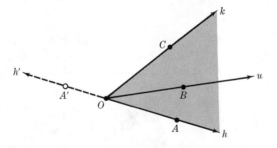

such that (AOA'). Since B does not lie on $\overleftrightarrow{OA}$ or $\overleftrightarrow{OC}$, then $u \neq h$, $u \neq k$, $u \neq h'$, and u is not opposite k. The relation (hkh') and Theorem 22.4(b) then yield the cases

$$(uhk), \qquad (huk), \qquad (kuh'), \qquad \text{and} \qquad (kh'u).$$

By the above comment the angles $\overline{ku}$ and $\overline{kh'}$ are nondegenerate and nonstraight. Therefore, the crossbar principle applies in each case (see Fig. 89):

(a)	(b)	(c)
(uhk)	(kuh')	$(kh'u)$
There is $E \in h$ with	There is $E \in u$ with	There is $E \in h'$ with
(BEC).	$(A'EC)$.	(BEC).
Then	Then by Theorem 24.4	Then by Theorem 24.4
$B \in H_2(C, \overleftrightarrow{OA})$,	$E \in H_1(A', \overleftrightarrow{OC})$,	$B \in H_1(E, \overleftrightarrow{OC})$,
a contradiction.	$E \in H_2(A, \overleftrightarrow{OC})$,	$B \in H_1(A', \overleftrightarrow{OC})$,
	$B \in H_2(A, \overleftrightarrow{OC})$,	$B \in H_2(A, \overleftrightarrow{OC})$,
	a contradiction.	a contradiction.

Figure 89

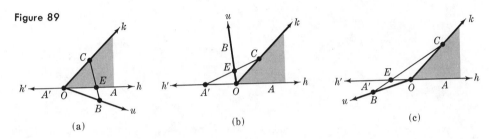

(a)   (b)   (c)

Hence, the only possibility remaining is (huk). ∎

EXERCISES

1. Prove that the interior of an angle is a convex set.

2. At some point O in the analytic plane the coordinates of the rays from O are redefined so that the coordinates of the rays $h[30]$ and $k[60]$ are interchanged (along with the coordinates of their opposite rays) as indicated in Fig. 90. Then Axiom 8 may be used to *define* angle measure at O. If all other concepts of analytic geometry are retained, with angle measure at other points as usual, show that the resulting "geometry" is a model for Axioms 1–9 with $\alpha = \infty$ but where Axiom 10 is denied. *Hint:* Consider the rays h, k, u as shown and the points A, B, and C. Does (ABC) imply (uhk)?

3. In the model described in Exercise 3 identify (a) the interior of $\angle AOC$ and (b) the set of interior points of the rays between $\overrightarrow{OA}$ and $\overrightarrow{OC}$.

4. Prove the following set equation (*Interior S* denotes *the set of interior points of* S):

$$\text{Interior } \overline{\angle ABC} = \{X: \; X \in \text{Interior } h \text{ and } (\overrightarrow{BA} \, h \, \overrightarrow{BC})\}. \qquad \textbf{(25.4)}$$

5. Prove that if $0 < x < 180$ and the rays $h[0]$ and $k[x]$ are determined relative to a coordinate system of some pencil of rays, the coordinate of the ray k' opposite k is $x - 180$.

6. Does the union of the interiors and sides of $\overline{\angle ABD}$ and $\overline{\angle DBC}$ give the interior and sides of $\overline{\angle ABC}$? Discuss.

7. Without referring to the text, write a direct argument of your own based on the axioms and theorems up to Axiom 10, of the following theorem: If (hku) then rays h and u lie

Figure 90

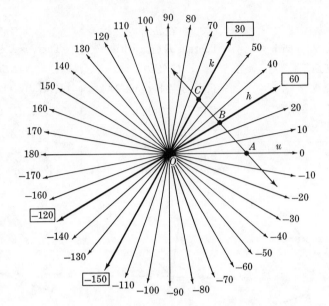

on opposite sides of the line l which contains k (meaning, explicitly, that each interior point of h and each interior point of u lie on opposite sides of line l).

★8. Write a definition for the term "angles about a point" so that the familiar principle *the sum of the measures of the angles about a point is* 360 may be proved. Then prove, using your definition.

★9. In Fig. 91, X is an interior point on segment $\overline{BC}$, with $x = BX$ and $\theta(x) = \angle BAX$. Thus, $\theta(x)$ is a real-valued function of x, $0 \le x \le BC$. Prove that $\theta(x)$ is an increasing, continuous function.

Figure 91

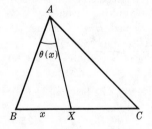

Figure 92

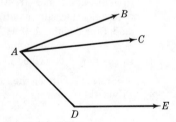

10. In Fig. 92 if $(\overrightarrow{AB}\ \overrightarrow{AC}\ \overrightarrow{AD})$, prove that $\overrightarrow{AC}$ must intersect $\overrightarrow{DE}$ if $\overrightarrow{AB}$ does.

11. Prove that if $A \in l$, $B \in l$, U lies on a given side of l, and $\angle XYZ$ is a given angle, there exists a unique ray $\overrightarrow{BC}$ on the same side of l as U such that $\angle ABC = \angle XYZ$.

PART III

ABSOLUTE GEOMETRY AND CONCEPTS OF PARALLELISM

4

The Triangle
and Its Properties

That part of Euclidean geometry which excludes any reference to parallel lines or notions directly related to them is known as *absolute geometry*. Traditionally this discipline excludes spherical geometry, but for us *absolute geometry is the study of the logical consequences of Axioms 1–10 and the additional axiom which is to be introduced in this chapter*. In such a development it is obvious that involvement with the Euclidean concept of parallelism is impossible since parallel lines do not exist on the sphere. (At this point the reader might enjoy the challenge of deciding whether Axioms 1–10, with $\alpha < \infty$, imply this property of the sphere.)

The development of the previous chapters will be carried further in this and the following chapter, and finally in Chapter 6 the implications of various assumptions regarding parallel lines will be studied. The present chapter in particular focuses attention on an object that is by far the best known and most studied of all geometric figures.

26. The Triangle: Correspondence and Congruence Between Two Triangles

The terminology normally associated with triangles will be introduced formally.

26.1 Definition: The union of the segments joining three distinct, noncollinear points is a **triangle**. The three points are called the **vertices,** while the three segments are called the **sides.** If A, B, and C are three noncollinear points, then it follows by a previous theorem (Theorem 24.5) that AB, BC, and AC are each less than α and segments $\overline{AB}$, $\overline{BC}$, and $\overline{AC}$ are uniquely determined. The triangle

$$\overline{AB} \cup \overline{BC} \cup \overline{CA}$$

is then defined and will be denoted

$$\triangle ABC.$$

The lines $\overleftrightarrow{AB}$, $\overleftrightarrow{BC}$, and $\overleftrightarrow{AC}$ are called the **extended sides** (sometimes also called **sides**). The **angles** of the triangle are $\overline{\angle BAC}$, $\overline{\angle ABC}$, and $\overline{\angle ACB}$; the angle at A and vertex A are each said to lie **opposite** side $\overline{BC}$, and similarly for the other two vertices and angles. The angle at A is said to be the **included angle** of sides $\overline{AB}$ and $\overline{AC}$, and side $\overline{AB}$ is said to be the **included side** of $\angle A$ and $\angle B$ (similarly with the other sides and angles). Finally, the **angle sum** of a triangle is the sum of the measures of its three angles,

$$\angle A + \angle B + \angle C.$$

The notion of the "congruence" of two triangles is the usual one:

26.2   Definition: Two triangles, distinct or not, are said to be **congruent** if and only if under some correspondence between their vertices, corresponding sides and corresponding angles have equal measures.

Figure 93

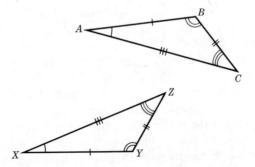

Thus, $\triangle ABC$ and $\triangle XYZ$ would be congruent (see Fig. 93) under the correspondence A to X, B to Y, and C to Z provided

$$AB = XY, \qquad BC = YZ, \qquad AC = XZ,$$
$$\angle A = \angle X, \qquad \angle B = \angle Y, \qquad \angle C = \angle Z.$$

This congruence will be denoted in the usual manner

$$\triangle ABC \cong \triangle XYZ. \tag{26.3}$$

A correspondence from $\triangle ABC$ to $\triangle XYZ$ may be written

$$ABC \leftrightarrow XYZ \tag{26.4}$$

to indicate the correspondence between the vertices in the pairs

$$(A, X), \qquad (B, Y), \qquad \text{and} \qquad (C, Z).$$

That is,

$$A \leftrightarrow X, \qquad B \leftrightarrow Y, \qquad \text{and} \qquad C \leftrightarrow Z.$$

Thus $ABC \leftrightarrow XZY$ would denote the different correspondence $A \leftrightarrow X$, $B \leftrightarrow Z$, and $C \leftrightarrow Y$.

Since each triangle has three vertices it is obvious that there are six distinctly different ways in which a correspondence between two triangles may be given, and a congruence under one correspondence need not be a congruence under a different one. As the correspondence in a congruence is so crucial, the above notation of congruent triangles will be used in the stronger sense

$$\triangle ABC \cong \triangle XYZ \qquad \text{only if} \qquad ABC \leftrightarrow XYZ. \qquad (26.5)$$

It is clear that in this connection one might have the *equality*

$$\triangle ABC = \triangle XYZ$$

(which is equivalent to the set equation $\overline{AB} \cup \overline{BC} \cup \overline{CA} = \overline{XY} \cup \overline{YZ} \cup \overline{ZX}$) without necessarily having the *congruence*

$$\triangle ABC \cong \triangle XYZ.$$

Observe that in Fig. 94, while $\triangle ABC = \triangle CBA$, it is not true that $\triangle ABC \cong \triangle CBA$.

It will prove convenient to extend the notion of congruence to segments and angles.

Figure 94

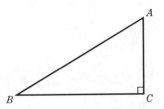

26.6 Definition: Two **segments** are **congruent** if their *lengths* are equal, and two **angles** are **congruent** if their *measures* are equal (the same symbol for congruence will be used as before).

Again there is a definite distinction between *congruence* and *equality*. If $\overline{AB} \cong \overline{XY}$, then by definition $AB = XY$, but we need not have $\overline{AB} = \overline{XY}$ (note that it does not make sense to write $AB \cong XY$). A similar comment about angles holds.

Definition 26.2 may now be restated to yield the principle (valid *by definition*):
Corresponding sides and corresponding angles of congruent triangles are congruent.

26.7 *Example:* The set equation $\triangle ABC = \triangle ACB$ always holds. Under what circumstances can one assert $\triangle ABC \cong \triangle ACB$?

Solution: By definition, the assertion holds only when $ABC \leftrightarrow ACB$ and $AB = AC$, $BC = CB$, $AC = AB$, $\angle A = \angle A$, $\angle B = \angle C$, and $\angle C = \angle B$. Eliminating the trivial and redundant information, it follows that

$$\triangle ABC \cong \triangle ACB \text{ if and only if } AB = AC \text{ and } \angle B = \angle C.$$

In Euclidean geometry, this can happen (and does happen) whenever $\triangle ABC$ is isosceles (Fig. 95). ∎

Figure 95

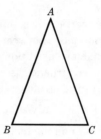

It may be helpful in our visualization of the concept of congruence to think of one triangle being "placed on top of" another with corresponding vertices in similar positions. The vertices of course "coincide" only when the triangles are congruent. (See Fig. 96.) To visualize the above situation for an isosceles triangle in Fig. 95, we might think of picking the triangle up, and, having marked the position it originally occupied, turning it over and placing it upside down on its original spot. We could in this manner observe the congruence we mentioned. However, it should be emphasized that this device is merely a mental crutch and does not have any basis in the present axiomatic development.

Figure 96

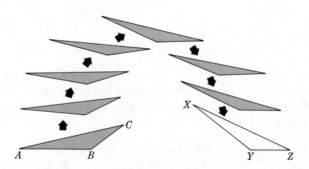

*27. A Model for Axioms 1–10: The Taxicab Geometry

It is probably not an exaggeration to say that every student of plane geometry has heard of the "side-angle-side" proposition: *If two sides and the included angle of one triangle are congruent, respectively, to two sides and the included angle of another, then the given triangles are congruent.* Euclid "proved" this proposition by his so-called method of *superposition*—the operation of placing one triangle over another to "see if they fit." But in the present development the proposition must be included as an axiom. It is the purpose of this section to construct a model to show the independence of this new axiom about to be added.

Consider the usual Cartesian plane with each point represented as a pair of coordinates (real numbers). Retain all the usual Euclidean concepts regarding points, lines, angles, and angle measure, but make the single following exception: *The distance from* $P(x_1, y_1)$ *to* $Q(x_2, y_2)$ *is defined by the formula*

$$\widetilde{PQ} = |x_1 - x_2| + |y_1 - y_2|. \tag{27.1}$$

That is, the "distance" from P to Q is the Euclidean distance *along the right-angle path PRQ*, as shown in Fig. 97. This "distance" could be called the "taxicab" distance from P to Q (for the obvious reasons). The resulting geometry is known by the more formal name *Minkowskian geometry*, after H. Minkowski (1864–1909).

Note that (27.1) implies $\widetilde{PQ} = \widetilde{QP}$. Furthermore, if $\widetilde{PQ} = 0$ then

$$|x_1 - x_2| + |y_1 - y_2| = 0,$$

which implies $x_1 = x_2$ and $y_1 = y_2$, or $P = Q$. Axioms 1–3 are therefore realized.

Figure 97

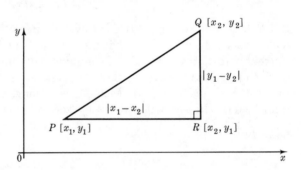

Since lines are ordinary Euclidean lines, Axiom 4 is valid with $\alpha = \infty$, and since Euclidean angle measure has not been altered, it is clear that Axioms 6–8 are satisfied. This leaves Axioms 5, 9, and 10. Let l be any line, with θ the angle it forms with the x axis (Fig. 98). If $P \in l$ and $Q \in l$, observe the Euclidean lengths a, b, and c of the sides of $\triangle PQR$. It follows that

$$\widetilde{PQ} = a + b = c \cos \theta + c \sin \theta.$$

That is, if $\lambda = \cos \theta + \sin \theta$ then

$$\widetilde{PQ} = c\lambda = PQ \cdot \lambda,$$

and we find that the Minkowski "distances" between points on l are constant multiples of Euclidean distances between those same points. It is then clear that Axiom 5 is operative, with the coordinate of each point on l λ times the original Euclidean coordinate. Consequently, it also follows that *Minkowski "betweenness" coincides with Euclidean betweenness*. This then implies that the remaining Axioms 9 and 10 also hold.

Figure 98

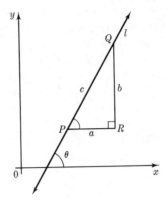

Minkowski geometry has many Euclidean features. But one striking departure from Euclidean geometry may be noticed in Fig. 98: Since $\widetilde{PR} = a$ and $\widetilde{RQ} = b$, it follows that

$$\widetilde{PR} + \widetilde{RQ} = \widetilde{PQ}.$$

But R does not lie on line $\overleftrightarrow{PQ}$! Thus, even though R is not "between" P and Q in the sense previously defined, its "distances" to P and Q behave as though it were. Furthermore, notice that there are a number of different "paths" leading from P to Q which presumably have the shortest lengths possible. For example, in Fig. 99, the routes P-R-Q and P-S-Q are each 14 units long, which is the taxicab "distance" from P to Q. (In this connection see Exercise 4 below.)

Figure 99

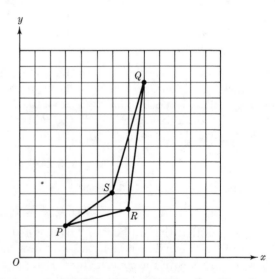

Another non-Euclidean property reveals itself in this example: Consider the two triangles shown in Fig. 100, $\triangle ABC$ and $\triangle XYZ$. They are both "right" triangles with right angles at B and Y, respectively. For the pairs of sides which include those angles we have $\overset{\frown}{AB} = \overset{\frown}{BC} = 6$, $\overset{\frown}{XY} = XM + MY = 6$, and $\overset{\frown}{YZ} = YM + MZ = 6$ so that $\angle B = \angle Y$, $\overset{\frown}{AB} = \overset{\frown}{XY}$, and $\overset{\frown}{BC} = \overset{\frown}{YZ}$. If one were operating in Euclidean geometry it could be asserted that $\triangle ABC \cong \triangle XYZ$ by "side-angle-side." But are they? Certainly under $ABC \leftrightarrow XYZ$ the corresponding angles are congruent. However, for the corresponding sides $\overline{AC}$ and $\overline{XZ}$, we have $\overset{\frown}{AC} = AB + BC = 12$ while $\overset{\frown}{XZ} = 6$, so $\overset{\frown}{AC} \neq \overset{\frown}{XZ}$! This provides the assurance that indeed the "side-angle-side" principle is not a logical consequence of Axioms 1–10.

Figure 100

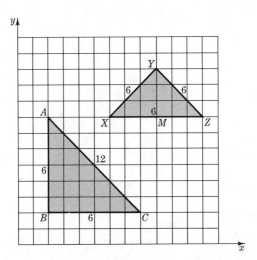

28. The Axiom of Congruence and Initial Consequences

The last axiom for absolute geometry will now be added. Since it is often referred to as the *side-angle-side* principle it will be convenient to designate it by the symbol SAS.

Axiom 11 (SAS Congruence Criterion for Triangles): *If in two triangles there is a correspondence in which two sides and the included angle of one are congruent, respectively, to the corresponding two sides and included angle of the other, the triangles are congruent.* (See Fig. 101.)

Figure 101

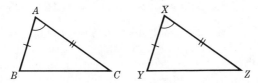

From the SAS axiom it is possible to infer:

28.1 ASA Congruence Criterion for Triangles: If in two triangles there is a correspondence in which two angles and the included side of one are congruent, respec-

tively, to the corresponding two angles and included side of the other, the triangles
are congruent.

PROOF: Let triangles ABC and XYZ be such that $\angle A = \angle X$, $\angle B = \angle Y$, and
$AB = XY$ (Fig. 102). If $BC = YZ$, then the triangles would already be congruent
by SAS. Hence, suppose $BC \neq YX$. It may be assumed without loss of generality
that $BC > YZ$. By the segment-construction theorem there is a point D on $\overleftrightarrow{BC}$ such
that $BD = YZ$ and (BDC). Consider the correspondence $ABD \leftrightarrow XYZ$ and note
that $AB = XY$, $\angle B = \angle Y$, and $BD = YZ$. By SAS, $\triangle ABD \cong \triangle XYZ$. Therefore,
$\angle BAD = \angle X$. But by hypothesis, $\angle X = \angle A = \angle BAC$. Therefore, $\angle BAD =$
$\angle BAC$. Since (BDC), Axiom 10 implies $(\overrightarrow{AB}\ \overrightarrow{AD}\ \overrightarrow{AC})$ and therefore, $\angle BAD <$
$\angle BAC$, a contradiction. Hence, $\triangle ABC \cong \triangle XYZ$. ∎

Figure 102

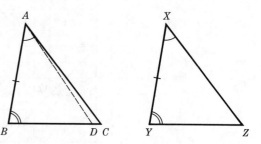

NOTE: It is premature to attempt a proof of the so-called *side-side-side* proposition—
the congruence criterion involving just the sides of two triangles. A proof could be given
at this time if $\alpha = \infty$, for then it would be known that for any three points A, B, and C on
a line, either (ABC), (BAC), or (ACB). (See Exercises 13 and 14 below.)

Now it is possible to obtain a most familiar proposition.

28.2   Theorem: If two sides of a triangle are congruent, the angles opposite those
sides are congruent, and conversely.

PROOF: Suppose $\triangle ABC$ has $AB = AC$ (Fig. 95). Under the correspondence
$ABC \leftrightarrow ACB$, $AB = AC$, $\angle A = \angle A$, and $AC = AB$, so SAS implies $\triangle ABC \cong$
$\triangle ACB$ and therefore, $\angle B = \angle C$. Conversely, suppose $\angle B = \angle C$. Then $\angle B = \angle C$,
$BC = CB$, and $\angle C = \angle B$ so ASA implies $\triangle ABC \cong \triangle ACB$. Therefore, $AB = AC$. ∎

28.3   Definition: The triangle mentioned in the previous theorem is called an
isosceles triangle; if a triangle has all three sides congruent it is called **equilateral,**
and if it has all three angles congruent it is called **equiangular.** If a triangle has
exactly one pair of congruent sides the remaining side is called the **base,** and the
angle opposite, the **vertex angle;** the vertex of the vertex angle of such a triangle is
called simply the **vertex.**

A corollary of Theorem 28.2 readily follows.

28.4   Corollary: An equilateral triangle is equiangular, and conversely.

EXERCISES

1. In how many ways can a triangle be congruent to itself? Under what circumstances do each of these congruences occur?
2. If $\triangle ABC \cong \triangle XYZ$, under what circumstances is $\triangle BCA \cong \triangle YZX$? $\triangle BAC \cong \triangle YZX$?
3. Prove that if $\overline{AB} = \overline{XY}$, then $\overline{AB} \cong \overline{XY}$, and if $\angle A = \angle X$, then $\angle A \cong \angle B$, but that the converses need not necessarily hold.
4. In the taxicab geometry described in Section 27, (a) identify all the points X which lie between $A(1, 3)$ and $B(5, 2)$, and those for which $AX + XB = AB$. (b) Identify those points X such that $AX = XB$, where A and B are as defined in (a). (c) Identify those points X such that $AX = 5$, if A is as defined in (a).
5. Prove that $AC \le AB + BC$ for any three points A, B, and C in the taxicab geometry. *Hint:* Use the inequality $|x + y| \le |x| + |y|$ in (27.1).
★6. **Non-Euclidean models for Axioms 1–10.** A variety of models for Axioms 1–10 may be constructed by starting with any differentiable function $f(x)$ having as domain some interval of real numbers and as range all real numbers, and with $f'(x) > 0$ for all x in the domain. (For example, consider $f(x) = \log x$ for $x > 0$.) Then one takes as "points" all points in the analytic plane, and as "lines" the *vertical Euclidean lines* together with *all curves of the form*

$$y = f(ax + b), \qquad a \text{ and } b \text{ constant,}$$

where x is such that $ax + b$ belongs to the domain of f. (a) Show that f^{-1} exists and that each pair of points $P(x_1, y_1)$, $Q(X_2, y_2)$ *lie on one and only one "line."* The "distance" $\widetilde{PQ}$ is then defined as the *arc length* of the "line" PQ from P to Q. Finally, define angle measure as in ordinary Euclidean geometry (as in the calculus). (b) Examine each of the Axioms 1–10 and find an argument to make each axiom plausible in this geometry, providing proofs whenever feasible. (c) Try to find a specific example [$f(x)$ is to be defined *explicitly*] where Axiom 11 fails. (d) What is the essential character of this geometry if $f(x) = x$? (See Exercise 10, Section 18.)
7. **Interior of a triangle.** The **interior** of a triangle ABC is defined as the set

$$H_1(A, \overleftrightarrow{BC}) \cap H_1(B, \overleftrightarrow{AC}) \cap H_1(C, \overleftrightarrow{AB}),$$

any point of which is called an **interior point** of the triangle. Prove (a) the interior of a triangle is a convex set, and (b) a line passing through an interior point intersects the triangle in *exactly two* points.
8. Prove (a) the interior of a triangle is the intersection of the interiors of any two of its angles, and (b) any line which contains exactly one interior point of a *side* of a triangle must also contain at least one interior point of the triangle.
9. In Fig. 103, it is given that (ABC) and (ADE) with $AC = AE$ and $BC = DE$. Prove that $CD = BE$.

Figure 103

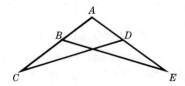

10. In Fig. 104, it is given that $\angle ACB = 90$ and C is the midpoint of BD. Prove that $AD = AB$.

11. In Fig. 105, it is given that $\angle ABC = 90$, $\angle CAB = \angle BAE$, (CBD), and $E \in H_1(D, \overleftrightarrow{AB})$. Prove that $\overrightarrow{AE}$ intersects $\overrightarrow{BD}$. (See Exercise 11, Section 25.)

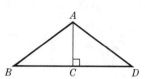
Figure 104

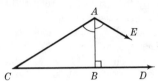
Figure 105

12. Show that the "angle-angle-side" hypothesis (two triangles have two angles and any side of one congruent to the corresponding parts in the other) does not imply congruency on the sphere.

13. Criticize the following argument for the SSS congruence proposition:

Suppose $AB = XY$, $BC = YZ$, and $AC = XZ$ (Fig. 106). Choose ray $h = \overrightarrow{AH}$ on the opposite side of $\overleftrightarrow{AC}$ as B such that $\angle CAH = \angle X$ (by the angle-construction theorem). Choose $D \in h$ such that $AD = XY$. Since $D \in H_2(B, \overleftrightarrow{AC})$, there is E on $\overleftrightarrow{AC}$ such that (BED). Now by SAS, $\triangle ADC \cong \triangle XYZ$ and therefore $AD = AB$, $DC = YZ = BC$, and $\angle ADC = \angle Y$. Thus $\triangle ABD$ is isosceles with $\angle ABD = \angle ADB$. Similarly, $\angle DBC = \angle BDC$. Therefore,

$$\angle ABC = \angle ABE + \angle EBC = \angle ADE + \angle EDC = \angle ADC.$$

That is, $\angle ABC = \angle Y$, so by SAS $\triangle ABC \cong \triangle XYZ$.

Figure 106

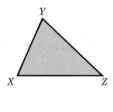
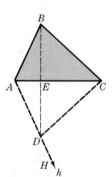

14. Complete the proof in Exercise 13 to make it valid in the case $\alpha = \infty$.

15. Can you prove the existence of an isosceles triangle on a given segment as base? An equilateral triangle? Discuss and give proofs where possible. What additional concepts appear to be needed?

29. Perpendicularity

The key to the definition of perpendicular lines is the following observation: *If two lines intersect, the four nonstraight angles thus formed determine four pairs of angles such that the sum of the measures of the angles in each pair is* 180. Thus in Fig. 107, if $h \cup h'$ and $k \cup k'$ are the intersecting lines, since Lemma 22.9 implies (hkh'), $(kh'k')$, $(h'k'h)$, and $(k'hk)$, it follows that

$$hk + kh' = kh' + h'k' = h'k' + k'h = k'h + hk = 180. \qquad (29.1)$$

Note that as a result,

$$hk = h'k', \qquad kh' = k'h. \qquad (29.2)$$

It follows that if the measure of *any one* of the four nonstraight angles mentioned is 90 then *all four* will have measure 90. Several definitions are now possible. (We shall use the terminology that lines $h \cup h'$ and $k \cup k'$ *determine* the four nonstraight angles mentioned above.)

Figure 107

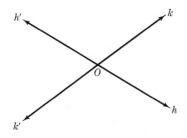

29.3 Definition: If the sum of the measures of two angles equals the measure of a straight angle, they are said to be **supplementary.** An angle is a **right angle** if its measure is 90. A pair of **complementary** angles is a pair of angles the sum of whose measures equals the measure of a right angle. An angle is said to be **acute** or **obtuse** according as its measure is less than, or more than, the measure of a right angle.

29.4 Definition: Two angles which have the property that the sides of one are, respectively, opposite those of the other are called **vertical angles.**

29.5 Definition: Two intersecting lines are said to be **perpendicular** at their point of intersection if they determine four right angles. If line l is perpendicular to line m, the relationship is denoted $l \perp m$.

Equations (29.1) and (29.2) yield two useful theorems (the first is merely a restatement of Lemma 22.9).

29.6 Theorem: Any ray whose origin is the vertex of a straight angle forms with its sides a pair of supplementary angles.

29.7 Theorem: Vertical angles are congruent.

The *existence* of perpendicular lines is of course obvious since right angles exist (Axiom 8). But in geometry it is important to be able to "construct" a perpendicular to any line, passing through an arbitrarily given point either lying on the given line or not. The first problem is almost trivial, while the second requires some effort.

29.8 *Theorem:* Through a given point on a given line there passes a unique line that is perpendicular to the given line.

Figure 108

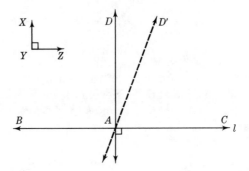

PROOF: As in Fig. 108, let A be any point on line l, and choose B and C on l such that (BAC). If $\angle XYZ$ is any angle whose measure is 90, the angle-construction theorem applied to angle BAC yields a ray $\overrightarrow{AD}$ such that $\angle DAB = \angle XYZ = 90$. Then $\overleftrightarrow{AD} \perp l$. To prove that $\overleftrightarrow{AD}$ is unique,[1] suppose $\overleftrightarrow{AD'} \perp l$ and $\overleftrightarrow{AD'} \neq \overleftrightarrow{AD}$. By Theorem 22.4(b) either $(\overrightarrow{AD'}\ \overrightarrow{AB}\ \overrightarrow{AD})$, $(\overrightarrow{AB}\ \overrightarrow{AD'}\ \overrightarrow{AD})$, $(\overrightarrow{AD}\ \overrightarrow{AD'}\ \overrightarrow{AC})$, or $(\overrightarrow{AD}\ \overrightarrow{AC}\ \overrightarrow{AD'})$. But each of these cases leads to the contradiction $\overleftrightarrow{AD'} = \overleftrightarrow{AD}$. For example, if $(\overrightarrow{AD'}\ \overrightarrow{AB}\ \overrightarrow{AD})$, then

$$\angle DAD' = \angle DAB + \angle D'AB = 90 + 90 = 180,$$

so $\overrightarrow{AD'}$ is opposite $\overrightarrow{AD}$ and therefore, $\overleftrightarrow{AD'} = \overleftrightarrow{AD}$. ∎

29.9 *Theorem:* Through a given point not on a given line there passes a line perpendicular to the given line. If the distance from the point of perpendicularity to the given point is not $\alpha/2$, the perpendicular is unique.

Figure 109

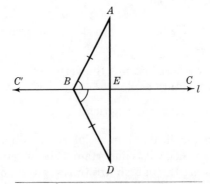

---

[1] That the question of uniqueness is not to be taken lightly can be seen in three-dimensional space where there are infinitely many perpendiculars to a line at a point on that line.

PROOF: (1) With A and l as shown in Fig. 109, choose $B \in l$. Since $A \not\subset l$, $\overleftrightarrow{AB} < \alpha$. If $\overleftrightarrow{AB} \perp l$, we are finished. Otherwise, if (CBC'), then by Theorem 29.6 either $\angle ABC'$ or $\angle ABC$, say $\angle ABC$, is less than 90. By the angle-doubling theorem there is a ray $\overrightarrow{BD}$ such that $(\overrightarrow{BA}\ \overrightarrow{BC}\ \overrightarrow{BD})$ and $\angle ABC = \angle CBD$, and by the segment-construction theorem it may be assumed that $AB = BD$. Since $\angle ABD < 2 \cdot 90 = 180$, and A and D are interior points on their respective rays, the crossbar theorem applies, and there is a point E on $\overrightarrow{BC}$ such that (AED). By SAS $\triangle BAE \cong \triangle BDE$ so that $\angle BEA = \angle BED$. By Theorem 29.6 $\angle BEA + \angle BED = 180$, so it follows that $\angle BEA = 90$ and therefore, $\overleftrightarrow{AE} \perp l$.

Figure 110

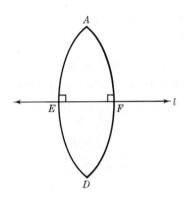

(2) If there were another perpendicular from A, say $\overleftrightarrow{AF}$ (Fig. 110), then by the coordinatization axiom there is D on $\overrightarrow{AE}$ such that $AE = ED$ [we do not necessarily have (AED) here]. As before, SAS implies $\triangle AEF \cong \triangle DEF$ and therefore, $\angle EFD = \angle EFA = 90$. Since it then follows that both $\overleftrightarrow{AF}$ and $\overleftrightarrow{DF}$ are perpendicular to l at F and by the preceding theorem that perpendicular is unique, $\overleftrightarrow{AF} = \overleftrightarrow{DF}$. Hence, $AD = \alpha$ (why?) and by the dual of Lemma 22.9,

$$(AED).$$

Hence from $AE = ED$ it follows that $AE = \alpha/2$, a contradiction. ∎

An illustration of Theorem 29.9 on the sphere is shown in Fig. 111, where $AE = \alpha/2$. Here there occur many perpendiculars to the equator l from pole A. Of course, as stated in Theorem 29.9, this *cannot* happen if $AE \neq \alpha/2$.

An important consequence of the last result is

29.10 Corollary: If the sides of a triangle are each of length less than $\alpha/2$, then the triangle can have at most one right angle.

The terminology commonly associated with a right triangle will be used.

Figure 111

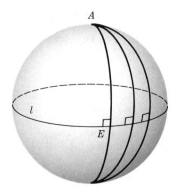

29.11   Definition: A triangle which has precisely one right angle is called a **right triangle.** The remaining two angles are called the **acute angles**[2] of the right triangle; the side opposite the right angle is called the **hypotenuse,** and the other two sides, the **legs.** A triangle having two right angles is said to be **birectangular,** and one having all three of its angles right angles is called **trirectangular.**

NOTE: The existence of birectangular and trirectangular triangles in the case $\alpha < \infty$ may be easily established. (See Exercise 7 below.)

EXERCISES

1. In Fig. 112 (assuming the betweenness properties from the figure) prove that if $\overleftrightarrow{AB} \perp \overleftrightarrow{BC}$, $\overleftrightarrow{CD} \perp \overleftrightarrow{BC}$, and $BE = EC$, then $AB = CD$. Mention in your proof those betweenness properties you actually used.

Figure 112

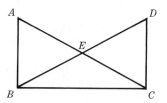

2. Using the figure as you did in Exercise 1, prove that if $\overleftrightarrow{AB} \perp \overleftrightarrow{BC}$, $\overleftrightarrow{CD} \perp \overleftrightarrow{BC}$, and $AB = CD$, then $BE = EC$ (Fig. 112).
3. (a) Prove that an isosceles triangle exists on a given segment as base. Establish all betweenness relations used. *(b) Prove that equilateral triangles exist. *Hint:* Do not start with the base (or any side). It will be established later that an equilateral triangle with given base can be constructed, but it is too early to attempt this now.
4. In this problem you are to establish all betweenness relations needed. Prove that if $\overleftrightarrow{AB}$ and $\overleftrightarrow{CD}$ bisect each other at M and $\overleftrightarrow{AC} \perp \overleftrightarrow{CD}$, then $\overleftrightarrow{BD} \perp \overleftrightarrow{CD}$ (Fig. 113).
5. In Fig. 114, (ABC) holds and D and E are on the same side of $\overleftrightarrow{AC}$ with $\angle ABD = \angle CBE$, $AB = BD$, and $BC = BE$. Prove $AE = CD$.
6. Prove Corollary 29.10.

---

[2] A purely technical term here, since the "acute angles" of a right triangle need not be acute in the sense of Definition 29.3. (See Exercise 8 below.)

Figure 113

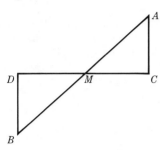

Figure 114

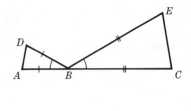

7. (a) Show how to "construct" via the segment- and angle-construction theorems a birectangular triangle in the case $\alpha < \infty$. This would then prove the existence of such triangles if $\alpha < \infty$. (b) Do birectangular triangles exist in the case $\alpha = \infty$? Cite the theorem(s) you use to arrive at your answer.

8. Make a sketch of a sphere on which there is pictured an isosceles right triangle which appears to have obtuse base angles. (Start with a right angle and proceed a distance greater than $\alpha/2$ along each side—this can be carried out axiomatically and succeeding sections will prove the properties intuitively inferred here.) Thus, one obtains a right triangle *both* of whose acute angles are *obtuse!*

9. In Fig. 115, suppose that $AB = BC = CD$ and $\angle ABC = \angle BCD$. If O is an interior point of both $\angle ABC$ and $\angle BCD$, equidistant from A, B, and C, prove that O is equidistant from A, B, C, and D by use of the SSS proposition (to be established later). Prove all betweenness relations used.

10. **Perpendicular bisectors.** Define the **perpendicular bisector** of a segment as the (unique) line which bisects that segment and is perpendicular to the line containing it. Prove that the perpendicular bisector of a segment is the set of all points equidistant from the endpoints of the segment. That is, in terms of set notation, if l is the perpendicular bisector of $\overline{AB}$ (Fig. 116) prove that

$$l = \{X: \ AX = BX\}.$$

A useful application of this is the following:

29.12 Theorem: The line passing through two points P and Q which are equidistant from the endpoints of a segment and such that $0 < PQ < \alpha$ is the perpendicular bisector of that segment.

Figure 115

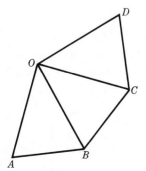

Figure 116

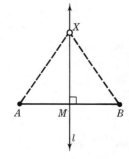

11. **Adjacent angles.** This and the following exercises will introduce the fundamental properties of adjacent angles. Define a pair of **adjacent angles** to be two angles having a side in common but whose interiors have no point in common. The sides which coincide are called the **adjacent sides.** Prove that if h, k, and u are concurrent rays such that (hku), then $\overline{hk}$ and $\overline{ku}$ are adjacent angles. Is the converse true? *Hint:* Use Corollary 25.3 See also (25.4).

12. A useful fact in working with adjacent angles is the following: If (hku), then

$$\text{Interior } \overline{hk} \subset \text{Interior } \overline{hu}.$$

Prove, using (25.4). Hence if $\overline{hk}$ and $\overline{hu}$ are adjacent angles (hku) does not hold.

13. Using Exercise 11 prove:

29.13   Theorem: An angle whose sides are the nonadjacent sides of a pair of supplementary adjacent angles is a straight angle.

Hint: Let $\overline{hk}$ and $\overline{ku}$ be the given angles and let h' be the ray opposite h. Apply Theorem 22.4(b).

14. Prove:

29.14   Theorem: If a line passes through the vertex of a pair of adjacent, supplementary angles and does not contain any of the sides of those angles, one of the two opposite rays lying on that line and having the given vertex as origin passes through the interior of one of the given angles.

(Actually this problem occurred previously in the proof of the crossbar principle; see Fig. 87.)

15. The bisectors of a pair of adjacent supplementary angles are perpendicular.

16. In Fig. 117 $\angle ABD = \angle CBE$, (ABC), and D and E lie in opposite half-planes. Prove that $\angle DBE$ is a straight angle.

Figure 117

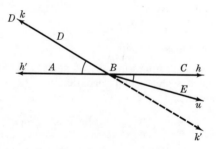

30.   The Exterior-Angle Theorem

One of the most important theorems in absolute geometry will now be considered. In Fig. 118 is shown a triangle with certain angles labeled 1, 2, and 3. Proposition 32 of Euclid's *Elements* (Book I) regarding an exterior angle of a triangle[3] implies that

$$\angle 1 = \angle 2 + \angle 3.$$

---

[3] T. L. Heath, *The Thirteen Books of Euclid's Elements* [14, p. 316].

Figure 118

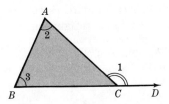

That this proposition is not possible in our development can best be illustrated in a trirectangular triangle on a sphere. In Fig. 119 such a triangle is shown, and here it is obvious that instead of

$$\angle ACD = \angle A + \angle B,$$

we have

$$\angle ACD < \angle A + \angle B!$$

In fact, in this case

$$\angle ACD = \angle A \quad \text{and} \quad \angle ACD = \angle B,$$

so even the weaker proposition $\angle ACD > \angle A$ and $\angle ACD > \angle B$ (often "proved" in elementary treatments) is not valid in general. However, this weaker proposition *can* be established for a triangle if its sides are small enough. First we define what is to be meant by an "exterior angle."

Figure 119

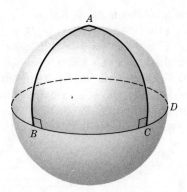

30.1 Definition: In triangle ABC, if D is any point such that (BCD), then $\overline{\angle ACD}$ is called an **exterior angle** of the triangle, and the angles at A and B are called the **opposite interior angles** corresponding to that exterior angle.

Before the weaker form of Euclid I, 32 can be proved two preliminary results must be established, which are useful by themselves.

30.2 Lemma: If $\overleftrightarrow{AB} \perp \overleftrightarrow{BC}$ and $AB = \alpha/2$, then $\overleftrightarrow{AC} \perp \overleftrightarrow{BC}$ and $AC = \alpha/2$.

PROOF: (See Fig. 120.) Since $\alpha < \infty$, suppose $\overleftrightarrow{AB}$ and $\overleftrightarrow{AC}$ meet at the additional point A' (thus $AA' = \alpha$). Then (ABA') and $BA' = \alpha - AB = \alpha/2 = AB$.

By definition of perpendicularity, $\angle ABC = \angle A'BC = 90$ and hence $\triangle ABC \cong \triangle A'BC$. Since (ACA'), $\overline{\angle ACA'}$ is a straight angle, so by Theorem 29.6 it follows that $\angle ACB = \angle BCA' = 90$ and therefore, $\overleftrightarrow{AC} \perp \overleftrightarrow{BC}$. Since $\triangle ABC$ has congruent base angles, $AC = AB = \alpha/2$.

Figure 120

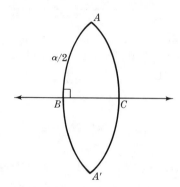

30.3 Lemma: If D is any interior point on side $\overline{BC}$ of $\triangle ABC$, $AB < \alpha/2$, and $AC \leqq \alpha/2$, then $AD < \alpha/2$.

PROOF: Suppose $AD \geqq \alpha/2$ (Fig. 121). Since $\overleftrightarrow{BC}$ does not pass through A, $AD < \alpha$. Locate E on $\overrightarrow{AD}$ such that $AE = \alpha/2$ and take $l \perp \overleftrightarrow{AD}$ at E. Since $AD \geqq AE$ it follows that either (AED) or $D = E$. In either case l does not cut $\overline{AB}$, for otherwise $F \in \overline{AB} \cap l$ implies $AB \geqq AF = \alpha/2$ (by the preceding lemma), a contradiction. In the case (AED), the postulate of Pasch applied to $\triangle ABD$ implies there is $G \in l$ such that (BGD). Then $(BGDC)$ and therefore (BGC). The case $D = E$ implies logically the same relationship: $D \in l$ and (BDC). Note Fig. 122. Since l cannot pass through A, B, or C, the postulate of Pasch again implies there is $H \in l$ such that (AHC). But again by Lemma 30.2 there is the contradiction

$$AC = AH + HC = \frac{\alpha}{2} + HC > \frac{\alpha}{2}.$$

Therefore, $AD < \alpha/2$. ∎

The following theorem is of major importance and will play a central role in our development.

Figure 121   **Figure 122**

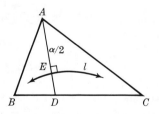
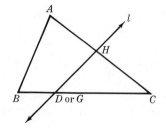

30.4 *Exterior-Angle Theorem for Absolute Geometry:* In any triangle whose sides are of length less than $\alpha/2$, an exterior angle has measure greater than the measure of either opposite interior angle.

PROOF: (1) Let $\triangle ABC$ have sides each of length less than $\alpha/2$ and let $\angle ACD$ be an exterior angle (Fig. 123). Let M be the midpoint of $\overline{AC}$. By Lemma

Figure 123

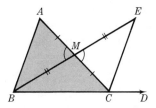

30.3, $BM < \alpha/2$. Hence the segment-doubling theorem applies, and segment $\overline{BM}$ may be extended its own length to E. Since (BME) and (AMC) imply that $(\overrightarrow{MA}, \overrightarrow{MC})$ and $(\overrightarrow{MB}, \overrightarrow{ME})$ are pairs of opposite rays, $\angle AMB$ and $\angle CME$ are vertical angles. Hence,

$$\triangle AMB \cong \triangle CME$$

by SAS and therefore,

$$\angle A = \angle MCE = \angle ACE.$$

To finish the argument it must be proved that $\overrightarrow{CE}$ lies between $\overrightarrow{CA}$ and $\overrightarrow{CD}$. Consider

$$\text{Interior } \overline{\angle ACD} = H_1(A, \overleftrightarrow{CD}) \cap H_1(D, \overleftrightarrow{CA}).$$

But (AMC) and (BME) imply by Theorem 24.4 that A and E both lie on the same side of $\overleftrightarrow{CD}$ as M. That is, A and E lie on the same side of $\overleftrightarrow{CD}$. Therefore,

$$E \in H_1(A, \overleftrightarrow{CD}).$$

Similarly, (BME) and (BCD) imply that D and E both lie on the opposite side of $\overleftrightarrow{CA}$ as B. Hence

$$E \in H_1(D, \overleftrightarrow{CA}).$$

Therefore, $E \in$ Interior $\overline{\angle ACD}$ and by Corollary 25.3

$$(\overrightarrow{CA}\ \overrightarrow{CE}\ \overrightarrow{CD}).$$

It then follows that

$$\angle ACD = \angle ACE + \angle ECD > \angle ACE,$$

and hence, the desired result

$$\angle ACD > \angle A.$$

(2) To prove $\angle ACD > \angle B$, let D' be such that (ACD') and consider exterior angle $\overline{\angle BCD'}$ (Fig. 124). By the preceding argument, $\angle BCD' > \angle B$. (The

figure illustrates the construction for this case.) But $\overline{\angle ACD}$ and $\overline{\angle BCD'}$ are vertical angles so it follows that

$$\angle ACD = \angle BCD' > \angle B. \ \blacksquare$$

It is to be observed that this theorem is the embodiment of practically all the previous axioms, since nearly all of them were employed in its proof. Of course, the betweenness relation for rays made the final step work—without it the proof breaks down completely. In this connection it is interesting to compare the above argument with the "classical" one, often found in elementary treatments:

Extend $\overline{BM}$ its own length to E and draw $\overline{CE}$ (Fig. 125). Then $AM = MC$, $\angle AMB = \angle EMC$, and $BM = ME$. By SAS $\triangle ABM \cong \triangle CME$ and $\angle A = \angle MCE$. But angle MCE is obviously *part of* angle ACD, so

$$\angle ACD > \angle MCE = \angle A. \quad \text{Q.E.D.}$$

Figure 124   **Figure 125**

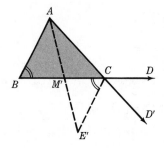
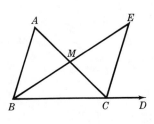

If no reference is made to the relation $(\overrightarrow{CA} \ \overrightarrow{CE} \ \overrightarrow{CD})$ in such a proof, then it is woefully inadequate. For, observe Fig. 126. A spherical right triangle ABC having congruent obtuse angles at B and C appears, with M the midpoint of $\overline{AC}$. But $BM > a/2$. Hence, if BM is "extended its own length" to E, E ends up *on the*

Figure 126

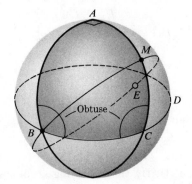

opposite side of $\overleftrightarrow{CD}$ as A. Instead of $\angle ACD > \angle MCE$, just the reverse is true: $\angle ACD < \angle MCE$!

It is an entertaining exercise to apply the exterior-angle theorem to one of the exterior angles of an isosceles triangle and deduce the following result:

30.5 Corollary: The base angles of any isosceles triangle whose sides are of length less than $\alpha/2$, are acute.

By applying the exterior-angle theorem to a right triangle, one may easily prove

30.6 Corollary: The acute angles of a right triangle whose sides are of length less than $\alpha/2$, are acute.

In view of Corollary 30.6 (see Fig. 127), Corollary 30.5 may be strengthened to give

30.5′ Corollary: The base angles of an isosceles triangle whose *legs* are of length less than $\alpha/2$, are acute.

Figure 127

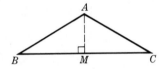

31.   Inequalities for Triangles

The exterior-angle theorem leads to several classical inequalities related to triangles. The first of these is fundamental. Usually known as the *triangle inequality*, it is the assertion that the sum of the lengths of any two sides of a triangle is greater than that of the third. The proof of a special case will lead to the general result.

31.1 Lemma: If in $\triangle ABC$, $AB < \alpha/2$ and $AC \leqq \alpha/2$, then $AB + AC > BC$.

PROOF: Assume that $AB + AC \leqq BC$. Then in particular, $AB < BC$ and by the segment-construction theorem point D may be located on $\overleftrightarrow{BC}$ such that $AB = BD$ and (BDC). Thus, $DC = BC - BD \geq AB + AC - BD = AC$, and there is E on $\overline{CD}$ such that $AC = EC$ and either $D = E$ or (DEC). The case $D = E$ may be resolved rather quickly, for then, as shown in Fig. 128, triangles

Figure 128

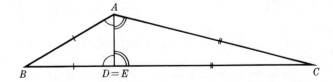

ABD and ACD are isosceles. Since $(\overrightarrow{AB}\ \overrightarrow{AD}\ \overrightarrow{AC})$ by Axiom 10 and $(\overrightarrow{DB}\ \overrightarrow{DA}\ \overrightarrow{DC})$ by Lemma 22.9, it follows that

$$\angle BAC = \angle BAD + \angle DAC = \angle ADB + \angle ADC = \angle BDC = 180,$$

a contradiction. If (DEC), then $(BDEC)$ follows (Fig. 129). Therefore, $AD < \alpha/2$ and $AE < \alpha/2$ by Lemma 30.3. Locate the midpoint M of $\overline{DE}$. Then $AM < \alpha/2$ and since $DE < BC < \alpha$, $DM = ME = \frac{1}{2}DE < \alpha/2$. Therefore, the exterior-angle theorem applies to both triangles ADM and AME. Observing the angles marked in the figure,

$$\angle 1 > \angle 2 = 180 - \angle 3 = 180 - \angle 4,$$
$$\angle 5 > \angle 6 = 180 - \angle 7 = 180 - \angle 8.$$

It follows by addition that

$$180 > 360 - \angle 4 - \angle 8,$$

which leads to the contradiction (by Axiom 10)

$$\angle BAC > \angle 4 + \angle 8 > 180.$$

Therefore,

$$AB + AC > BC. \ \blacksquare$$

Figure 129

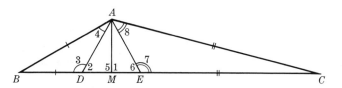

31.2 Theorem: In any triangle, the sum of the lengths of any two sides is greater than the length of the third side.

PROOF: With the notation of the preceding theorem it suffices to prove the single inequality $AB + AC > BC$. If both $AB \geq \alpha/2$ and $AC \geq \alpha/2$, then $AB + AC \geq \alpha/2 + \alpha/2 = \alpha > BC$, and the inequality already follows. Therefore, assume that one of the segments AB or AC is less than $\alpha/2$, say $AB < \alpha/2$.[4] Then if $AC \leq \alpha/2$, the lemma implies $AB + AC > BC$. If $AC > \alpha/2$, then the inequality would again follow trivially if, in addition, $BC \leq \alpha/2$. Hence, the remaining case is $AC > \alpha/2$ and $BC > \alpha/2$. Since this involves the hypothesis $\alpha < \infty$, let $\overleftrightarrow{AC}$ and $\overleftrightarrow{BC}$ meet at C', with $CC' = \alpha$ (Fig. 130). It follows that $C'B = \alpha - BC < \alpha/2$, and since $AB < \alpha/2$ Lemma 31.1 applies to $\triangle BAC'$:

$$BC' + AB > AC',$$

or

$$\alpha - BC + AB > \alpha - AC$$

---

[4] By interchanging B and C in the argument which follows, one could obtain an explicit proof for the case $AC < \alpha/2$.

which reduces to

$$AB + AC > BC,$$

the desired inequality. This completes the proof in all cases. ▌

A very important property of distance is an easy extension of the preceding theorem:

31.3 Triangle Inequality: If A, B, and C be any three distinct points in the plane,

$$AB + BC \geq AC,$$

with equality only if (ABC). (See Fig. 131.)

An immediate corollary is a formulation of the ancient geometric precept:

A line is the shortest path between two points.

Figure 130

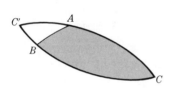

Figure 131

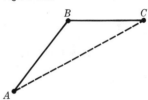

31.4 Polygonal Inequality: For any positive integer $n \geq 3$, if P_1, P_2, $\cdots$, P_n are any n points in the plane (Figure 132), then

$$P_1P_2 + P_2P_3 + P_3P_4 + \cdots + P_{n-1}P_n \geq P_1P_n.$$

PROOF: By Theorem 31.3,

$$P_1P_2 + P_2P_3 \geq P_1P_3,$$
$$P_1P_2 + P_2P_3 + P_3P_4 \geq P_1P_3 + P_3P_4 \geq P_1P_4,$$
$$P_1P_2 + P_2P_3 + P_3P_4 + P_4P_5 \geq P_1P_4 + P_4P_5 \geq P_1P_5,$$

and so on, the process ending (by mathematical induction) in the step

$$P_1P_2 + P_2P_3 + P_3P_4 + \cdots + P_{n-1}P_n \geq P_1P_n. ▌$$

One of Euclid's propositions is concisely put:

"In any triangle the greater side subtends the greater angle." [5]

Figure 132

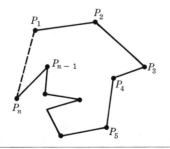

---

[5] Euclid I, 18: Heath [14, p. 283].

In order that we may use this rather efficient language, let the order relation $<$ for real numbers be extended to *segments* and *angles* in the following way: If $\overline{AB}$ and $\overline{XY}$ are any two segments and $\angle A$ and $\angle X$ are any two angles, define

$$\overline{AB} < \overline{XY} \quad \text{if and only if} \quad AB < XY,$$

and

$$\angle A < \angle X \quad \text{if and only if} \quad \angle A < \angle X.$$

Further, if x is any real number,

$$\overline{AB} < x \text{ (or } \angle A < x) \quad \text{if and only if} \quad AB < x \text{ (or } \angle A < x).$$

Thus, the definition of "acute angle" becomes "an angle less than a right angle" or "an angle less than 90."

31.5 Theorem: If one angle of a triangle is greater than a second, then the side opposite the first is greater than the side opposite the second.

PROOF: In Fig. 133, let $\angle B > \angle C$. Then there exists a ray $\overrightarrow{BX}$ such that $\angle XBC = \angle C$ and $(\overrightarrow{BA}\ \overrightarrow{BX}\ \overrightarrow{BC})$, by the angle-construction theorem. $\overrightarrow{BX}$ will meet $\overline{AC}$ in a point D such that (ADC). The triangle inequality then implies

$$AB < AD + DB$$
$$= AD + DC$$
$$= AC. \ \blacksquare$$

There are two immediate corollaries; the first is actually the converse of Theorem 31.5.

31.6 Corollary: If one side of a triangle is greater than a second, then the angle opposite the first is greater than the angle opposite the second.

PROOF: Let $\triangle ABC$ be such that $AB > AC$ (Fig. 134). Now either $\angle B > \angle C$, $\angle B = \angle C$, or $\angle B < \angle C$. If $\angle B > \angle C$, then by Theorem 31.5, $AC > AB$, which contradicts $AB > AC$. If $\angle B = \angle C$, then $AB = AC$, again contradicting $AB > AC$. Therefore, $\angle B < \angle C$. $\blacksquare$

Figure 133   **Figure 134**

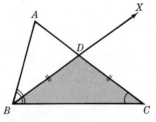

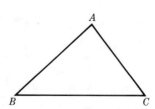

In view of Corollary 30.6 the preceding theorem implies the following important result.

31.7 Corollary: If the sides of a right triangle are each less than $\alpha/2$, the hypotenuse is the longest side.

For the case $\alpha = \infty$ Corollary 31.7 applies to all right triangles, but if $\alpha < \infty$ there are triangles to which it does not. It is of importance later to be able to remove part of that restriction and to specialize Corollary 31.7 in a particular way.

31.7′ Corollary: If a right triangle has one of its legs less than $\alpha/2$, then the angle opposite that leg is acute, and the hypotenuse is greater than that leg.

PROOF: Let $\triangle ABC$ have a right angle at C (Fig. 135) and suppose $AC < \alpha/2$. If $AB \geqq \alpha/2$, then $AB > AC$ and by Corollary 31.6 $\angle B < \angle C = 90$. If on the other hand $AB < \alpha/2$, locate M, the midpoint of $\overline{BC}$. Then $\triangle ABM$ and $\triangle AMC$ are two triangles each having sides less than $\alpha/2$. By Corollary 31.7, $\angle AMC < 90$ and by the exterior-angle theorem,

$$\angle B < \angle AMC < 90.$$

Figure 135

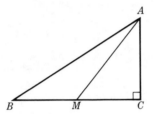

Then, $\angle B < \angle C$ and by Theorem 31.5

$$AB > AC. \quad \blacksquare$$

Another important inequality for triangles is one which involves two *different* triangles:

31.8 Theorem: If two triangles have two sides of one congruent, respectively, to two sides of the other, but the included angle of the first is greater than that of the second, then the side opposite the first angle is greater than the side opposite the second, and conversely. (See Fig. 136.)

PROOF: If $\angle A > \angle X$ there is a ray $\overrightarrow{AD}$ such that $\angle BAD = \angle X$ and $(\overrightarrow{AB}\ \overrightarrow{AD}\ \overrightarrow{AC})$; assume D is such that $AD = XZ$. Then $\triangle ABD \cong \triangle XYZ$, and

Figure 136

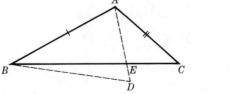
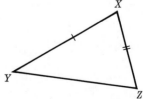

$BD = YZ$. It suffices to prove $BC > BD$. By the crossbar principle there is an interior point E on $\overrightarrow{AD}$ such that (BEC). Then there are three cases: (AED), $E = D$, and (ADE). We observe that the case $E = D$ is trivial, however, since then (BDC) would already imply $BC > BD$, so we proceed to the other two.

(1)   Assume (AED) as shown in Fig. 137. By Axiom 10, (BEC) implies $(\overrightarrow{DB}\ \overrightarrow{DE}\ \overrightarrow{DC})$ and (AED) implies $(\overrightarrow{CA}\ \overrightarrow{CE}\ \overrightarrow{CD})$. Since $\triangle ACD$ is isosceles, $\angle ADC = \angle ACD$. If the betweenness relations are now applied, we may infer (reading the angles from the figure)

$$\angle 1 + \angle 2 > \angle 2 = \angle 3 + \angle 4 > \angle 3.$$

Hence,

$$\angle BDC > \angle BCD,$$

and by Theorem 31.5

$$BC > BD.$$

Figure 137

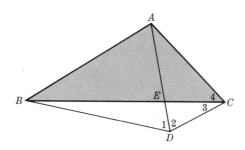

(2)   Assume (ADE) as shown in Fig. 138. Again Axiom 10 and the properties of opposite rays imply the betweenness relations $(\overrightarrow{DB}\ \overrightarrow{DE}\ \overrightarrow{DC})$ and $(\overrightarrow{CA}\ \overrightarrow{CD}\ \overrightarrow{CE}\ \overrightarrow{CF})$, where F is any point such that (ACF). Again reading the angles from the figure, we observe that

$$\angle 1 + \angle 2 > \angle 2 = 180 - \angle 3 = 180 - \angle 4 = \angle 5 + \angle 6 > \angle 5.$$

That is,

$$\angle BDC > \angle BCD,$$

so $BC > BD$. ▍

Figure 138

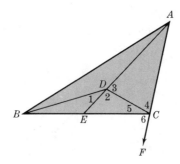

EXERCISES

1. When does equality occur in Theorem 31.4?
2. Prove Corollary 30.5.
3. Prove Corollaries 30.6 and 30.5'. (See Exercise 4.)
4. Illustrate on the sphere that M is not necessarily the midpoint of $\overline{BC}$ if merely $\overleftrightarrow{AM} \perp \overleftrightarrow{BC}$ at M and $AB = AC$.
5. Suppose $\overleftrightarrow{AD} \perp \overleftrightarrow{BC}$ at D and $AD < \alpha/2$ (Fig. 139). Show that (BDC) holds if it is known that both $\angle B$ and $\angle C$ are acute angles. *Hint:* Let M be the midpoint of $\overline{BC}$. By Theorem 19.6, if (BDC) does *not* hold then since $B \neq D$ and $C \neq D$, either (DBM) or (DCM). Apply Corollary 31.7'.

Figure 139

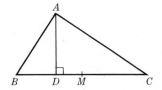

6. If A, B, C, and D be any four points, then $AC + BD \leq AB + BC + CD + AD$. Prove.
7. Prove by the triangle inequality that if $AB \leq r$, $AC \leq r$, and X is any point on segment BC then $AX \leq 2r$. Do you think this inequality could be improved to, say, $AX \leq 3r/2$?
8. Prove that if a, b, and c are fixed positive reals, $a \geq b \geq c$, such that for each positive integer n the numbers a^n, b^n, c^n are the lengths of the sides of some triangle, then these triangles must all be isosceles. *Hint:* Use the factoring law $a^{n+1} - b^{n+1} = (a - b)(a^n + a^{n-1}b + a^{n-2}b^2 + \cdots + b^n)$. (*Hungarian Problem Book II*, Random House: New York, 1963, p. 16; used by permission.)
9. Prove that there exists an isosceles triangle each of whose sides are less than any given real $a > 0$.
10. In Fig. 140 assume that the sides of the two triangles are each less than $\alpha/2$ and that $\angle A = \angle X$, $AB = XY$, but that $\angle B < \angle Y$. Prove $\angle C > \angle Z$.
11. Again using Fig. 140, suppose that $AB = XY$, $\angle A = \angle X$, but $AC < XZ$. Prove that $\angle C > \angle Z$.

Figure 140

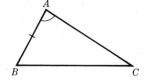 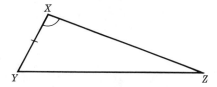

12. In triangles ABC and XYZ, each having sides less than $\alpha/2$, it is given that $\angle A = \angle X$, $\angle B = \angle Y$, but $AB < XY$. Prove that $AC < XZ$.
13. (a) Assuming $\alpha > 1$, what is the most that the lengths of two sides of a triangle can differ if the length of the third side is $1/10$? $1/100$? (b) Referring to Fig. 141 prove that

if $X_0X_1 = 1$, $X_0X_2 = \frac{1}{2}$, $X_0X_3 = \frac{1}{3}$, $\cdots$, $X_0X_n = 1/n$, $\cdots$, then $\lim\limits_{n\to\infty} |PX_0 - PX_n| = 0$,

or, equivalently, $\lim\limits_{n\to\infty} PX_n = PX_0$.

Figure 141

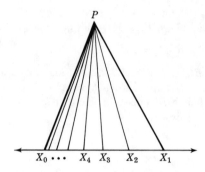

14. (a) In Fig. 142, with M the midpoint of $\overline{AB}$ and $\overline{CD}$, use the triangle inequality for $\triangle BCD$ to prove the inequality

$$2d < a + b.$$

   (b) Prove the proposition: The median to any side of a triangle whose sides are less than $\alpha/2$ is less than the arithmetic mean of the lengths of the other two sides.

   (c) Is the proposition in (b) true on the sphere for all triangles? *Hint:* Study Fig. 143; what do you know about a', b', and d'?

Figure 142

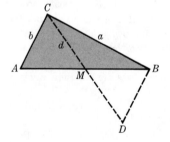

Figure 143

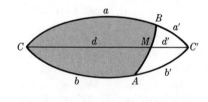

15. (a) Let a and b denote the lengths of sides $\overline{BC}$ and $\overline{AC}$, respectively, of $\triangle ABC$ and consider the segments joining C and the points on $\overline{AB}$ which divide $\overline{AB}$ into eight congruent segments (as shown in Fig. 144). Using Exercise 14(b), prove the following three inequalities:

$$d_{\frac{1}{2}} < \tfrac{1}{2}a + \tfrac{1}{2}b,$$
$$d_{\frac{1}{4}} < \tfrac{1}{4}a + \tfrac{3}{4}b, \qquad\qquad (31.9)$$
$$d_{\frac{3}{4}} < \tfrac{3}{4}a + \tfrac{1}{4}b.$$

   (b) Prove the further group of inequalities

$$d_p < pa + qb, \qquad\qquad (31.10)$$

where $d_p = CX$, $p = AX/AB$, $q = 1 - p = XB/AB$, and p is of the form $n/2^3$ for $n = 1, 2, \cdots, 7$, from the preceding inequalities (31.9) and Exercise 14(b).

Figure 144

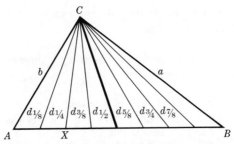

$$d_{1/8} \quad d_{1/4} \quad d_{3/8} \quad d_{1/2} \quad d_{5/8} \quad d_{3/4} \quad d_{7/8}$$

★16. A number of the form $n/2^k$, where k and n are non-negative integers with $0 \leqq n \leqq 2^k$ is called a *diadic rational of order* k. By induction on k, prove that with the notation of Exercise 13

$$d_p \leqq pa + qb. \tag{31.11}$$

for any diadic rational p.

NOTE: From the fact that there exists a sequence of diadic rationals converging to every real number p, $0 \leqq p \leqq 1$, and from the continuity of d_p as a function of p (essentially the result of Exercise 13) it can be established that inequality (31.11) prevails for all real p, $0 \leqq p \leqq 1$. It is interesting to compare this with the Euclidean formula for d_p which we shall prove later:

$$d_p{}^2 = pa^2 + qb^2 - pqc^2,$$

where a, b, and c are the lengths of the sides of the triangle. Therefore, in Euclidean geometry one obtains the further inequality

$$d_p{}^2 \leqq pa^2 + qb^2. \tag{31.12}$$

17. Exercise 19 at the end of Chapter 1 involves the use of the triangle inequality in Euclidean geometry. Solve it if you have not already done so. (Use the fact that if $a \leqq b \leqq c$ and $a + b > c$ there exists a triangle whose sides are of length a, b, and c.)

32. Further Congruence Criteria

The exterior-angle theorem and the resulting inequalities we have established for triangles provide easy proofs of some of the well-known congruence theorems for triangles. The first is the so-called "side-side-side" theorem:

32.1 SSS Congruence Criterion for Triangles: If under some correspondence two triangles have their corresponding sides congruent, the triangles are congruent.

PROOF: Suppose that in triangles ABC and XYZ, $AB = XY$, $AC = XZ$, and $BC = YZ$ (Fig. 145). By Theorem 31.8, if $\angle A > \angle X$, then $BC > YZ$, and if

Figure 145

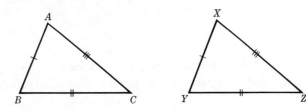

$\angle A < \angle X$, then $BC < YZ$, which are both impossible. Therefore, $\angle A = \angle X$ and the triangles are congruent by SAS. ∎

A familiar chain of propositions in elementary geometry culminates in a much used congruence criterion for triangles:

(a) The measure of an exterior angle of a triangle equals the sum of the measures of the two opposite interior angles. [$\angle 1 = \angle 2 + \angle 3$ in Fig. 146(a).]

(b) The angle sum of any triangle is 180. [$\angle 1 + \angle 2 + \angle 3 = 180$ in Fig. 146(b).]

(c) If two triangles have two angles of one congruent, respectively, to two angles of the other, then the remaining angles are congruent. [If $\angle 1 = \angle 1'$, $\angle 2 = \angle 2'$, then $\angle 3 = \angle 3'$ in Fig. 146(c).]

(d) Two triangles are congruent if they have a pair of corresponding angles and *any* pair of corresponding sides congruent.

Figure 146

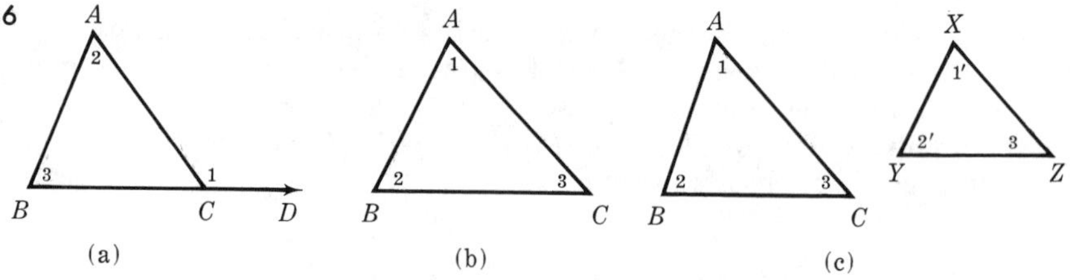

(a) (b) (c)

We have already seen why the first of these propositions cannot be true in our development. But a single example will show simultaneously that none of them are valid. Consider the spherical triangles ABC and XYZ in Fig. 147, where line $\overleftrightarrow{BC}$ is the "equator," A is the "north pole," X is the "south pole," and $\angle A > \angle X$. The angles at B, C, X, and Y will then be right angles, with $AB = XY = \alpha/2$ and

Figure 147

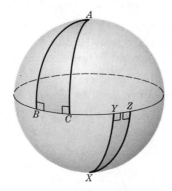

$AC = XZ = \alpha/2$. This example denies each of the propositions (a)–(d) mentioned above! Note, however, that the sides of the triangles in this example are not less

than $\alpha/2$, a requirement which has previously produced Euclidean-like results. While (a), (b), and (c) are still not true with this requirement, proposition (d) can be saved.

32.2 AAS Congruence Criterion for Triangles: Two triangles are congruent if their sides are each less than $\alpha/2$ and if under some correspondence between their vertices two angles and a side not included by those angles in one triangle are congruent, respectively, to the corresponding two angles and side in the other.

PROOF: In Fig. 148 let $\angle A = \angle X$, $\angle B = \angle Y$, and $BC = YZ$ (logically equivalent to the case $AC = XZ$). Assume $AB > XY$. There is D on $\overline{AB}$ such that (ADB) and $BD = XY$, so by SAS $\triangle DBC \cong \triangle XYZ$. Therefore,

$$\angle BDC = \angle X = \angle A.$$

Since the sides of $\triangle ABC$ are each less than $\alpha/2$, $CD < \alpha/2$ so the sides of $\triangle ADC$ are each less than $\alpha/2$ as well. But then the exterior-angle theorem implies

$$\angle BDC > \angle A,$$

a contradiction. In a similar manner, the assumption $AB < XY$ leads to a contradiction. Therefore, $AB = XY$ and the conclusion follows. $\blacksquare$

An immediate corollary may be stated.

Figure 148

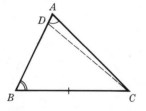 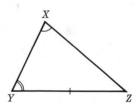

32.3 Corollary: Two right triangles are congruent if their sides are each less than $\alpha/2$ and if an acute angle and hypotenuse of one are congruent, respectively, to an acute angle and hypotenuse of the other.

The possible combinations of three corresponding parts of two triangles may be easily listed and examined for their possible validity as congruence criteria. So far, the following criteria have been considered, SAS (Axiom 10), ASA (Theorem 28.1), SSS (Theorem 32.1), and AAS (Theorem 32.2). This leaves as possible criteria those indicated by SSA and AAA. These two criteria are more complicated: The first is valid in both Euclidean and spherical geometry for *right triangles only*, while the second, even though obviously not valid in Euclidean geometry, holds *in general*, remarkably enough, in spherical geometry. We postpone further considera-

tion of the latter for now. As for the former, the reader should be able to show
that the restriction to right triangles is necessary if it is to hold, even for triangles
whose sides are less than $\alpha/2$ (study Fig. 149). Further, Fig. 147 shows that the
restriction to triangles whose sides are less than $\alpha/2$ is necessary.

Figure 149

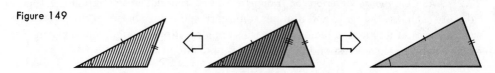

32.4   SSA Congruence Criterion for Right Triangles: Two right triangles are congruent if their sides are each less than $\alpha/2$, and a leg and hypotenuse of one are
congruent, respectively, to a leg and hypotenuse of the other.

   PROOF: Let $\triangle ABC$ and $\triangle XYZ$ be given, with sides each less than $\alpha/2$,
$\angle C = \angle Z = 90$, $AB = XY$, and $AC = XZ$ (Fig. 150). Then since $BC + YZ < \alpha$,
there is a segment $\overline{UV}$ whose length is $BC + YZ$, and we may accordingly locate D on $\overset{\leftrightarrow}{BC}$ such that $BD = UV$ and (BCD). Then $BC + CD = BC + YZ$, or
$CD = YZ$. Hence, by SAS $\triangle ADC \cong \triangle XYZ$ and $AD = XY = AB$. Therefore
$\triangle ABD$ is isosceles, with $\angle B = \angle D$. Since $\angle D = \angle Y$, we have $\angle B = \angle Y$. Hence
by AAS the given triangles are congruent. ∎

   NOTE: It can be observed on the sphere that the above theorem is false if AB and AC
are allowed to assume the value $\alpha/2$. Where then was the assumption $AB < \alpha/2$ and
$AC < \alpha/2$ used in the above proof?

Figure 150

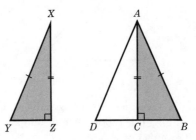

33.   The Distance from a Point to a Line

   In Euclidean geometry the distance from a point A to a line l is defined to be
the distance from A to the point where the perpendicular from A meets l. But this
would not make a good definition in spherical geometry: Even though the perpendicular to the given line may be unique, there is more than one point of intersection
between the perpendicular and the line. To eliminate a possible source of confusion
in this connection, consider two lines l and m which are perpendicular at A (Fig.
151). If $\alpha < \infty$, then l and m have a further point B in common, and $AB = \alpha$.

Locate C and D on l and m such that $AC = AD = \alpha/2$. Since (ACD) and (ADB), it follows that $CB = DB = \alpha/2$. By SSS $\triangle ACD \cong \triangle BCD$, and therefore, $\angle B$ is a right angle. Hence: *If line* l *is perpendicular to line* m *at* A *and* l *meets* m *at a second point* B, *then* l *is perpendicular to* m *also at* B.

Figure 151

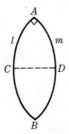

Now consider point A and line l, and suppose we undertake the task of finding a point X on l that is *nearest* to A (that is, $AX \le AX'$ for all $X' \in l$)—if such exists. The case $A \in l$ is trivial since we need only set $X = A$ for the solution. Assume $A \notin l$. By Theorem 29.9 there is at least one point B on l such that $\overleftrightarrow{AB} \perp l$ at B. There are then three cases:

(1) $AB < \alpha/2$. In this case it suffices to take $X = B$. For, suppose C is any other point on l. In the case in which A, B, and C are collinear, see Fig. 152(a), then, since $C \ne B$, $BC = \alpha$ and (BAC) implies $AC = \alpha - BA > \alpha/2 > AB$. Otherwise, triangle ABC is a right triangle with leg $AB < \alpha/2$. See Fig. 152(b). By Corollary 31.7′, again $AB < AC$. Thus, in this case AB *is the unique minimum of the set of distances from* A *to points of* l.

Figure 152

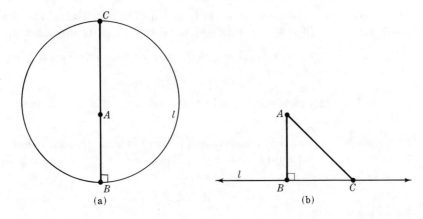

(a)   (b)

(2) $AB = \alpha/2$. Then $\alpha < \infty$ and by Lemma 30.2, $AX = AB = \alpha/2$ and $\overleftrightarrow{AX} \perp l$ for *every* point $X \in l$ (Fig. 153). In this case, for any $B \in l$ chosen at random. AB *is minimal among all the distances from* A *to points of* l.

Figure 153

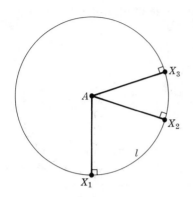

Figure 154

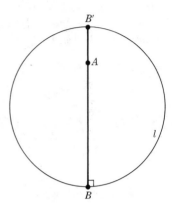

(3) $AB > \alpha/2$. Again $\alpha < \infty$ and by Corollary 24.6 $\overleftrightarrow{AB}$ meets l at a second point B' such that $BB' = \alpha$ (Fig. 154). Then it follows that $\overleftrightarrow{AB'} \perp l$ at B' with $AB' < \alpha/2$. In this case, AB' is the unique minimum sought for by applying the argument in Case 1 to B'.

We have then proved

33.1   Theorem: If A is any point and l is any line not passing through it, there exists a point B on l such that $\overleftrightarrow{AB} \perp l$ and $AB \leq \alpha/2$. The number $d = AB$ is minimal among the distances from A to all the points on line l: If $d = \alpha/2$, then $d = AX$ for each point X on l; if $d < \alpha/2$, then $d < AX$ for each point X on l, except B.

33.2   Definition: The number d mentioned in Theorem 33.1 is the **distance** from point A to line l (if $A \in l$ d is defined to be zero). If B is any point on l such that $\overleftrightarrow{AB} \perp l$ and $AB \leq \alpha/2$, then B is called a **foot of the perpendicular from** A **to line** l, or simply, a **foot of** A **on** l.

Theorem 33.1 partly anticipates the result that if $d < \alpha/2$, then the foot of A on l is unique.

33.3   Theorem: Suppose A is some point and l is some line such that $AP \neq \alpha/2$ for at least one point P on l, with $A \notin l$. Then the foot B of A on l is unique, and it is the only point on l having the property

$$AB < AX$$

for all points X on l, $X \neq B$.

PROOF: Let B be any foot of A on l. By Theorem 33.1, if $AB = \alpha/2$, then $AP = \alpha/2$, a contradiction. Hence, $AB < \alpha/2$. If B_1 is any other foot of A on l, Theorem 33.1 implies

$$AB < AB_1$$

if $B_1 \neq B$. But Theorem 33.1 also implies

$$AB_1 \leqq AB$$

which is impossible. Therefore, the foot of B is unique. Now suppose there were another point B' on l with the property that

$$AB' < AX$$

for all X on l, $X \neq B'$. If $B \neq B'$, then set $X = B$, and the preceding inequality gives

$$AB' < AB,$$

a contradiction. Therefore, $B' = B$. ∎

34. Special Segments Associated with Triangles

It is the purpose of this section to introduce formally a few of the segments associated with triangles which are familiar to most students of elementary geometry.

34.1 Definition: In triangle ABC let D, E, and F be the *unique feet* (if they exist) of the respective vertices A, B, and C on the opposite sides; L, M, and N the respective *midpoints of the sides* opposite A, B, and C; and X, Y, and Z the respective *points of intersection of the angle bisectors of* $\angle A$, $\angle B$, and $\angle C$ with the opposite sides $\overline{BC}$, $\overline{CA}$, and $\overline{AB}$ (Fig. 155). Then the segments

$$\overline{AD}, \quad \overline{BE}, \quad \overline{CF}$$

are called the **altitudes** of the triangle,

$$\overline{AL}, \quad \overline{BM}, \quad \overline{CN}$$

are the **medians**, and

$$\overline{AX}, \quad \overline{BY}, \quad \overline{CZ}$$

are the **angle bisectors**.

Figure 155

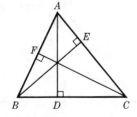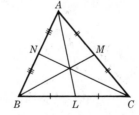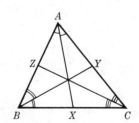

In Euclidean geometry each of the triples of segments defined here are *concurrent* (or, more accurately, determine *lines* which are concurrent). However, in absolute geometry this is true only for the angle bisectors and the medians—the reason for this will appear later. It is premature to attempt a proof that the medians of a triangle are concurrent. However, the fact that the angle bisectors are concurrent is the result of the following "locus" property, which depends mainly on the ideas of the preceding two sections:

34.2 Theorem: The locus of points[6] equidistant from the sides of an angle and lying in its interior is the interior of the bisector of that angle.

We shall indicate a proof of Theorem 34.2 for the case $\alpha = \infty$ as an illustration of the greater simplicity sometimes afforded by this assumption, the case $\alpha < \infty$ being left to the reader. Let u be the bisector of angle $\overline{hk}$, and take X any interior point of u (Fig. 156). Since $XA \neq \alpha/2$, the unique feet Y and Z of X on the lines containing h and k exist. Because $\overline{hk}$ is assumed to be nonstraight, angles $\overline{hu}$ and $\overline{uk}$ are acute and it follows that $Y \in h$ and $Z \in k$. (Why?) Then in right triangles XAY and XAZ, $\angle XAY = \angle XAZ$ and $AX = AX$. By Corollary 32.3 $\triangle XAY \cong \triangle XAZ$ and hence,

$$XY = XZ.$$

Therefore, X is equidistant from h and k.

Now it must be proved that if X is equidistant from h and k and lies in the interior of $\overline{hk}$, then $X \in u$. Again let Y and Z be the feet of X on h and k, respectively (Fig. 156). Then in right triangles XAY and XAZ,

$$XY = XZ, \qquad AX = AX.$$

Theorem 32.4 then implies the congruence of those triangles, and therefore,

$$\angle XAY = \angle XAZ.$$

Since $\overrightarrow{AX}$ lies between h and k, it follows that $\overrightarrow{AX}$ is the bisector of the given angle. ∎

Figure 156

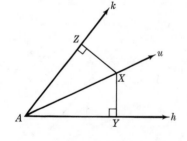

Figure 157

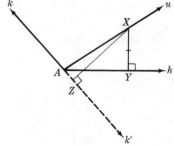

It should be pointed out that in the latter argument, like most elementary presentations of this problem, several details were lacking. The fact that Y and Z were interior points on rays h and k was not justified, and until it is justified, a case

---

[6] The term "locus of points" as used in geometry refers to a *set* of points all having some property in common. Although the term is probably obsolete, avoiding its use would seem most unnatural.

like that illustrated in Fig. 157 is logically possible. It was also tacitly assumed that $Y \neq A$ and $Z \neq A$. But these difficulties are easily handled and will be left to the reader.

As indicated earlier, Theorem 34.2 is the basis for

34.3 Theorem: The angle bisectors of a triangle are concurrent.

Figure 158

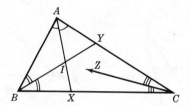

PROOF: Let the angle bisectors of $\triangle ABC$ be $\overline{AX}$, $\overline{BY}$, and $\overline{CZ}$ (Fig. 158). Consider the first two. By the crossbar principle $\overrightarrow{BY}$ cuts $\overline{AX}$ at a point I such that (AIX). Therefore, by Theorem 34.2 established above, I is equidistant from $\overleftrightarrow{AB}$ and $\overleftrightarrow{AC}$ and also from $\overleftrightarrow{AB}$ and $\overleftrightarrow{BC}$. Hence I is equidistant from $\overleftrightarrow{AC}$ and $\overleftrightarrow{BC}$. Furthermore,

$$I \in \text{Interior } \overline{ZBAC} \cap \text{Interior } \overline{ZABC}$$
$$= \text{Interior } \triangle ABC \subset \text{Interior } \overline{ZACB}.$$

Therefore, again by Theorem 34.2, $I \in \overrightarrow{CZ}$. That is, $\overrightarrow{AX}$, $\overrightarrow{BY}$, and $\overrightarrow{CZ}$ all pass through I. ∎

34.4 Definition: The point I of concurrency of the three angle bisectors of a triangle is called the **incenter** of the triangle.

EXERCISES

1. In Fig. 159, it is given that $\angle A = \angle X$, $AB = XY$, and $BC = YZ$. Prove that either $\triangle ABC \cong \triangle XYZ$ or $\overline{ZC}$ and $\overline{ZZ}$ are supplementary.

Figure 159

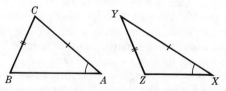

2. Using Exercise 1 show that for triangles having sides of length less than $\alpha/2$, the only time a line can bisect both a side of a triangle and the angle opposite is when the triangle is isosceles.

3. (a) Using Exercise 2 show that the following construction is inadequate (commonly attempted by high school students): To trisect an angle lay off segments $\overline{AB}$ and $\overline{AC}$

on the sides such that $AB = AC < \alpha/2$ (Fig. 160). Trisect the base $\overline{BC}$ of the resulting isosceles triangle by points D and E. Then $\overrightarrow{AD}$ and $\overrightarrow{AE}$ are the desired trisectors (!)
(b) Does the construction in (a) *ever* work in absolute geometry if we do not restrict $AB < \alpha/2$?

Figure 160

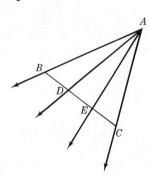

Figure 161

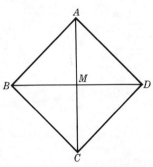

4. In the cross shown in Fig. 161, M is the midpoint of congruent segments $\overline{AC}$ and $\overline{BD}$, $\overleftrightarrow{AC} \perp \overleftrightarrow{BD}$. Prove that the four-sided figure $ABCD$ is equilateral and equiangular.
5. An equilateral triangle is equiangular. Do the angles measure 60?
6. Prove that the angle sum for triangles is 180 if and only if it is a constant for all triangles. *Hint:* Study Fig. 162 to obtain two independent expressions involving angles 1, 4, 5, and 6 and the assumed constant angle sum γ (C. R. Wylie, Jr., *Foundations of Geometry* [31, p. 107]; used by permission).

Figure 162

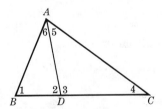

7. Is the line passing through the midpoints of two sides of a triangle parallel to the third side? Does its length equal one half that of the third side? Illustrate by sketches.
8. Assuming proposition (a) in Section 32 concerning the measure of an exterior angle of a triangle, deduce the remaining propositions (b), (c), and (d) listed there.
9. Prove the following assertion, the converse of Corollary 30.5′: *If the base angles of an isosceles triangle are acute, the legs are each less than $\alpha/2$.*
10. The medians of any isosceles triangle are concurrent.
11. For triangles whose sides are less than $\alpha/2$ and whose angles are acute, the altitudes of an isosceles triangle are concurrent.
12. For triangles whose sides are less than $\alpha/2$ and whose angles are acute, two altitudes of a triangle are congruent if and only if the triangle is isosceles.
13. The medians to the legs of an isosceles triangle are congruent. (The converse holds, but is more difficult.)

14. The bisectors of the base angles of an isosceles triangle are congruent.

★15. *The Steiner-Lehmus Theorem for absolute geometry.* The converse of the proposition in Exercise 14 is known as the *Steiner-Lehmus Theorem* and has an interesting history [see V. Thebault, "The Theorem of Lehmus," *Scripta Mathematica*, **15** (1949), pp. 87–88]. It is a corollary to the following result (we restrict ourselves to triangles whose sides are less than $\alpha/2$): *If one side of a triangle is greater than a second side the angle bisector to the smaller side is the greater bisector.* Prove, using the result of Exercise 12, Section 31. *Hint:* In Fig. 163, locate D on $\overline{AY}$ such that $\angle DBY = \angle ACZ$ (justify this), then apply the exercise mentioned to triangles BDY and CFD.

16. *The circumcenter of a triangle.* Prove

   34.5 Theorem: The perpendicular bisectors of the sides of a triangle are concurrent if any two of them intersect. Moreover, the point of concurrency is equidistant from the vertices of the triangle.

   The point of concurrency of Theorem 34.5 (if it exists) is called the **circumcenter** of the given triangle. To anticipate a later discussion of circles, it follows that the circumcenter is the center of a circle which passes through the three vertices of the triangle, called its **circumcircle.**

17. The feet of the incenter of a triangle on the three sides are unique interior points on those sides. (See Exercise 5, Section 31.)

Figure 163

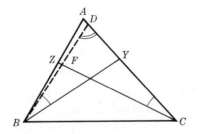

5

Quadrilaterals, Polygons, and Circles

The fundamental properties of polygons and circles commonly studied in elementary geometry may be easily derived axiomatically. The purpose of the present chapter is to explore that derivation briefly. The polygon has an important special case in the quadrilateral—the object of discussion for the opening sections.

35. Quadrilaterals

If four points A, B, C, and D are chosen in the plane, one may describe a total of six distinct "shortest paths" beginning and ending at any one of those four points, say at A, and passing through each of the other points precisely once. Each such path must accordingly be the union of four segments, illustrated in Figs. 164 and 165. It is to be noticed that these "paths" occur in *pairs* whose members consist of the *same point sets* but having *opposite directions*. For example, the path A–B–D–C–A shown in Fig. 164(b) is the same as the path A–C–D–B–A shown in Fig. 164(d), except for its direction.

Figure 164

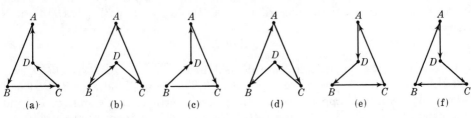

(a)   (b)   (c)   (d)   (e)   (f)

Figure 165

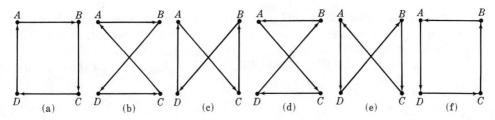

(a)   (b)   (c)   (d)   (e)   (f)

One important difference is obvious in the paths shown in the two figures, however. All those shown in Fig. 164 and in (a) and (f) of Fig. 165 are *simple*[1] paths, while those in (b)–(d) (Fig. 165) are not. This distinction is closely related to intuitive notions concerning *quadrilaterals*. Since a quadrilateral is a four-sided "figure" which somehow encloses a "region," one must evidently rule out figures like (b), (c), (d), and (e) of Fig. 165 as "quadrilaterals." For, the paths illustrated there are "self-intersecting" and thus do not enclose a region in the usual sense. This must be taken into account in formulating the definition.

35.1   Definition: Let P_0, P_1, P_2, P_3, and P_4 be points in the plane, distinct or not, such that for all $i, j = 0, 1, 2, 3, 4$, $P_i P_j < \alpha$. The set union

$$\overline{P_0 P_1} \cup \overline{P_1 P_2} \cup \overline{P_2 P_3} \cup \overline{P_3 P_4}$$

is called a **polygonal path of order four,** denoted

$$[P_0 P_1 P_2 P_3 P_4],$$

with the given points as **vertices** and the named segments as **sides.**[2] The polygonal path is **closed** if $P_0 = P_4$, and **simple** if the vertices are distinct, with the single possible exception $P_0 = P_4$, and if the sides intersect at endpoints or not at all. A

---

[1] A curve in the plane is called *simple* if it has the property that as a point proceeds along the curve it does not occupy the same position twice, except possibly for its initial and terminal positions.

[2] Note that this definition does not include the concept of "direction" on a polygonal path. Thus, for example $[P_0 P_1 P_2 P_3 P_4] = [P_4 P_3 P_2 P_1 P_0]$. If $P_0 = P_4$, then other relations are possible, such as $[P_0 P_1 P_2 P_3 P_0] = [P_2 P_3 P_0 P_1 P_2]$. (See Exercise 1 below, in this connection.)

quadrilateral is then a simple, closed polygonal path of order four, no three of whose vertices are collinear. If the quadrilateral is defined by the path $[ABCDA]$, it is denoted

$$\Diamond ABCD.$$

The **angles** of the quadrilateral are defined to be $\overline{\angle DAB}$, $\overline{\angle ABC}$, $\overline{\angle BCD}$, and $\overline{\angle CDA}$, while the **diagonals** are $\overline{AC}$ and $\overline{BD}$. The pairs of sides $(\overline{AB}, \overline{CD})$ and $(\overline{BC}, \overline{DA})$ are termed **opposite,** as are the pairs of vertices (A, C) and (B, D), and the pairs of angles $(\overline{\angle A}, \overline{\angle C})$ and $(\overline{\angle B}, \overline{\angle D})$. All other pairs of vertices, sides, or angles are called **adjacent** or **consecutive.**

It is obvious from the examples shown in Figs. 164 and 165 that sometimes four points are the vertices of more than one quadrilateral, in contrast with the situation for triangles. Thus, in Fig. 166 the points A, B, C, and D are in such a position that the paths $[ABDCA]$, $[ACBDA]$, and $[ABCDA]$ are all three quadri-

Figure 166

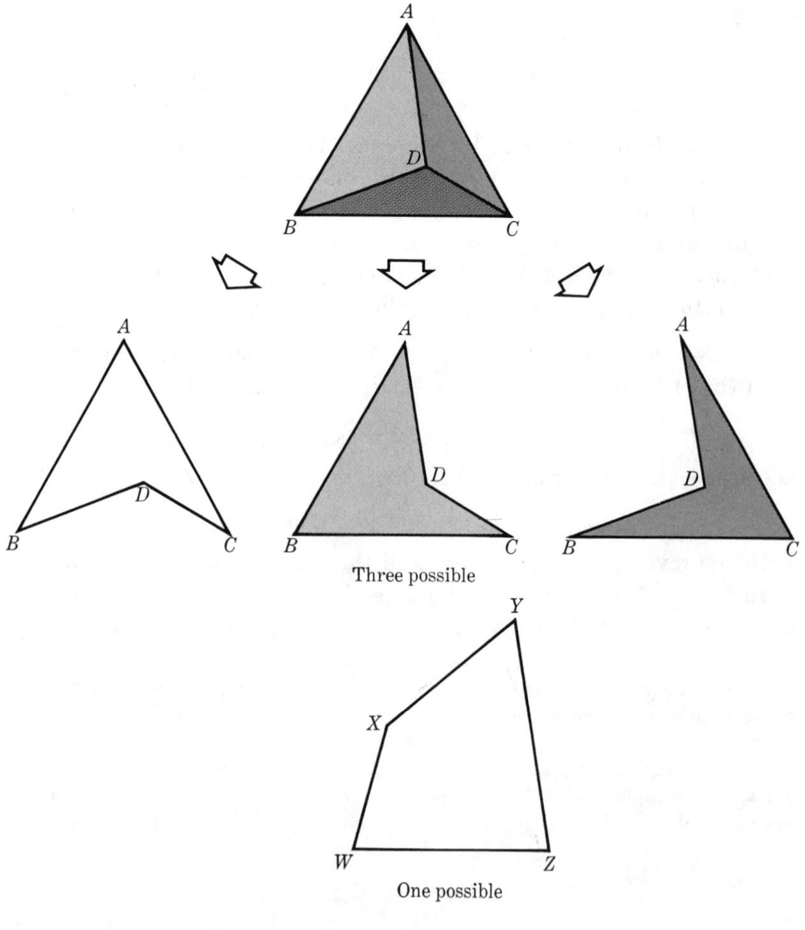
A
D
B
C
A
D
B
C
A
D
B
C
A
D
B
C
Three possible
Y
X
W
Z
One possible

laterals. On the other hand, the points X, Y, Z, and W shown in that figure determine several *polygonal paths* but only the one *quadrilateral* $\Diamond XYZW$. (For example, $[XZYWX]$ is not a quadrilateral because $\overline{XZ}$ meets $\overline{YW}$.) This phenomenon is of more significance than one might at first suspect.

36. Convex Quadrilaterals

Most applications of the properties of triangles to the study of quadrilaterals make use of the fact that certain quadrilaterals have diagonals which *intersect*. This in turn may be derived from a more fundamental property, namely, that *if ℓ is any one of the (extended) sides of the quadrilateral, then the rest of the quadrilateral lies entirely on one side of ℓ.* This property has bearing on the question of whether a quadrilateral and its interior determine a convex set. For this reason, a quadrilateral having the above property will be termed "convex" (and not because the quadrilateral by itself is a convex set—thus, in connection with quadrilaterals this meaning of the term "convex" differs from its previous meaning in the phrase "convex set").

It will be convenient to state the definition of convex quadrilaterals in a slightly different manner. Note that $\Diamond ABCD$ lies completely on one side of the line $\overset{\leftrightarrow}{AB}$ (except for the points on $\overline{AB}$) if and only if line $\overset{\leftrightarrow}{AB}$ does not cut side $\overline{CD}$ (symbolically, $\overset{\leftrightarrow}{AB} \cap \overline{CD} = \varnothing$—see Fig. 167).

Figure 167

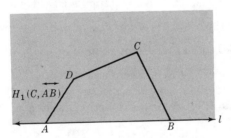

36.1 Definition: A quadrilateral is **convex** if it has the property that no side is intersected by the line determined by the opposite side. In symbols, $\Diamond ABCD$ is convex if and only if

$$\overset{\leftrightarrow}{AB} \cap \overline{CD} = \overset{\leftrightarrow}{BC} \cap \overline{DA} = \overset{\leftrightarrow}{CD} \cap \overline{AB} = \overset{\leftrightarrow}{DA} \cap \overline{BC} = \varnothing.$$

The reader should test this definition on the quadrilaterals shown in Fig. 166 to discover whether it agrees with his preconceived notion of a "convex quadrilateral." The property of the diagonals mentioned earlier can now be easily established for convex quadrilaterals. It is a corollary of the following result:

36.2 Lemma: If $\Diamond ABCD$ is a quadrilateral, the two conditions

(a) $\overset{\leftrightarrow}{AB} \cap \overline{CD} = \overset{\leftrightarrow}{BC} \cap \overline{DA} = \overset{\leftrightarrow}{CD} \cap \overline{AB} = \varnothing,$

(b) $\overline{AC} \cap \overline{BD} \neq \varnothing$

are equivalent.

PROOF: Suppose (a) holds. Since C and D do not lie on $\overleftrightarrow{AB}$, and A and D do not lie on $\overleftrightarrow{BC}$, the first two relations in (a) imply $D \in H_1(C, \overleftrightarrow{BA})$ and $D \in H_1(A, \overleftrightarrow{BC})$. Hence,

$$D \in \text{Interior } \overline{\angle ABC}$$

(Fig. 168). By Corollary 25.3 and the crossbar principle, there is P on $\overrightarrow{BD}$ such that (APC). But the second and third relations in (a) imply $A \in H_1(D, \overleftrightarrow{CB}) \cap H_1(B, \overleftrightarrow{CD})$. Therefore,

$$A \in \text{Interior } \overline{\angle BCD},$$

and again, Corollary 25.3 implies

$$(\overrightarrow{CB}\ \overrightarrow{CA}\ \overrightarrow{CD}).$$

By Axiom 10,

$$(BPD).$$

Hence, P is an interior point on both $\overline{AC}$ and $\overline{BD}$, and (b) follows.

Conversely, suppose (b) holds, with $P \in \overline{AC} \cap \overline{BD}$. Since the definition of $\Diamond ABCD$ involves $P \neq A$, $P \neq B$, $P \neq C$, and $P \neq D$, it follows that P is an interior point on each diagonal. Now, if $\overleftrightarrow{AB}$ cuts CD, it must do so at an interior point E (Fig. 169). By the postulate of Pasch, since $BP < \alpha$ implies that $\overleftrightarrow{AB}$ cannot pass through P, line $\overleftrightarrow{AB}$ cuts $\overline{PC}$ or $\overline{PD}$ at an interior point. If $\overleftrightarrow{AB} \cap \overline{PC} = F$, then (APC) and (PFC) implies $(APFC)$. Therefore, $\overleftrightarrow{AB} = \overleftrightarrow{AF}$, and hence, A, B, and C are collinear, a contradiction. The other case is similar. It is obvious that the remaining conditions in (a) may be easily derived in like manner. ∎

The above argument showing that $\overleftrightarrow{AB} \cap \overline{CD} = \emptyset$ may be easily adapted to prove also the fourth relation in Definition 36.1

$$\overleftrightarrow{DA} \cap \overline{BC} = \emptyset.$$

Figure 168   **Figure 169**

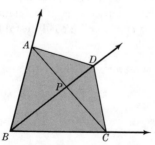

This proves the equivalence of (b) and the convexity of $\Diamond ABCD$, or

(c) $\overleftrightarrow{AB} \cap \overline{CD} = \overleftrightarrow{BC} \cap \overline{DA} = \overleftrightarrow{CD} \cap \overline{AB} = \overleftrightarrow{DA} \cap \overline{BC} = \emptyset.$

Since, therefore, (a) and (c) are both equivalent to (b), they are equivalent to each other. The following useful theorem is thereby inferred.

36.3 Theorem: A quadrilateral is convex if and only if it has the property that no *three* sides are intersected by the lines containing the sides opposite.

The equivalence of (b) and (c) imply

36.4 Theorem: A quadrilateral is convex if and only if its diagonals intersect at a point which is interior to each diagonal.

36.5 Corollary: With either endpoint of a diagonal of a convex quadrilateral as origin, the diagonal determines a ray which lies between the sides of the angle of the quadrilateral whose vertex is the given endpoint.

PROOF: Axiom 10. (See Fig. 168.) ▌

37. Congruence Theorems for Convex Quadrilaterals

As in triangles, the notation $ABCD \leftrightarrow XYZW$ is used to establish the correspondence

$$A \leftrightarrow X \qquad C \leftrightarrow Z$$
$$B \leftrightarrow Y \qquad D \leftrightarrow W$$

between the vertices of the quadrilaterals $ABCD$ and $XYZW$. Two quadrilaterals are **congruent** if and only if under some such correspondence the pairs of corresponding sides and angles are congruent. The notation

$$\Diamond ABCD \cong \Diamond XYZW$$

will be used to assert that quadrilaterals ABCD *and* XYZW *are congruent under the correspondence* ABCD $\leftrightarrow$ XYZW.

Just as the case with triangles, congruence among certain parts of a pair of quadrilaterals leads to the congruence of all pairs and thus of the quadrilaterals. Only one of those "congruence criteria" will be proved here (the one that is used in the next section); the others are left to the reader (see Exercises 5 and 6 below).

37.1 SASAS Congruence Criterion for Quadrilaterals: Two convex quadrilaterals are congruent if, under some correspondence, three sides and the two adjacent angles included by those sides in one quadrilateral are congruent, respectively, to the corresponding three sides and two adjacent angles of the other.

PROOF: In Fig. 170 let $AB = XY$, $\angle B = \angle Y$, $BC = YZ$, $\angle C = \angle Z$, and $CD = ZW$. By SAS,

$$\triangle ABC \cong \triangle XYZ.$$

Figure 170

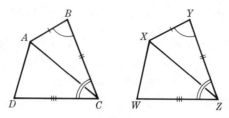

Therefore, $AC = XZ$ and $\angle ACB = \angle XZY$. Since the quadrilaterals are convex, Corollary 36.5 implies

$$(\overrightarrow{CB} \ \overrightarrow{CA} \ \overrightarrow{CD})$$

and

$$(\overrightarrow{ZY} \ \overrightarrow{ZX} \ \overrightarrow{ZW}).$$

Therefore, in triangles ACD and XZW, $AC = XZ$, $\angle ACD = \angle BCD - \angle ACB = \angle YZW - \angle XZY = \angle XZW$, and $CD = ZW$. SAS then implies

$$\triangle ACD \cong \triangle XZW.$$

It then follows that $AD = XW$, $\angle D = \angle W$, and (by Corollary 36.5)

$$\angle BAD = \angle BAC + \angle CAD = \angle YXZ + \angle ZXW = \angle YXW.$$

Therefore,

$$\Diamond ABCD \cong \Diamond XYZW. \ \blacksquare$$

Figure 171

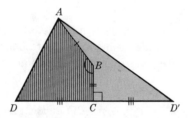

Fig. 171 shows the necessity of the restriction to convex quadrilaterals in Theorem 37.1: Obviously, $\Diamond ABCD \not\cong \Diamond ABCD'$, but under $ABCD \leftrightarrow ABCD'$ one has $AB = AB$, $\angle ABC = \angle ABC$, $BC = BC$, $\angle BCD = \angle BCD' = 90$, and $CD = CD'$.

EXERCISES

1. Since there are 24 ways of ordering the letters A, B, C, and D, there are allegedly 24 distinct closed polygonal paths possible having the four points A, B, C, and D as vertices. Classify these permutations according to the paths they represent and show that each one is set-theoretically equal to either $[ABCDA]$, $[ABDCA]$, or $[ACBDA]$ (see Fig. 164).

2. If four points determine a convex quadrilateral, then no other quadrilateral can have those four points as vertices. Prove. (See Exercise 1.)

3. If four points determine a quadrilateral which is not convex, then there are precisely three quadrilaterals having those four points as vertices. Prove. (See Exercise 1.)

4. Write an explicit proof for Corollary 36.5.

5. Consider each of the congruence criteria for convex quadrilaterals suggested by the following symbols: (a) SASAA, (b) SAASA, (c) ASASA, and (d) SAAAA. Decide which of these propositions are valid and prove those which are (by assuming the sides and diagonals are $< \alpha/2$ if necessary) and provide counterexamples in Euclidean geometry for the rest.

6. Repeat Exercise 5 for: (a) SSSSA, (b) SSASA, and (c) SASSA.

7. Congruent quadrilaterals have their corresponding diagonals congruent.

8. Assuming that the quadrilaterals formed are convex, prove that in two congruent, convex quadrilaterals a segment joining the midpoints of a pair of opposite sides in one is congruent to the segment joining the midpoints of the two corresponding sides in the other.

9. An **isosceles quadrilateral** may be defined as a convex quadrilateral having a pair of opposite sides congruent (the **legs**) and a pair of consecutive angles opposite those sides congruent (the **base angles**). Prove: (a) The consecutive angles opposite the base angles of an isosceles quadrilateral are congruent and the diagonals are congruent, and ⋆(b) if a convex quadrilateral has one pair of consecutive angles congruent, respectively, to the other pair, the quadrilateral is isosceles.

10. Let the **interior** of a convex quadrilateral be defined as the intersection of the half planes determined by the sides which contain the sides opposite. Prove:

    (a) Interior $\Diamond ABCD$ is a convex set.

    (b) $\Diamond ABCD \cup$ Interior $\Diamond ABCD$ is a convex set.

    (c) Interior $\Diamond ABCD =$ Interior $\angle A \cap$ Interior $\angle C$.

11. Prove that a convex quadrilateral is *not* convex.

12. Prove that if a line passes through an interior point of any convex quadrilateral, then it must intersect that quadrilateral in precisely two points. (See Exercise 7, Section 28.)

38. The Saccheri Quadrilateral

Let A and B be any two points, $AB < \alpha$, and erect perpendiculars to $\overleftrightarrow{AB}$ at A and B, respectively. On the perpendicular at A, select any point A' such that $AA' < \alpha/2$, and on the perpendicular at B select a second point B' lying on the same side of $\overleftrightarrow{AB}$ as A' and such that

$$BB' = AA'.$$

(See Fig. 172.)

Figure 172

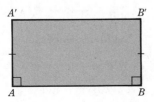

By choosing $B' \in H_1(A', \overleftrightarrow{AB})$, we guarantee that $\overleftrightarrow{AB} \cap \overline{A'B'} = \varnothing$, and $\overleftrightarrow{AA'} \cap \overline{BB'} = \varnothing$ follows from Theorem 29.9. It follows that A, B, B', and A' are the vertices of a quadrilateral $ABB'A'$, called a **Saccheri quadrilateral,** after the Italian geometer Girolamo Saccheri (1667–1733), who used it in a vain attempt to prove the parallel postulate.

Certain special terms are commonly used in connection with the Saccheri quadrilateral: side $\overline{AB}$ is called the **base,** sides $\overline{AA'}$ and $\overline{BB'}$, the **legs,** and side $\overline{A'B'}$, the **summit.** The angles at A and B are the **base angles,** while those at A' and B' are the **summit angles.**

Before going further, the reader should be cautioned not to infer Euclidean properties from the Saccheri quadrilateral which may not necessarily follow from our set of axioms. In Euclidean geometry, the Saccheri quadrilateral would surely be a rectangle, with $A'B' = AB$ and $\angle A' = \angle B' = 90$. But this cannot be inferred from the axioms thus far introduced because of the existence of spherical geometry as a model. In Fig. 173, a Saccheri quadrilateral which has $\angle A' = \angle B' > 90$ is shown. Indeed, by increasing the size of AB, AA', and BB' *these angles can be made to assume values arbitrarily close to* 180!

Figure 173

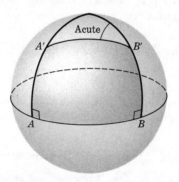

In the preceding example, the fact that $\angle A' = \angle B'$ was no accident. This is one of the easily proved properties of the Saccheri quadrilateral. First, however, the following result must be obtained:

38.1   Theorem: A Saccheri quadrilateral is convex.

PROOF: The same observations used to show that $[ABB'A'A]$ was a quadrilateral also shows it is convex: $\overleftrightarrow{AB}$ does not cut $\overline{A'B'}$ by construction, so therefore,

$$\overleftrightarrow{AB} \cap \overline{A'B'} = \varnothing.$$

By the uniqueness of the perpendiculars (since $AA' = BB' < \alpha/2$), it follows that

$$\overleftrightarrow{A'A} \cap \overline{BB'} = \overleftrightarrow{BB'} \cap \overline{A'A} = \varnothing.$$

Therefore by Theorem 36.3, $\Diamond A'ABB'$ is convex. ∎

38.2 *Theorem:* The summit angles of a Saccheri quadrilateral are congruent.

PROOF: Under the correspondence $A'ABB' \leftrightarrow B'BAA'$ (Fig. 174), note that:
$A'A = B'B$, $\angle A = \angle B$, $AB = BA$, $\angle B = \angle A$, and $BB' = AA'$.
Therefore, by SASAS

$$\Diamond A'ABB' \cong \Diamond B'BAA'.$$

Hence, $\angle A' = \angle B'$. ∎

Figure 174

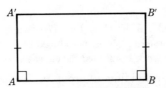

A property of the Saccheri quadrilateral to be used next is somewhat interesting by itself: *If points* M *and* M′ *be selected at random on the base and summit of a Saccheri quadrilateral, two further convex quadrilaterals are formed.* (See Fig. 175.) This property holds more generally for any convex quadrilateral, where M and M' lie on a pair of opposite sides. The more general property will be left as an exercise for the reader (Exercise 9, below, which then establishes the above.)

Figure 175

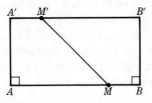

38.3 *Theorem:* The line passing through each of the midpoints of the base and summit of a Saccheri quadrilateral is their common perpendicular.

PROOF: In Fig. 176, let M be the midpoint of base $\overline{AB}$ and M' the midpoint of summit $\overline{A'B'}$. Then,

$$M'A' = M'B', \quad \angle A' = \angle B' \ (38.2), \quad A'A = B'B,$$
$$\angle A = \angle B, \quad \text{and} \quad AM = MB.$$

By SASAS,

$$\Diamond M'A'AM \cong \Diamond M'B'BM,$$

and therefore,

$$\angle AMM' = \angle M'MB, \quad \angle A'M'M = \angle B'M'M.$$

Since these pairs of angles are also supplementary, it follows that

$$\angle AMM' = \angle MM'A' = 90. \;\blacksquare$$

Figure 176

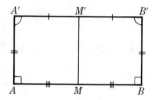

NOTE: The quadrilateral $AMM'A'$ in Fig. 176 is a quadrilateral having *three* right angles, called a **Lambert quadrilateral**. It was introduced by Johann H. Lambert (1728–1777), a German investigator whose work resembled Saccheri's.

Saccheri dreamed of providing a proof of Euclid's parallel postulate. His method was to introduce the quadrilateral named after him and investigate systematically the three possible cases (Fig. 174):

(a) $\angle A' > 90$—*the hypothesis of the obtuse angle*,
(b) $\angle A' = 90$—*the hypothesis of the right angle*,
(c) $\angle A' < 90$—*the hypothesis of the acute angle*.

He dismissed the hypothesis of the obtuse angle in short order (we shall see the possibility of doing this later). But the hypothesis of the acute angle caused him considerable difficulty, and in final desperation he begged the question by an appeal to intuition, claiming to have settled the issue for all time. It is ironic that in this effort Saccheri believed he had rendered Euclid's *Elements* more perfect, for as it turned out, had Saccheri actually succeeded, his work would have dealt a terrible blow to Euclid—it would have meant ultimately that *Euclidean geometry is inconsistent*. Misguided though he was, Saccheri very nearly discovered hyperbolic geometry—and his work anticipated that discovery by about 100 years.

39.   Polygons

Much of the discussion concerning quadrilaterals may be generalized. Only a brief development of an introductory nature will be undertaken here, however, primarily to indicate the validity of such ideas in non-Euclidean geometry.

For each integer $n > 1$, define a **polygonal path of order** n as a set of points of the form

$$[P_0P_1P_2 \cdots P_n] = \overline{P_0P_1} \cup \overline{P_1P_2} \cup \cdots \cup \overline{P_{n-1}P_n},$$

with the points P_0, P_1, P_2, $\cdots$, P_n as **vertices** and the named segments as **sides**. The path is **closed** if $P_0 = P_n$, and **simple** if the vertices are distinct, with the possible exception $P_0 = P_n$, and if the sides intersect at endpoints or not at all. The polygonal path $[P_0P_1P_2 \cdots P_n]$ is called a **polygon**, denoted

$$\bigcirc P_1P_2 \cdots P_n,$$

if it is simple and closed and if no three of its vertices lie on a line. (See Fig. 177.) The terminology regarding *angles* will be adopted as before, except that here the term "opposite" has no meaning; **consecutive vertices** are defined as a sequence of vertices of the form P_i, P_{i+1}, P_{i+2}, $\cdots$, P_{i+k} (where $i + j$ is understood to be reduced modulo n^3 for each j). Consecutive angles and consecutive sides may be defined in an analogous manner. Finally, a **diagonal** is a segment joining any pair of nonconsecutive vertices.

Figure 177

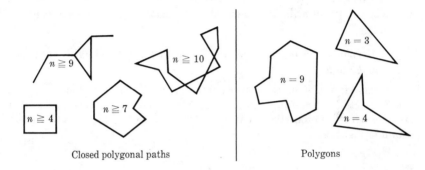

| Closed polygonal paths | Polygons |

As in the case of quadrilaterals, an important class of polygons are the **convex** polygons, defined as those having the property that no side is intersected in an interior point by the line determined by any other side. It may be easily shown by the postulate of Pasch that the ordering of the vertices of a convex polygon determines the betweeness relations of the rays having one vertex as origin and passing through the others. That is, *if* $\bigcirc P_1 P_2 \cdots P_n$ *is a convex polygon and* $1 < r < s < t$, *then* $(\overrightarrow{P_1P_r}, \overrightarrow{P_1P_s}, \overrightarrow{P_1P_t})$ (Fig. 178). Most work with convex polygons depends on this property.

Figure 178

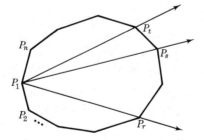

The *regular polygons* in turn form an important subclass of convex polygons, and a very familiar one. The Pythagoreans identified themselves as members of their society by wearing a five-pointed star—formed by the diagonals of a regular

---

[3] Basically this means to subtract a multiple of n from $i + j$ so that the result lies between 1 and n.

pentagon. The study of regular polygons led to unsolved problems regarding their construction, and they subsequently became the subject of many probing investigations in number theory and algebra. Euclid's compass, straightedge construction of the regular pentagon which involves the golden ratio suggests the difficulty of the construction of the general polygon.[4] In fact, we now know that this construction is impossible for certain values of n (see H. Eves, *A Survey of Geometry* [9, pp. 217–224] for more details).

The formal definition of the regular polygon leaves its existence a matter to be settled; the proof will be indicated later in the exercises (Exercises 10, 11, Section 42).

39.1 *Definition:*   **A regular polygon** is a convex polygon having congruent sides and congruent angles.

NOTE: It might be conjectured that the condition of convexity imposed in Definition 39.1 need not be assumed but, rather, may be derived as a consequence of the seemingly strong congruence properties. That this is false, however, is provided by the simple example shown in Fig. 179.

Figure 179

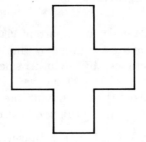

A characteristic property of the regular polygon is introduced in the next theorem.

39.2 *Theorem:*   The angle bisectors of a regular polygon are concurrent at a point which is equidistant from the vertices and sides of that polygon. Moreover, the distance from that point to each of the vertices and sides is less than $a/2$, and the feet of the perpendiculars from the point to the sides are the midpoints of the sides.

PROOF: The reader will be led through an instructive betweenness argument to prove the *existence* of the point (Exercises 7 and 8 below), so we assume that the bisectors of the angles at B and C meet at O and that O lies interior to the angles at B, C, and D (Fig. 180). Now

$$\angle 1 = \tfrac{1}{2}\angle B = \tfrac{1}{2}\angle C = \angle 2$$

by definition. Hence, $\triangle OBC$ is isosceles, and since $\angle B = \angle C < 180$, it follows that the base angles of $\triangle OBC$ are acute. By a previous exercise (Exercise 9, Section 34)

---

[4] In this connection the reader should return to Exercise 28, Chapter 1, and attempt its solution at this time if he has not already done so (some Euclidean ideas will be needed).

Figure 180

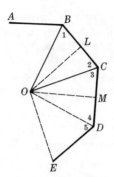

$OB = OC < \alpha/2$. Since $\angle 2 = \angle 3$ and $BC = CD$, $\triangle OCB \cong \triangle OCD$, and therefore,

$$\angle 4 = \angle 1 = \tfrac{1}{2}\angle B = \tfrac{1}{2}\angle D.$$

Since $(\overrightarrow{DC}\ \overrightarrow{DO}\ \overrightarrow{DE})$,

$$\angle 5 = \angle D - \angle 4 = \angle D - \tfrac{1}{2}\angle D = \tfrac{1}{2}\angle D = \angle 4.$$

Therefore, $\overrightarrow{DO}$ bisects $\overline{\angle D}$, and it follows in addition that $OC = OD$. It is obvious that the process continues, proving that the remaining bisectors pass through O and that O is equidistant from the vertices. To finish the proof, let L and M be the feet of O on $\overleftrightarrow{BC}$ and $\overleftrightarrow{CD}$. Since $OL \leq OB < \alpha/2$, it follows by another previous exercise (Exercise 5, Section 31) that (BLC). Similarly, (CMD). Since the triangles OBL, OLC, OCM, and OMD are congruent, L and M are midpoints, and $OL = OM$. ∎

EXERCISES

1. *The triangle and its associated Saccheri quadrilateral.* (a) In Fig. 181, M and N are midpoints and A', B', and C' are the feet of the perpendiculars to $\overleftrightarrow{MN}$ from A, B, and C. (a) Using the betweenness properties that are apparent from the figure, prove that $\lozenge B'CCC'$ is a Saccheri quadrilateral with base $\overline{B'C'}$ (called the **Saccheri quadrilateral associated with triangle** ABC), and that

$$\text{Angle-sum } (\triangle ABC) = \angle B'BC + \angle BCC' \qquad \text{and} \qquad MN = \tfrac{1}{2}B'C'.$$

(b) What do Saccheri's three hypotheses imply with regard to the angle sum of triangles?

Figure 181

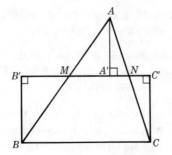

2. Assume the result of Exercise 1(b) for all triangles. (a) Prove that under the hypothesis of the acute angle the summit of a Saccheri quadrilateral is greater than the base. *Hint:* In Fig. 182, show that $\angle 1 > \angle 3$ and apply Theorem 31.8. (b) Prove the analogous fact that under the hypothesis of the obtuse angle the summit is less than the base. (c) How does the length of segment $\overline{MN}$ joining the midpoints of $\overline{AB}$ and $\overline{AC}$ compare with that of segment BC under the three hypotheses?

Figure 182

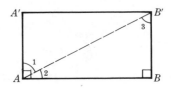

3. Prove that if M and M' lie on opposite sides of a convex quadrilateral two other convex quadrilaterals are formed. (See Fig. 175.)
4. (a) How many polygonal paths of order n are possible? (b) How many of these paths are distinguishable as sets of points?
5. How many diagonals does an n-sided polygon have?
6. Establish the SASASASASASASAS congruence criterion for convex nonagons. Generalize.
7. Let $\bigcirc P_1 P_2 P_3 \cdots P_{2k}$ $(k \geq 2)$ be a regular polygon having an even number of sides. (a) Show that the bisector of the angle at P_1 passes through P_{k+1} and is collinear with the angle-bisector at P_{k+1}. (b) Thus it follows that the bisector of the angle at P_2 passes through P_{k+2}. Prove that in this case the two bisectors meet at a point O which is interior to all the angles of the polygon. *Hint:* For (a) consider the diagonals from P_1 and make use of the italicized statement in Section 39 (Fig. 178). For (b) use the italicized statement, Axiom 10, and the crossbar principle.
★8. Let $\bigcirc P_1 P_2 P_3 \cdots P_{2k-1}$ $(k \geq 2)$ be a regular polygon having an odd number of sides. (a) Show that the angle-bisector at P_1 is the perpendicular bisector of $\overline{P_k P_{k+1}}$. (b) As in Exercise 7 show that the bisectors at P_1 and P_2 meet at a point O which is interior to all the angles of the polygon. *Hint:* Same as for Exercise 7.
9. Give a suitable definition for the *interior* of a convex polygon and prove it is a convex set.

40. Circles

Many familiar properties of circles carry over to non-Euclidean geometry. The next definition will introduce the subject formally.

40.1 Definition: A circle is the set of all points which lie at a positive, fixed distance r away from some fixed point O. Equivalently, a circle is any set of the form

$$\{X: \ OX = r\}.$$

The number r is called the **radius** and the point O, the **center** of the circle. (It is customary to use the term "radius" also for any segment which joins a point of the circle and the center; it will be clear by context which meaning is intended.)

The usual terminology associated with circles will be employed. For the convenience of the reader, a pictorial glossary is provided in Fig. 183; any definition illustrated there may be easily stated in terms of the concepts which have already been introduced. For example, an **arc** of a circle is the intersection of that circle with a line and one of its half planes.

Figure 183

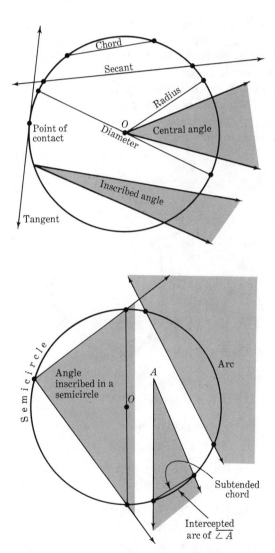

In spherical geometry we know that a circle can have *two* centers and *two* radii (Fig. 184). And from the axiomatic standpoint we know that if $\alpha < \infty$ and ω is the circle defined by

$$\omega = \{X: \ OX = r\},$$

then for each X in ω,

$$O'X = OO' - OX = \alpha - r,$$

where O' is the extremal of O. Conversely, if $O'X = \alpha - r$ then $OX = r$ and $X \in \omega$. Hence,

$$\omega = \{X: \quad O'X = \alpha - r\}.$$

Then ω also has the center O' and radius $r' = \alpha - r$. It follows, therefore, that *the center of any circle may be chosen in such a manner that the radius of the circle with respect to that center is not greater than* $\alpha/2$. Furthermore, if the radius of a circle is different from $\alpha/2$, *then there is exactly one center such that the corresponding radius is less than* $\alpha/2$. From now on the convention of choosing this point as *the* center will be followed, whenever meaningful. With this convention then, every circle whose radius is not $\alpha/2$ *has radius less than* $\alpha/2$.

A few elementary theorems regarding circles follow, the proofs being left to the reader.

Figure 184

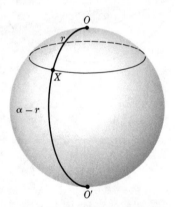

40.2 Theorem: The center of a circle is the midpoint of any diameter.

40.3 Theorem: The perpendicular bisector of any chord of a circle passes through the center.

40.4 Theorem: The line passing through the center of a circle and the midpoint of a chord is perpendicular to that chord, provided the chord is not a diameter.

40.5 Theorem: The perpendicular from the center of a circle to any chord, if it is unique, bisects the given chord.

40.6 Theorem: Two equal central angles subtend equal chords, and conversely.

The above theorems show that there is considerable agreement between Euclidean and non-Euclidean geometry with regard to circles. But there are some Euclidean properties that do *not* hold in absolute geometry. One of those is the principle: *The product of the lengths of the segments formed on each of two intersecting chords* are equal. (That is, in Fig. 185, $ab = cd$.) A counterexample to this is outlined in Exercise 14 below.

Figure 185

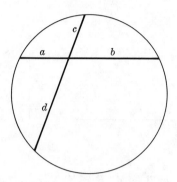

41. The Secant Theorem

If a line passes through an "interior" point of a circle, it seems obvious that it must intersect that circle at two distinct points. It is now our purpose to show how this fact may be established from the previous development. First, a definition is needed.

41.1 Definition: The **interior** of any circle whose radius is less than $\alpha/2$ is the set of all points P such that the distance from P to the center is less than the radius. The **exterior** of a circle is the set of points that are neither interior points nor lie on the circle.

NOTE: In view of the convention regarding the center of a circle, the radius of any circle may be assumed to be not greater than $\alpha/2$. Hence, the only circles for which the interior is not defined are those whose radii *equal* $\alpha/2$.

In order to prove the property mentioned above, a geometric function considered previously (Exercise 13, Section 31) will be used. Let O be any point on one of the sides of right angle ABC, say $O \in \overrightarrow{BA}$ (O is allowed to coincide with B). Consider OX as a function of the distance BX for $X \in \overrightarrow{BC}$. Set

$$d(x) = OX, \qquad \text{where } x = BX, \quad X \in \overrightarrow{BC}.$$

See Fig. 186(a).

An easy argument shows that the function d is continuous. The triangle inequality applied to the points O, X, and X_0 shows that

$$OX_0 \leqq OX + X_0X \qquad \text{and} \qquad OX \leqq OX_0 + X_0X.$$

See Fig. 186(b). Therefore,

$$-X_0X \leqq OX_0 - OX \leqq X_0X,$$

or, equivalently,

$$|OX_0 - OX| \leqq X_0X.[5] \tag{41.2}$$

---

[5] See Appendix 1.

Figure 186

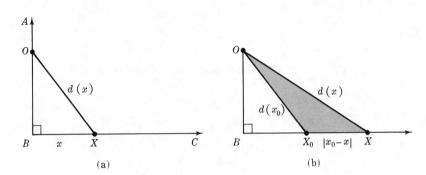

(a)   (b)

If $x_0 = BX_0$, then since either $B = X_0$, $B = X$, $X_0 = X$, (BX_0X), or (BXX_0) we have

$$X_0X = \pm(BX_0 - BX) = |x_0 - x|.$$

Then (41.2) becomes

$$|d(x_0) - d(x)| \leq |x_0 - x|. \tag{41.3}$$

It is obvious that (41.3) implies the continuity of d at x_0, for if $|x_0 - x|$ is small then $|d(x_0) - d(x)|$ must also be small. ∎

It is also to be noticed that if $OB < a/2$ the function d is *strictly increasing*. For, if $0 < x_1 < x_2$ and $BX_1 = x_1$, $BX_2 = x_2$ (Fig. 187), then (BX_1X_2). By Corollary 31.7′ both angles BX_1O and BX_2O are acute. Hence

$$\angle OX_2X_1 < 90 < 180 - \angle OX_1B = \angle OX_1X_2,$$

and by Theorem 31.5,

$$OX_1 < OX_2.$$

That is, if $x_1 < x_2$, then $d(x_1) < d(x_2)$.

Figure 187

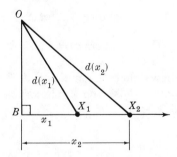

41.4 Secant Theorem: If a line passes through an interior point of a circle, it intersects the circle in exactly two points and is therefore a secant of the circle.

PROOF: Suppose line l passes through an interior point A of a circle centered at O, with radius $r < \alpha/2$ (Fig. 188). If O is not on l let B be the foot of the perpendicular from O on l; if $O \in l$, take $B = O$ (B may coincide with A). In either case, $OB \leqq OA < r$, and the function $d(x) = OX$, $x = BX$, defined as before for $X \in h$ where h is the ray $\overrightarrow{BA}$ (or one of the rays on l from B), will be strictly increasing. It is obvious that there is C on h such that $OC > r$ (by simply taking $r < BC < \alpha/2$). Set $a = BA$ and $c = BC$. Then $d(a) < r$ (by hypothesis) and $d(c) > r$. By the intermediate-value theorem[6] for continuous functions there is b, $a < b < c$, such that $d(b) = r$. Hence, if D is a point on h such that $BD = b$, then $OD = r$ and thus h meets the circle at D. By the increasing property of $d(x)$, D is the *only* point in common between h and circle O. By repeating the argument for the ray h' opposite h, it follows that l intersects the circle in precisely two points. ∎

Figure 188

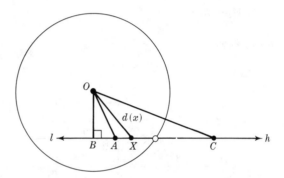

The function d used in the above proof may be used to prove other theorems in geometry (see for example Exercises 6 and 7 below). This represents an important step in our development. The fact that certain geometrically-defined functions are continuous makes available the tools of analysis for the solution of certain problems in geometry—which might be difficult or impossible by other means.

42. The Tangent Theorem

The theorem of this section will establish a characteristic property of tangents in absolute geometry. We define a **tangent** to a circle as any line which has precisely one point in common with the circle. The secant theorem just proved immediately implies that if a line is tangent to a circle having radius less than $\alpha/2$, *all points on that line different from the point of contact are exterior points of the circle.*

42.1 Theorem: If a circle centered at O has radius $\overline{OA} < \alpha/2$[7] and line l passes through A, then l is tangent to the circle at A if and only if l is perpendicular to $\overleftrightarrow{OA}$ at A.

---

[6] See any elementary calculus text.
[7] See Section 31.

PROOF: (1) Suppose first that l is tangent to circle O at A, and that l is not perpendicular to $\overleftrightarrow{OA}$. Let A' be the foot of the perpendicular from O to l. [See Fig. 189(a).] Then

$$OA' < OA.$$

But by the definition of tangent and our above observation,

$$OA' > OA,$$

providing a contradiction. Therefore, l must be perpendicular to OA at A.

(2) Conversely, suppose $l \perp \overleftrightarrow{OA}$ at A and $X \in l$, $X \neq A$. [See Fig. 189(b).] Since $OA < \alpha/2$, Corollary 31.7' applies to right triangle OAX and hence

$$OX > OA.$$

Therefore, A is the only point in common between l and circle O, and by definition, line l is tangent at A. ∎

Figure 189

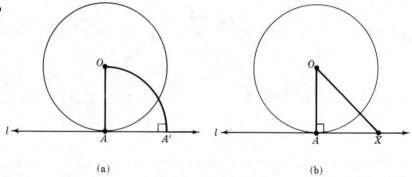

(a) (b)

EXERCISES

1. Prove Theorems 40.2, 40.3, and 40.4.
2. Prove Theorems 40.5 and 40.6.
3. Prove:

   42.2 Theorem: Each pair of tangents to a circle from a common external point form congruent angles with the line joining that point and the center of the circle, and the external point is equidistant from the two points of contact.

4. Use Theorem 42.2 to prove there can be at most two tangents to a circle from a common external point, and if there is one such tangent there will necessarily be two.
5. *Incircle of a triangle.* The incircle of a triangle is the circle having the incenter (Section 34) as center and tangent to each of the sides (the radius of this circle is called the inradius). (a) Prove that the incircle of every triangle exists. (b) Using Theorem 42.2 and the notation provided in Fig. 190, deduce the formulas

   $$x = s - a, \qquad y = s - b, \qquad z = s - c,$$

   where a, b, c, and s are as in standard notation ($2s = a + b + c$).

Figure 190

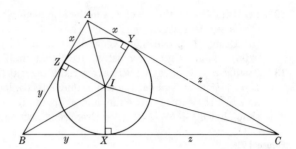

6. (a) Prove that an equilateral triangle exists with a given segment (less than $\alpha/2$) as base. (b) What is the maximal length of a side of an equilateral triangle on a sphere of radius α/π?

7. Use the continuity of the function $d(x)$ defined in Section 41 to prove that if P is a given external point of a circle at a distance less than $\alpha/2$ from its center, there exists a tangent to the circle which passes through the given point. *Hint:* In Fig. 191 select A, any point of the circle, and use the function d to show that there is $B \in t$ such that $OB = OP$.

Figure 191

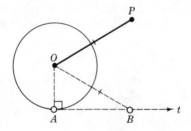

8. Prove:

   42.3 Theorem: All interior points of a segment whose endpoints lie either on or interior to a circle are themselves interior points of the circle. Thus, the interior of a circle is convex.

   Hint: Use the secant theorem, Section 41. A more direct proof may be had by applying inequality (31.11). [The function d is also used in establishing (31.11) for all real p, $0 \leq p \leq 1$; see *Note* following Exercise 16, Section 31.]

9. Consider the analytic plane with all the familiar concepts regarding "distance" and "angle," but with "points" as those pairs (x, y) such that either x or y (or both) are *rational*. (a) Which of our axioms are satisfied in this geometry? (b) Find an example of a circle and line which denies the secant theorem.

10. A polygon is **inscribed** in a circle if its vertices lie on the circle. Prove that a polygon inscribed in a circle of radius less than $\alpha/2$ is convex. (Use Theorem 42.3, Exercise 8.)

11. Prove that for each integer $n \geq 3$, a regular polygon having n sides may be inscribed in a given circle of radius $< \alpha/2$. (Theorem 39.2 implies the converse proposition, so this is a characteristic property of regular polygons. Note that this finally establishes the *existence* of regular polygons having a prescribed number of sides.)

12. (a) Using the continuity of d as in the argument which proved the secant theorem show that a segment which joins an interior and exterior point of a circle crosses the circle. (b) Using (a) prove more generally that a polygonal path joining an interior and exterior point of a circle must intersect that circle at least once.

13. Assume $\alpha < \infty$. Let $\triangle ABC$ have a right angle at C and sides less than $\alpha/2$ (Fig. 192). If $a^2 + b^2 = c^2$ (standard notation understood), locate L and M, the midpoints of $\overline{AC}$ and $\overline{BC}$, respectively. If a', b', and c' denote the lengths of the sides of $\triangle CML$, then show that $a'^2 + b'^2 < c'^2$. (See Exercise 2, Section 39.)

Figure 192

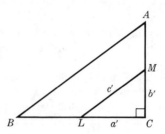

14. Using Exercise 13 above, again assuming that $\alpha < \infty$, construct an example of two intersecting chords of a circle such that $ab \neq cd$, where a, b, c, and d are the lengths of the segments on those chords. (Fig. 193 should provide ample guidance; assume that $r^2 \neq a^2 + s^2$.)

Figure 193

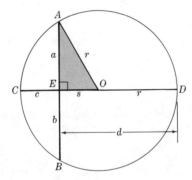

6

Three Geometries

Having carefully set down those axioms which reflected our intuitive knowledge of Euclidean and spherical geometry, we undertook in the preceding chapters a modest development of geometry from those axioms. In that development it was tacitly assumed that if α (the least upper bound of distances) is finite, spherical geometry is obtained, and if α is infinite, Euclidean geometry results. It may come as a surprise to the reader to learn that along with Euclidean and spherical geometry, he was actually studying a *third* geometry just as important and just as consistent. This "hidden" geometry was the one which early nineteenth century mathematicians did not believe existed. It arises quite naturally by considering certain possibilities regarding parallel lines, as we shall see.

Before proceeding further it is desirable to state precisely what the concept of parallelism shall be.

Definition: Two lines are said to be **parallel** if they have no point in common. If line l is parallel to line m, we write $l \parallel m$, as customary.

Now suppose point P does not lie on line l. There are clearly three mutually exclusive cases:

(a)   There exist no lines through P parallel to l—the **elliptic hypothesis.**
(b)   There exists exactly one line through P parallel to l—the **parabolic hypothesis.**
(c)   There exist at least two lines through P parallel to l—the **hyperbolic hypothesis.**
These three hypotheses give rise to three independent geometries each having their own interesting properties. They are called **elliptic, parabolic,** and **hyperbolic** geometry, respectively. Parabolic geometry is the geometry with which we are perhaps

most familiar. It is better known as *Euclidean geometry*. The elliptic and hyperbolic geometries are sometimes also referred to as *Riemannian* and *Lobachevskian geometry*, in that order.

Figure 194

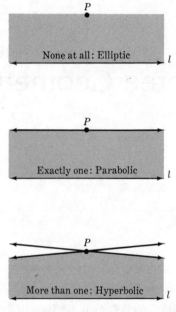

NOTE: The use of the names of special conic sections to designate the three geometries is purely traditional and is not intended to imply any immediate connection with the conics themselves. There is an analogy which might be noted at this time, however. Recall from analytic geometry that the equation

$$ax^2 + bxy + cy^2 + dx + ey + f = 0$$

represents a general conic section in the xy plane. It represents an *ellipse, parabola,* or *hyperbola* as the discriminate $\triangle = b^2 - 4ac$ is *negative, zero,* or *positive*—that is, as the quadratic equation

$$ax^2 + bx + c = 0$$

has *no real roots, one real root,* or *at least two real roots.*

A brief development of each of the three geometries will be taken up in order. But first, several key results concerning the value of α must be obtained.

43. The Influence of α on Parallelism

It is interesting that the boundedness and unboundedness of α has definite bearing on the existence of parallel lines. If $\alpha < \infty$, one feels that the previous axioms depict spherical geometry and that under this assumption, therefore, parallel lines should not exist. This is indeed the case.

43.1 Theorem: If $\alpha < \infty$, each pair of lines intersect.

Figure 195

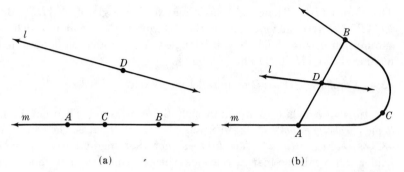

(a)   (b)

PROOF: Let l and m be any two lines. Choose A, B, and C any three points on m, but with $AB = \alpha$, and let D be any point on l. See Fig. 195(a). Then (ACB) and (ADB), and $AC < \alpha$, $BC < \alpha$ imply that $\overleftrightarrow{AC} = \overleftrightarrow{BC} = m$. If l passes through A, B, or C, we are finished. Otherwise, the postulate of Pasch implies that in Fig. 195(b), l passes through either a point E such that (AEC) or a point F such that (BFC). In either case, however, l intersects m. ∎

The next theorem will require some new terminology.

43.2 Definition: If a line intersects two other lines, it is called a **transversal.** Further, let line t intersect lines l and m at points A and B (Fig. 196) and choose C, D, E, F, G, and H on l, m, and t as indicated in the figure, with the betweenness relations as shown. Then the pairs $(\angle GAC, \angle ABE)$, $(\angle CAB, \angle EBH)$, $(\angle GAD, \angle ABF)$, and $(\angle DAB, \angle FBH)$ are **corresponding angles,** and the pairs $(\angle CAB, \angle ABF)$ and $(\angle EBA, \angle BAD)$ are **alternate interior angles** with respect to the transversal t. (The remaining pairs are sometimes referred to as **interior** and **exterior angles on the same side of the transversal,** respectively.)

Figure 196   **Figure 197**

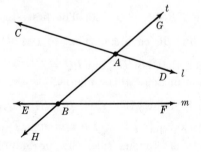   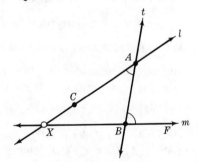

43.3 Theorem: If $\alpha = \infty$ and two lines are cut by a transversal so that a pair of alternate interior angles are congruent, the lines are parallel.

PROOF: In Fig. 197, suppose that $\angle CAB = \angle ABF$ and that l and m meet at X. If $X \in H_1(C,t)$, then since $F \in H_2(C,t)$ we cannot have (XFB) or (BXF).

Therefore (XBF). Hence $\overline{\angle ABF}$ is an exterior angle of $\triangle ABX$. But $\angle XAB = \angle CAB = \angle ABF$ in violation of the exterior-angle theorem which, owing to the assumption $\alpha = \infty$, holds for all triangles. A contradiction is obtained in a similar manner if $X \in H_2(C,t)$. Thus, $l \parallel m$. ∎

43.4 Corollary: If $\alpha = \infty$, parallel lines exist.

It is worthwhile to mention that the *converse* of Theorem 43.3 is *equivalent to the parabolic hypothesis* (see Exercise 4, below). That is, the converse of Theorem 43.3 is *characteristically Euclidean*—a fact that might be easily overlooked. The reader will recall the many occasions when a geometric fact and its converse were almost automatic companions. It is therefore quite natural to regard Theorem 43.3 and its converse as being of similar disposition and, having observed the easy proof of Theorem 43.3, to attempt its converse. But because of the validity of hyperbolic geometry any such attempt is doomed to failure. This was of course the pitfall of those eighteenth century mathematicians (and others) who felt impelled to provide a proof of Euclid's fifth postulate. It is somewhat remarkable that in his own writings Euclid preferred to avoid the issue by simply stating his immortal postulate. One wonders to what extent Euclid realized the impossibility of proving the postulate of parallels and whether its appearance was more than just a matter of convenience.

44. The Influence of α on the Angle Sum of Triangles

Saccheri was able to eliminate with ease the hypothesis of the obtuse angle.[1] His proof made the tacit assumption that the line was "unbounded," or in the terms being used here, that $\alpha = \infty$. The fact was later verified by Legendre in the following theorem due to him (we employ Legendre's own ingenious argument):

44.1 Legendre's First Theorem: If $\alpha = \infty$, the angle sum of any triangle is less than or equal to the measure of a straight angle.

PROOF: Suppose triangle ABC has $\angle A + \angle B + \angle C > 180$. The first step is to construct a sequence of triangles on base line $\overleftrightarrow{BC}$ each congruent to $\triangle ABC$. In Fig. 198, consider ray $\overrightarrow{BC}$ and D on $\overrightarrow{BC}$ such that (BCD) and $BC = CD = a$ (since $\alpha = \infty$). Since $\angle ACD > \angle B$ by the exterior angle theorem, there is a ray $\overrightarrow{CE}$ such that $(\overrightarrow{CA}\ \overrightarrow{CE}\ \overrightarrow{CD})$ and $\angle ECD = \angle B$. It follows that since $(\overrightarrow{CB}\ \overrightarrow{CA}\ \overrightarrow{CD})$, then also $(\overrightarrow{CB}\ \overrightarrow{CA}\ \overrightarrow{CE}\ \overrightarrow{CD})$. Choose F on $\overrightarrow{CE}$ such that $CF = AB = c$. Thus by SAS $\triangle ABC \cong \triangle FCD$, and $AC = FD = b$. It is obvious that this procedure leads by mathematical induction to two sequences of congruent triangles (Fig. 199)

$$\triangle A_0B_0B_1 \cong \triangle A_1B_1B_2 \cong \triangle A_2B_2B_3 \cong \cdots \cong \triangle A_{n-1}B_{n-1}B_n,$$

and

$$\triangle A_0B_1A_1 \cong \triangle A_1B_2A_2 \cong \triangle A_2B_3A_3 \cong \cdots \cong \triangle A_{n-1}B_nA_n,$$

---

[1] See Section 38.

where $A_0, A_1, \cdots, A_n$ lie on the same side of $\overleftrightarrow{BC}$, and n is any integer greater than 1. (The points $A_0, A_1, \cdots, A_n$ need not be collinear due to the validity of hyperbolic geometry—see Exercise 5 below.)

Figure 198

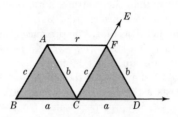

Figure 199

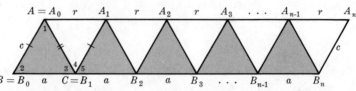

Now with the notation of Fig. 199, $\angle 2 = \angle 5$. Since $\angle 1 + \angle 2 + \angle 3 > 180$, then $\angle 1 + \angle 5 + \angle 3 > 180$. That is,

$$\angle 1 + \angle 5 + \angle 3 > \angle 3 + \angle 4 + \angle 5,$$

and therefore,

$$\angle 1 > \angle 4.$$

By Theorem 31.8,

$$a > r.$$

By the polygonal inequality,

$$B_0B_n \leq B_0A_0 + A_0A_1 + A_1A_2 + \cdots + A_{n-1}A_n + A_nB_n.$$

That is,

$$na \leq 2c + nr.$$

Since $a > r$, set $\epsilon = a - r > 0$. Then

$$n(\epsilon + r) \leq 2c + nr,$$

and from this may be derived

$$\epsilon \leq \frac{2c}{n}.$$

But the limit, as n becomes infinite, yields the contradiction

$$\epsilon \leq 0.$$

Therefore, it must be concluded that

$$\angle A + \angle B + \angle C \leq 180. \quad \blacksquare$$

EXERCISES

1. Prove that if $\alpha = \infty$ and two lines cut a transversal making a pair of corresponding angles congruent, the lines are parallel.

2. If $\alpha = \infty$, two lines perpendicular to the same line are parallel. Prove this in two different ways: (a) As a corollary to Theorem 43.3, and (b) as a corollary to Theorem 29.9.
3. The measure of an exterior angle of a triangle is less than, equal to, or greater than the sum of the measures of the opposite interior angles as the angle sum of the triangle is greater than, equal to, or less than the measure of a straight angle. Prove.
4. The parabolic hypothesis is sometimes called *Playfair's postulate*, after John Playfair (1748–1819), a Scottish mathematician. In the presence of Axioms 1–11 prove Playfair's postulate from Theorem 43.3. Since Theorem 43.3 can be proved from Playfair's postulate, it follows that the two are equivalent assumptions.
5. In the proof of Theorem 44.1 show that if the points A_0, A_1, A_2, . . . , A_n are collinear (Fig. 199), then there exists a Saccheri quadrilateral satisfying the hypothesis of the right angle (Section 38). *Hint:* Drop perpendiculars from A_0, A_1, and A_2.
6. Prove that the statement of Theorem 43.3 is equivalent to Euclid's fifth postulate. (See Appendix 3 for a statement of the fifth postulate.)

45. Elliptic Geometry: The Pole–Polar Theory

The axiom which serves to characterize elliptic geometry is

Axiom 12:   *If a point and a line not passing through it be given, there exists no line passing through the given point which is parallel to the given line.*

Simply stated, Axiom 12 disallows parallel lines. In view of Corollary 43.4 one must conclude that $\alpha < \infty$. With Axioms 1–12 and with any specific value of α, the *plane*—defined earlier as simply the *set of all points*—will be called the **elliptic plane**. It follows that there is an elliptic plane corresponding to each value of α. For convenience, the "canonical" value $\alpha = \pi$ will be assumed here (this value for α would be realized on a sphere of unit radius).

A previously established result (Theorem 24.6) can be combined with Axiom 12 and stated in the following form:

45.1   Theorem: Each pair of lines in the elliptic plane intersect at two points lying at a distance π from each other.

Corollary 24.7 and the definition of "extreme" points leads to the statement

45.2   Theorem: To each point in the elliptic plane there corresponds one and only one extreme point whose distance from the given point is π.

One of the important properties of extreme points to be recalled is that every line passing through one of such a pair also passes through the other.

The fact that each pair of lines in the elliptic plane meet at a pair of extreme points gives rise to a "pole-polar" theory. Begin with a line l and the family of lines *perpendicular to l*. Any two of these perpendiculars meet at a pair of extreme points P and P', and since congruent triangles are formed by ASA (Fig. 200), it immediately follows that P and P' each lie at a (perpendicular) distance $\pi/2$ from line l. Since P and P' are unique with this property, and any two perpendiculars were selected for the argument, this proves

45.3   Theorem: The perpendiculars to any line in the elliptic plane meet at a pair of extreme points P and P' which are each a distance $\pi/2$ from the line. Moreover, the distance from P (or P') to each point on that line is $\pi/2$.

Figure 200

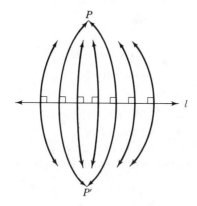

45.4   Corollary: Each line in elliptic geometry is a circle of radius $\pi/2$.

The points P and P' mentioned in the previous theorem are called the **poles** of l.[2] Conversely, suppose P and P' are a pair of extreme points. One may locate a point M on any line through P and P' such that $MP = MP' = \pi/2$ and determine the perpendicular l to that line at M. Then line l has P and P' as its poles and is uniquely determined in this fashion. Line l is called the **polar** of the points P and P'.

An easily proved fundamental property of poles and polars is the following result:

45.5   Theorem: If P is either pole of line l, then the polar of any point on l passes through P, and dually, any line through P has its poles on l.

PROOF: Let X be any point on l (Fig. 201). Then $PX = \pi/2$, so if x is the perpendicular to $\overleftrightarrow{PX}$ at P, it follows that X is a pole of x and, consequently, x is the polar of X. Conversely, let x be any line through P and consider the perpendicular m to x at P. It must meet l in two extreme points X, X'. But $PX = PX' = \pi/2$, since P is a pole of l, and hence, X and X' are the poles of x. ∎

---

[2] The fact that each line possesses *two* poles gives rise to the term *double* elliptic geometry. By identifying extreme points one obtains *single* elliptic geometry, with one pole for each line. Of course this necessitates a thorough revision of the previous axioms. The difficulties inherent with a study of single elliptic geometry are reflected by the fact that the pseudosphere discussed earlier is a model for this geometry.

Figure 201

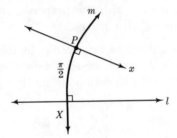

Now consider a triangle ABC. Let A', B', and C' be one of the poles of each of the sides $\overleftrightarrow{BC}$, $\overleftrightarrow{AC}$, and $\overleftrightarrow{AB}$, respectively (Fig. 202). Triangle $A'B'C'$ is called a **polar triangle** of triangle ABC. By Theorem 45.5 it follows that A, B, and C are poles of $\overleftrightarrow{B'C'}$, $\overleftrightarrow{A'C'}$, and $\overleftrightarrow{A'B'}$, respectively. Hence: *If $\triangle A'B'C'$ is a polar triangle of $\triangle ABC$, then $\triangle ABC$ is a polar triangle of $\triangle A'B'C'$.*

There is an important relation between the measures of the angles of a triangle and the lengths of the sides of one of its polar triangles, which leads to the angle-sum theorem for elliptic geometry.

Figure 202

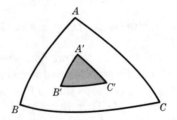

46.   A Formula in Elliptic Geometry

Let the polar triangle $A'B'C'$ of triangle ABC be so determined that

$$A' \in H_1(A, \overleftrightarrow{BC}), \qquad B' \in H_1(B, \overleftrightarrow{AC}), \qquad C' \in H_1(C, \overleftrightarrow{AB}).$$

Two fundamental relations will be derived here (stated only for one vertex):

$$\angle B'AC' + \angle BAC = 180, \tag{46.1}$$

$$\frac{\angle B'AC'}{180} = \frac{B'C'}{\pi}. \tag{46.2}$$

(See Fig. 203.) Thus may be obtained formulas which relate the *angles* of triangle ABC to the *sides* of one of its polar triangles, $A'B'C'$.

First observe that if $p = \overrightarrow{AB'}$, $q = \overrightarrow{AB}$, $r = \overrightarrow{AC}$, and $s = \overrightarrow{AC'}$ (Fig. 204), the rays p and q lie on the same side of line $\overleftrightarrow{AC}$ (since B and B' do so), and thus either $p = q$, (pqr), or (qpr).[3] Similarly, since C and C' lie on the same side of $\overset{\leftrightarrow}{AB}$ either $r = s$ (qrs) or (qsr). But B' is a pole of $\overset{\leftrightarrow}{AC}$, and C' is a pole of $\overset{\leftrightarrow}{AB}$. Hence,

$$pr = \angle B'AC = 90,$$
$$qs = \angle BAC' = 90. \tag{46.3}$$

It can now be proved that (pqr) holds if and only if (qrs): Suppose (pqr) and $r = s$. Then (pqs) and $pq + qs = ps = pr$ implies, by (46.3),

$$pq + 90 = 90.$$

That is, $pq = 0$, which contradicts (pqr). Next, suppose that (pqr) and (qsr) both hold. Then $(pqsr)$, and hence,

$$pq + qs + sr = pr$$
$$pq + 90 + sr = 90$$
$$pq + sr = 0,$$

a contradiction. Hence, if (pqr), then (qrs). The converse obviously follows in like manner. Further, it can be shown that (qpr) implies (psr).

Thus we have the three cases:

(a) $p = q$ and $r = s$,
(b) (pqr) and (qrs); therefore $(pqrs)$ by the dual of Theorem 19.3,
(c) (qpr) and (psr); therefore $(qpsr)$.

Figure 203   **Figure 204**

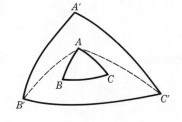
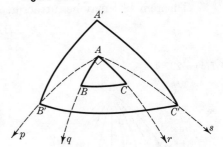

In (a) the equation (46.1) is trivial, for then

$$\angle B'AC' + \angle BAC = ps + qr = pr + qs = 180,$$

by (46.3). In (b), Fig. 204 shows that

$$\angle B'AC' + \angle BAC = \angle B'AC + \angle CAC' + \angle BAC = 90 + \angle BAC' = 180.$$

And in (c) we have (observing Fig. 205)

$$\angle B'AC' + \angle BAC = \angle B'AC' + \angle BAB' + \angle B'AC = \angle BAC' + 90 = 180. \blacksquare$$

The property stated in (46.2) is ultimately a result of the observation: *If*

---

[3] See Exercise 7, Section 25.

Figure 205

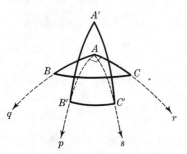

Figure 206

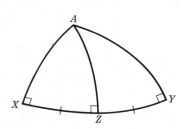

A *is a pole of* $\overleftrightarrow{XY}$ *and* Z *is the midpoint of* $\overline{XY}$, *then* $\overrightarrow{AZ}$ *bisects* $\angle XAY$ (Fig. 206). This is an obvious property of poles, for $\overleftrightarrow{AX}$, $\overleftrightarrow{AY}$, and $\overleftrightarrow{AZ}$ are each perpendicular to $\overleftrightarrow{XY}$, and therefore, by ASA, $\triangle AXZ \cong \triangle AZY$ and $\angle XAZ = \angle ZAY$.

The following more general property immediately follows:

46.4   Theorem: If A is a pole of line $\overleftrightarrow{BC}$ and

$$B = B_0, B_1, B_2, \cdots, B_n = C$$

are distinct points on $\overline{BC}$ which determine congruent subsegments on $\overline{BC}$, then the corresponding angles formed at A are congruent (Fig. 207).

Theorem 46.4 may be further generalized to give (Fig. 208):

$$\frac{\angle BAC}{\angle BAD} = \frac{BC}{BD}. \qquad\qquad (46.5)$$

Figure 207

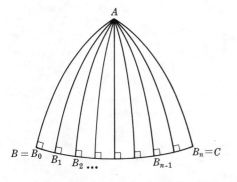

Figure 208

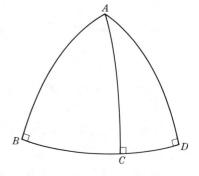

The method of proof is based on a pattern akin to Archimedes so-called "method of exhaustion" which was commonly used prior to the invention of the calculus to solve problems involving limits. The reader should pay particular attention here as the argument will be used again in the development of parabolic geometry in an almost identical fashion.

In Fig. 209 choose points $B = B_0, B_1, B_2, \cdots, B_n = D$ on $\overline{BD}$, with the betweenness relations as indicated by the figure, such that

$$B_0B_1 = B_1B_2 = \cdots = B_{n-1}B_n = r_n.$$

By Theorem 46.4,

$$\angle B_0AB_1 = \angle B_1AB_2 = \cdots = \angle B_{n-1}AB_n = \theta_n.$$

Figure 209

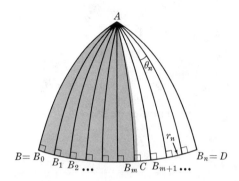

Now for some m, C must lie on segment $\overline{B_mB_{m+1}}$, and hence ray $\overrightarrow{AC}$ will lie between $\overrightarrow{AB_m}$ and $\overrightarrow{AB_{m+1}}$. Set $s_n = B_mC$ and $\phi_n = \angle B_mAC$. Thus

$$0 \leq s_n \leq r_n \quad \text{and} \quad 0 \leq \phi_n \leq \theta_n.$$

It follows that

$$\angle BAC = m\theta_n + \phi_n, \quad BC = mr_n + s_n, \qquad (46.6)$$
$$\angle BAD = n\theta_n, \quad BD = nr_n.$$

Dividing yields

$$\frac{\angle BAC}{\angle BAD} = \frac{m}{n} + \frac{\phi_n}{n\theta_n}, \quad \frac{BC}{BD} = \frac{m}{n} + \frac{s_n}{nr_n}. \qquad (46.7)$$

Set $\epsilon_n = \phi_n/\theta_n \leq 1$ and $\delta_n = s_n/r_n \leq 1$, and subtract in (46.7):

$$\frac{\angle BAC}{\angle BAD} - \frac{BC}{BD} = \frac{\epsilon_n}{n} - \frac{\delta_n}{n}. \qquad (46.8)$$

The quantity in the right member of (46.8) is independent of n. Since ϵ_n and δ_n are bounded, $\epsilon_n/n \to 0$ and $\delta_n/n \to 0$ as $n \to \infty$.[4] Thus, the limit as $n \to \infty$ in (46.8) yields

$$\frac{\angle BAC}{\angle BAD} - \frac{BC}{BD} = 0,$$

the desired result. ∎

46.9 Theorem: If A is a pole of $\overleftrightarrow{BC}$, then $\dfrac{\angle BAC}{180} = \dfrac{BC}{\pi}$.

---

[4] Compare these assertions with such limits commonly introduced in the calculus as $\lim_{n \to \infty} (-1)^n/n = 0$ and $\lim_{n \to \infty} (\sin n)/n = 0$.

PROOF: In equation (46.5) let $\overline{\angle BAD}$ be a straight angle. Then B and D are extreme points and thus $BD = \pi$. Upon substitution, (46.5) becomes the desired relation. ▌

It is clear that since A is a pole of $\overleftrightarrow{B'C'}$, the previous theorem finally establishes equation (46.2). If we write a', b', and c' for $B'C'$, $A'C'$, and $A'B'$ (46.2) takes on the form

$$\frac{\angle B'AC'}{180} = \frac{a'}{\pi}.$$

Substitution into (46.1) then yields the relation

$$\frac{\angle A}{180} + \frac{a'}{\pi} = 1 \qquad (\angle A = \angle BAC)$$

which, when applied to all three angles of triangle ABC, implies

$$\boxed{\frac{\angle A}{180} + \frac{a'}{\pi} = \frac{\angle B}{180} + \frac{b'}{\pi} = \frac{\angle C}{180} + \frac{c'}{\pi} = 1} \qquad \textbf{(46.10)}$$

(see Fig. 210).

Figure 210

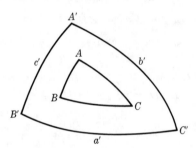

47.   The Angle-Sum Theorem for Elliptic Geometry

With the result (46.10), a surprisingly simple fact is all that is needed to establish the relationship concerning the sum of the measures of the angles of a triangle.

47.1   Theorem: The sum of the lengths of the sides of any triangle in elliptic geometry is less than 2π.

PROOF: If C' is the extreme point of C (Fig. 211), then, by the triangle inequality for $\triangle ABC'$,

$$AB < C'A + C'B.$$
Then,

$$AB + BC + AC < C'A + AC + C'B + BC = C'C + C'C = 2\pi. \ ▌$$

Figure 211

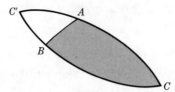

Now let triangle ABC be given, and let the polar triangle $A'B'C'$ be determined as in Section 46. By (46.10)

$$\frac{\angle A + \angle B + \angle C}{180} + \frac{a' + b' + c'}{\pi} = 3.$$

But the theorem just proved implies

$$\frac{a' + b' + c'}{\pi} < 2.$$

Hence,

$$\frac{\angle A + \angle B + \angle C}{180} > 1.$$

Thus we have proved

47.2   Angle-Sum Theorem for Elliptic Geometry: The angle sum of any triangle in the elliptic plane is greater than a straight angle.

48.   Area in Elliptic Geometry

It is an interesting fact that in elliptic geometry the angle sum of a triangle *varies with its size.* It can be shown that the smaller the sides of a triangle, the closer its angle sum is to the measure of a straight angle. For example, in $\triangle ABC$ where A is a pole of $\overleftrightarrow{BC}$ (Fig. 212), if $BC = x$, then by Theorem 46.9

$$\frac{\angle A}{180} = \frac{x}{\pi}$$

and hence,

$$\angle A + \angle B + \angle C = \frac{180}{\pi} \cdot x + 90 + 90 = \frac{180}{\pi} \cdot x + 180.$$

Figure 212

In this case as $x \to 0$, $\angle A + \angle B + \angle C \to 180$. This suggests that the difference

$$\angle A + \angle B + \angle C - 180,$$

called the **excess** of triangle ABC, measures the *area* of the triangle.

This is made even more credible by the following result:

48.1   Theorem: If D is any interior point on side $\overline{BC}$ of triangle ABC, the excess of triangle ABC equals the sum of the excesses of triangles ABD and ADC.

   PROOF: If ϵ is the excess of $\triangle ABC$ (Fig. 213), then

$$\begin{aligned}
\epsilon &= \angle A + \angle B + \angle C - 180 \\
&= \angle A + \angle B + \angle C + \angle 3 + \angle 4 - 360 \\
&= (\angle 1 + \angle B + \angle 3 - 180) + (\angle 2 + \angle 4 + \angle C - 180) \\
&= \epsilon_1 + \epsilon_2. \quad \blacksquare
\end{aligned}$$

Figure 213

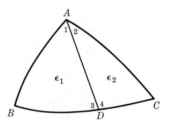

The preceding theorem leads to the more general notion of **excess for convex polygons,** defined for each polygon as the *angle sum diminished by* $(n - 2)\,180$, *where* n *is the number of sides.* The additive property of excess then extends to (convex) polygons in the following manner: *If the union of the polygons,* Π_1, Π_2, $\ldots$, Π_n *and their interiors is the polygon* Π *and its interior, and no two interiors of* Π_1, Π_2, $\ldots$, Π_n *intersect, then the sum of the excesses of* Π_1, Π_2, $\ldots$, Π_n *equals that of* Π. (The ambitious reader would do well to consult E. Moise, *Elementary Geometry from an Advanced Standpoint* [21, pp. 324–346], which discusses the general theory of area.) Thus, it is not altogether unreasonable to state:

48.2   Definition: The **area** of triangle ABC in elliptic geometry is its excess, $\angle A + \angle B + \angle C - 180$.

   NOTE: The preceding ideas would be trivial in Euclidean geometry where the "excess" of a triangle is always zero. Thus a different definition for area must be used there. But while the formal definitions for area in elliptic and Euclidean geometry differ, the resulting theory is the same. Indeed, a formula for the area of an elliptic triangle in terms of its sides is possible, much like the one in Euclidean geometry (see Chapter 10).

   A corollary of Theorem 48.1 clinches the non-Euclidean character of elliptic geometry. First, a definition is needed.

48.3   Definition: Two triangles are **similar** if under some correspondence between their vertices the lengths of corresponding sides are in the same ratio, and corresponding angles are congruent.

If $\triangle ABC$ is similar to $\triangle XYZ$ under the correspondence $ABC \leftrightarrow XYZ$, we shall use the notation

$$\triangle ABC \sim \triangle XYZ,$$

following the convention established earlier for congruence. By definition, therefore, $\triangle ABC \sim \triangle XYZ$ if and only if

$$\frac{AB}{XY} = \frac{AC}{XZ} = \frac{BC}{YZ}, \qquad \angle A = \angle X, \qquad \angle B = \angle Y, \qquad \text{and} \qquad \angle C = \angle Z.$$

It follows that *congruent triangles are similar*.

48.4 Theorem: There exist no similar, noncongruent triangles in elliptic geometry.

PROOF: Suppose $\triangle ABC \sim \triangle XYZ$ and $\triangle ABC \ncong \triangle XYZ$. Since $\angle A = \angle X$ it may be assumed without loss of generality that the triangles are so arranged that $A = X$, $Y \in \overline{AB}$, and $Z \in \overline{AC}$, as indicated in Fig. 214, with the areas of the various triangles as indicated, and with ϵ and ϵ_4 the respective areas of triangles ABC and ABZ. Then by Theorem 48.1

$$\epsilon = \epsilon_3 + \epsilon_4 = \epsilon_3 + (\epsilon_1 + \epsilon_2) > \epsilon_1.$$

But by hypothesis, $\angle A = \angle X$, $\angle B = \angle Y$, and $\angle C = \angle Z$ so that $\epsilon = \epsilon_1$, a contradiction. ∎

Figure 214

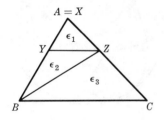

Figure 215

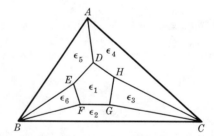

EXERCISES

1. If a triangle coincides with one of its polar triangles, it is called **self polar**. Prove that there exist self-polar triangles in the elliptic plane. What are the main properties of such triangles?
2. Do self-polar *quadrilaterals* exist? (Refer to Exercise 1.)
3. What is the maximal area of a right triangle having an acute angle of 60?
4. Using Exercise 8, Section 25, show that in Fig. 215, with ϵ the excess of $\triangle ABC$,

$$\epsilon = \epsilon_1 + \epsilon_2 + \epsilon_3 + \epsilon_4 + \epsilon_5 + \epsilon_6.$$

5. Show that the entire elliptic plane may be expressed as the union of eight triangles and their interiors, with no two interiors having any points in common. Thus show that 720 is a reasonable value for the *area of the elliptic plane*.

6. If K stands for the area of $\triangle ABC$ and a', b', and c' denote the sides of its polar triangle defined as in Section 46, prove that

$$K = \frac{180}{\pi}(2\pi - a' - b' - c').$$

7. Identify the set of points on the same side of a line and equidistant from that line. Is it another line as it is in Euclidean geometry?

8. Prove that each pair of lines in the elliptic plane have a common perpendicular.

9. It is interesting to compare the concept of area introduced above to the one with which we are already familiar from, say, the theory of integration. Thus, we know that the area of a sphere of unit radius is 4π, and that the area of a **lune of angle** A (Fig. 216) is therefore given by the formula $L = \pi \angle A/90$. (a) Prove this formula from an examination of Fig. 216 and the "familiar" concept of area. *Hint:* What part of a hemisphere would a lune of $\overline{\angle A}$ be? (b) In Fig. 217 we have indicated the extension of the sides of $\triangle ABC$ to obtain the great circles shown, meeting at A', B', and C'. It is clear (?) that $\triangle ABC \cong \triangle A'B'C'$. Therefore, $K = K'$. The region $K \cup I \cup II \cup III$ is a hemisphere while $K \cup I$, $K' \cup II$, and $K \cup III$ are each lunes. Applying the formula proved in (a) prove that

$$K = \frac{\pi}{180}(\angle A + \angle B + \angle C - 180).$$

Figure 216

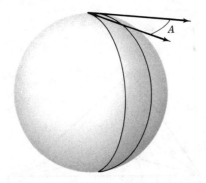

Figure 217

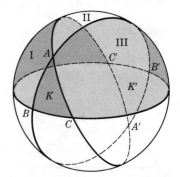

10. In elliptic geometry the only convex set which contains two distinct lines is the entire plane. (See in connection Exercise 10, Section 55 and Exercise 6, Section 87.)

49.   Hyperbolic Geometry: Limit Parallels and Hyperparallels

The object of the next several sections is to study the effect which the following axiom has on absolute geometry:

Axiom 12':   *If a point and a line not passing through it be given, there exist at least two lines which pass through the given point parallel to the given line.*

Axioms 1–11 and 12' provide a formal basis for the geometry known as hyperbolic geometry. The "plane" with this set of axioms is called the **hyperbolic plane,** following the precedent set earlier for elliptic geometry. It is not immediately clear that Axiom 12′ is consistent with the rest (this consistency will be dealt with in a later chapter). But granting this for the moment, it follows that all previously proved theorems in absolute geometry are valid and may be freely used. It is to be emphasized that Axiom 12′ does not *contradict* anything proved from Axioms 1–11, but rather it *specializes* the development in a particular direction.

In the presence of Axiom 12′, Theorem 43.1 implies $\alpha = \infty$. This simplifies many previous theorems. For instance, two distinct points *always* determine a line, three points on a line *always* satisfy a unique betweenness relation, and so on. In addition, Theorem 43.3 and its corollaries are valid.

Figure 218   **Figure 219**

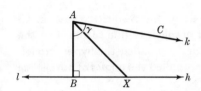   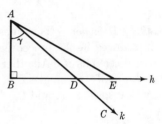

Suppose in Fig. 218 $\overleftrightarrow{AB} \perp l$, and that X lies on one of the rays h on l from B. As BX becomes large $\angle BAX$ increases (by Axiom 10) but remains less than 90 (since $\overline{\angle BAX}$ is one of the acute angles of right triangle ABX). Thus

$$\gamma = \sup \ \{\angle BAX \colon \ X \in h\}$$

exists and is not greater than 90 (see Appendix 2).

On the same side of $\overleftrightarrow{AB}$ as h, take $\overrightarrow{AC} = k$ such that

$$\angle BAC = \gamma.$$

Since k is the "limiting position" of rays intersecting h it might seem surprising at first that k itself *does not meet* h. But of course this is impossible, for if $k = \overrightarrow{AD}$ for some $D \in h$ (Fig. 219), then there exists a point E on h such that (BDE) and hence, by definition of γ,

$$\angle BAD < \angle BAE \leqq \gamma,$$

a contradiction since $\angle BAD = \gamma$.

Another property of k is easily established: *If* $u = \overrightarrow{AF}$ *is any ray from* A *between* $\overrightarrow{AB}$ *and* k, *then* u *meets* h. For if $\theta = \angle BAF$ then $\theta < \gamma$ and there is $X \in h$ such that, as shown in Fig. 220,

$$\theta < \angle BAX < \gamma$$

(otherwise θ is an upper bound). Thus the crossbar principle implies that u meets segment $\overline{BX}$ and thus ray h. ∎

From this discussion it is apparent that ray k may be regarded as the "first" ray from A that is parallel to (does not meet) h. The procedure is carried out for the

other side of $\overleftrightarrow{AB}$, resulting in the "first" ray $k^* = \overrightarrow{AC}^*$ parallel to the ray h'
opposite h, where the measure γ^* of $\angle BAC^*$ is defined by

$$\gamma^* = \sup\ \{\angle BAX':\ \ X' \in h'\}.$$

Figure 220

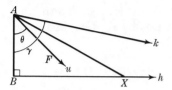

Figure 221

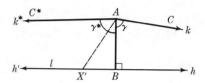

49.1   Definition: The lines m and m^* determined by the respective rays k, k^*
introduced in the preceding discussion are the **limit parallels** to line l from A in the
two directions of l. All other lines parallel to l through A are called the **hyperparallels**
to l. Finally, the angles BAC and BAC^* in Fig. 221 are called the **angles of parallelism**
with respect to A and l.

A very important fact for hyperbolic geometry is the following:

49.2   Theorem: The angles of parallelism BAC and BAC^* are congruent.

PROOF: Recall that

$$\gamma = \sup\ \{\angle BAX:\ X \in h\}, \qquad \gamma^* = \sup\ \{\angle BAX':\ X' \in h'\}.$$

In Fig. 222, let $BX = BX'$. Then

$$\triangle ABX \cong \triangle ABX',$$

and therefore,

$$\angle BAX = \angle BAX'.$$

This shows that the two sets of real numbers defining γ and γ^* are identical and
hence, $\gamma = \gamma^*$. ∎

Figure 222

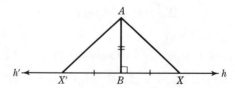

The preceding analysis was accomplished without the explicit use of Axiom 12′
(only the fact $\alpha = \infty$ was needed). The full power of that axiom will now be used
for the first time.

49.3   Theorem: The angles of parallelism BAC and BAC^* are acute.

PROOF: It is already known that $\angle BAC = \angle BAC^* \leqq 90$. Suppose that $\angle BAC = \angle BAC^* = 90$. Then by Theorem 29.13 k and k^* are opposite rays and $m = k \cup k^*$ is a line parallel to l (Fig 223). If m' is any other line through A, by Theorem 29.14 it enters the interior of one of the angles BAC or BAC^* and must, therefore, intersect l. Hence, m is the only line through A parallel to l, a denial of Axiom 12'. Therefore, $\angle BAC = \angle BAC^* < 90$. ∎

Figure 223

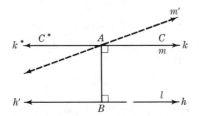

49.4 Corollary: The perpendicular to $\overleftrightarrow{AB}$ at A is hyperparallel to l at A.

It follows that the lines through A fall into three classes: The *intersectors* (those lines which intersect l), the *limit parallels* (the two "first" parallels to l), and the *hyperparallels* (parallels which are not limit-parallel to l, of which there are infinitely many). (See Fig. 224.)

One further theorem on parallelism can be stated. Since it is not actually used in the development the proof will be left as an exercise.

Figure 224

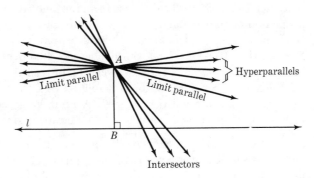

49.5 Theorem: If the distances from each of two points to two lines are equal, the angles of parallelism at the given points with respect to the given lines are congruent, and conversely.

*50.   The Theory of Parallelism in Hyperbolic Geometry

The results of the preceding section may be used to develop a theory for limit parallelism that is reminiscent of Euclidean parallelism. It is our aim merely to introduce the reader to these ideas informally, the proofs being postponed until a model for hyperbolic geometry is developed. The reader will discover at that time that all the concepts and theorems being discussed here are portrayed in the model in a beautifully simple and completely revealing manner.

The first fact is that if m is limit-parallel to l at A, and B is any other point on m, then m is limit-parallel to l at B in the same direction. Thus it is meaningful to say simply that m *is limit-parallel to* l, and we write

$$m \mid\mid\mid l.$$

The symmetric and transitive laws for limit-parallelism hold just as they do for Euclidean parallelism. That is, for any three distinct lines l_1, l_2, and l_3, and with a particular direction understood,

$$l_1 \mid\mid\mid l_2 \text{ implies } l_2 \mid\mid\mid l_1$$

and

$$l_1 \mid\mid\mid l_2 \text{ and } l_2 \mid\mid\mid l_3 \text{ imply } l_1 \mid\mid\mid l_3.$$

A similar development for hyperparallels is valid, except that here the transitive law does not hold. One of the very interesting theorems concerning hyperparallelism is: *Two lines are hyperparallel if and only if they have a common perpendicular.* Though somewhat difficult to prove from the axiomatic approach,[5] it can be viewed with ease in the model to be given later. It then follows that there are *intersecting* lines which are hyperparallel to the *same* line, a rather emphatic denial of the transitive law. To see this merely drop perpendiculars from some point to the legs of a Saccheri quadrilateral (Section 38).

Theorem 49.5 shows that the measure of the angle of parallelism at each point depends only on the distance from the point to the given line. Thus it becomes a function of that distance: If x is the distance from point A to line l, denote by $\Pi(x)$ the measure of the angle of parallelism at A. It can readily be shown that if $x_1 < x_2$, then $\Pi(x_1) \geqq \Pi(x_2)$. Since Theorem 49.5 denies equality, $\Pi(x)$ *is a strictly decreasing function of* x. Indeed, it can be proved that as x increases from 0 ta ∞, $\Pi(x)$ *decreases from* 90 *to* 0. Thus it follows that the function Π has an inverse Π^{-1} defined on the interval $0 < \theta < 90$. A remarkable conclusion is that if θ is the measure of a given acute angle $\angle ABC$, however small, $\Pi^{-1}(\theta)$ gives the distance x such that if $AB = x$, then $\overleftrightarrow{BC}$ is limit-parallel to the perpendicular to $\overleftrightarrow{BA}$ at A. Indeed, a *synthetic construction* of the segment $\overline{AB}$ is known which does not make use of the function Π^{-1} at all. (See H. E. Wolfe, *Non-Euclidean Geometry* [30, pp. 123–124].)

51. The Angle-Sum Theorem for Hyperbolic Geometry

In view of Theorem 44.1 any triangle ABC in the hyperbolic plane has

$$\angle A + \angle B + \angle C \leqq 180.$$

---

[5] For a proof that is reasonably easy to follow, see H. Eves, *A Survey of Geometry* [9, pp. 355–356].

The methods of Legendre will be used to show that the strict inequality is maintained. The following result will be needed:

51.1 Lemma: If the angle sum of triangle ABC is 180 and D is any point between B and C, then the angle sum of each of the triangles ABD and ADC is 180.

PROOF: We know that

$$\angle B + \angle 1 + \angle 3 \leqq 180 \quad \text{and} \quad \angle 2 + \angle C + \angle 4 \leqq 180$$

(Fig. 225). If either inequality is strict, then

$$(\angle B + \angle 1 + \angle 3) + (\angle 2 + \angle C + \angle 4) < 360.$$

Since $\angle 1 + \angle 2 = 180$ and $\angle 3 + \angle 4 = \angle A$, this simplifies to

$$\angle A + \angle B + \angle C < 180,$$

a contradiction. Therefore, $\angle B + \angle 1 + \angle 3 = 180$ and $\angle 2 + \angle C + \angle 4 = 180$. ▌

Figure 225

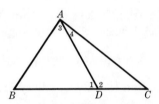

Now suppose that in Fig. 225 the acute angles of triangle ABC are at B and C (each triangle in hyperbolic geometry can have at most one obtuse angle since the exterior-angle theorem applies to all triangles). Then the altitude from A meets $\overline{BC}$ at an interior point D. Thus, the existence of a triangle having angle sum 180 implies the existence of a *right triangle* having angle sum 180. By "putting two congruent right triangles together" the existence of a quadrilateral having four right angles and opposite sides congruent, that is, a **rectangle,** is made clear (Fig. 226). Hence:

51.2 Theorem: If a single triangle has angle sum 180, then rectangles exist.

Figure 226

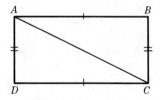

Using the one rectangle known to exist, others may be constructed. Given rectangle AA_1B_1B, points $A_2, A_3, \cdots, A_n$ and $B_2, B_3, \cdots, B_n$ are chosen on $\overrightarrow{AA_1}$ and $\overrightarrow{BB_1}$ such that $AA_1 = A_1A_2 = \cdots = A_{n-1}A_n = BB_1 = B_1B_2 = \cdots = B_{n-1}B_n,$

with the betweenness relations as depicted in Fig. 227. It is clear that the resulting quadrilaterals $A_1A_2B_2B_1$, $A_2A_3B_3B_2$, $\cdots$, $A_{n-1}A_nB_nB_{n-1}$ are congruent. Hence, $\angle A_n = \angle B_n = 90$, $A_nB_n = AB$, and $AA_n = BB_n$. Thus, AA_nB_nB is a rectangle with $AA_n = nAA_1$. Since the above procedure may be repeated for rectangle AA_nB_nB along the sides $\overline{AB}$ and $\overline{A_nB_n}$, the argument proves:

51.3 Theorem: If a single triangle has angle sum 180, then a rectangle having sides of arbitrarily large lengths exists.

Figure 227

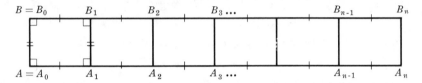

By considering a diagonal of the rectangle of Theorem 51.3 we conclude:

51.4 Corollary: If a single triangle has angle sum 180, then a right triangle having legs of arbitrarily large lengths exists with angle sum 180.

Now let right triangle ABC be given, still under the assumption that there exists a triangle having angle sum 180. By Corollary 51.4 there is A' on $\overrightarrow{CA}$ and B' on $\overrightarrow{CB}$ such that $CA' > CA$, $CB' > CB$, and such that the angle sum of right triangle $A'B'C$ is 180 (Fig. 228). By Lemma 51.1 the angle sum of each of the triangles $A'B'B$, $AA'B$, and ABC is 180. Thus, if a single triangle has angle sum 180 then *all right triangles have angle sum* 180. This obviously proves

51.5 Legendre's Second Theorem: If a single triangle has angle sum 180, then all triangles have angle sum 180.

NOTE: It should be pointed out that Axiom 12' was not used in proving the last theorem. Thus it is actually a theorem of absolute geometry, with $\alpha = \infty$.

Figure 228

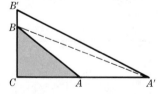

51.6 Angle-Sum Theorem for Hyperbolic Geometry: The angle sum of any triangle in hyperbolic geometry is less than a straight angle.

PROOF: Suppose the theorem is false. Then Legendre's Theorem shows that the angle sum of every triangle is 180. Let l be a given line with A any point not on it (Fig. 229). Take $\overleftrightarrow{AB} \perp l$ and let $\overline{\angle BAC}$ be one of the angles of parallelism at A.

Locate points $B = B_0, B_1, \cdots, B_n, \cdots$ on l on the same side of $\overleftrightarrow{AB}$ as C such that $BB_1 = BA$, $B_1B_2 = B_1A$, $B_2B_3 = B_2A$, $\cdots$, $B_nB_{n+1} = B_nA$, $\cdots$, with betweenness relations as shown in the figure. For convenience let $\angle B_{n-1}AB_n = \theta_n$ for $n = 1$, 2, $\cdots$. Since the triangles involved are isosceles, and since the angle sum of all triangles is 180, it follows that

$$\theta_1 = 180 - \angle AB_1B_2 = 2\theta_2,$$
$$\theta_2 = 180 - \angle AB_2B_3 = 2\theta_3.$$

It is clear (by mathematical induction) that in general

$$\theta_n = 2\theta_{n+1}.$$

Thus,

$$\theta_2 = \frac{\theta_1}{2}, \qquad \theta_3 = \frac{\theta_2}{2} = \frac{\theta_1}{4}, \qquad \cdots \qquad \theta_{n+1} = \frac{\theta_n}{2} = \frac{\theta_1}{2^n}, \qquad \cdots .$$

By definition of γ,

$$\angle BAC = \gamma > \angle BAB_n = \theta_1 + \theta_2 + \theta_3 + \cdots + \theta_n = \theta_1\left(1 + \tfrac{1}{2} + \tfrac{1}{4} + \cdots + \frac{1}{2^{n-1}}\right)$$

$$= 45 \cdot \frac{1 - \tfrac{1}{2}^n}{1 - \tfrac{1}{2}}$$

$$= 90\left(1 - \tfrac{1}{2}^n\right).$$

As n becomes infinite we find

$$\gamma \geqq 90,$$

a contradiction. ∎

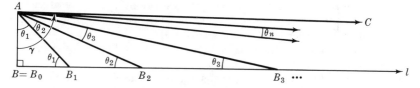

52. Area in Hyperbolic Geometry

In view of the angle-sum theorem the theory of area in the hyperbolic plane may be based on the **defect** of triangles—the amount by which the measure of a straight angle exceeds the angle sum.

As in elliptic geometry, *defect is additive:*

52.1   Theorem: If D is any interior point on side $\overline{BC}$ of triangle ABC, the defect of triangle ABC equals the sum of the defects of triangles ABD and ADC.

PROOF: If δ is the defect of $\triangle ABC$ (Fig. 230), then

$$\delta = 180 - \angle A - \angle B - \angle C$$
$$= 360 - \angle 3 - \angle 4 - \angle A - \angle B - \angle C$$
$$= (180 - \angle 1 - \angle B - \angle 3) + (180 - \angle 2 - \angle 4 - \angle C)$$
$$= \delta_1 + \delta_2. \quad∎$$

Figure 230

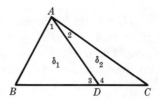

Following the procedure used in elliptic geometry we state:

52.2 Definition: The **area** of a triangle in hyperbolic geometry is its defect.

Analogous to the situation in elliptic geometry, Theorem 52.1 implies the non-Euclidean result:

52.3 Theorem: There exist no similar noncongruent triangles in hyperbolic geometry.

EXERCISES

1. Prove, as asserted in Section 51, that if right triangle ADC has angle sum 180 and $\triangle ADC \cong \triangle CEA$, then $ADCE$ is a rectangle (Fig. 226).
2. Using the theorem that lines are hyperparallel if and only if they have a common perpendicular, prove that if two lines are cut by a transversal such that a pair of corresponding angles are congruent, the lines are hyperparallel. *Hint:* Drop perpendiculars from some point on the transversal.
3. Decide whether the following proposition is true in hyperbolic geometry: *Two parallel lines are everywhere equidistant.*
4. Deduce that the summit angles of a Saccheri quadrilateral are acute.
5. Define "defect" for convex polygons and work Exercise 4, Section 48, for hyperbolic geometry.
6. Decide whether in hyperbolic geometry an angle inscribed in a semicircle is a right angle. *Hint:* Draw $\overline{OB}$ as shown in Fig. 231.
7. The segment joining the midpoints of the base and summit of a Saccheri quadrilateral is less than either leg.
8. In the notation of the statement made in Section 50 concerning the function $\Pi(x)$, prove that if $x < y$, then $\Pi(x) \geqq \Pi(y)$.
★9. As in the previous exercise, prove that if $x < y$ then $\Pi(x) \neq \Pi(y)$, thus completing the proof of Theorem 49.5.
10. If line m is limit-parallel to line l with respect to point A on m, then it is limit-parallel to l with respect to any other point A' on m (Fig. 232). You may use the betweenness relations evident from the figure. *Hint:* Use the fact that every ray between $\overrightarrow{AD}$ and $\overrightarrow{AB}$ intersects l while ray $\overrightarrow{AD}$ itself does not. Establish this property at A'.

Figure 231

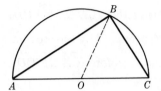

Figure 232

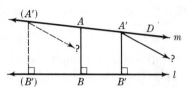

53. Parabolic Geometry: The Euclidean Angle-Sum Theorem

A foundation for Euclidean geometry is provided by Axioms 1–11 and the following postulate for parallelism:

Axiom 12″: *If a point and a line not passing through it be given, there exists one and only one line which passes through the given point parallel to the given line.*

The plane with these axioms will be called the **Euclidean plane.** As in hyperbolic geometry, it follows that $\alpha = \infty$ and certain simplifications of previous theorems result. The development begins with the *converse* of Theorem 43.3:

53.1 Theorem: If two parallel lines cut a transversal, each pair of alternate interior angles thus formed are congruent.

PROOF: Suppose $l \parallel m$ and let $\angle CAB$ and $\angle ABF$ be a pair of alternate interior angles (Fig. 233). If $\angle CAB \neq \angle ABF$, then there exists C' on the same side of $\overleftrightarrow{AB}$ as C such that

$$\angle C'AB = \angle ABF,$$

and by Theorem 43.3, $\overleftrightarrow{AC'} \parallel l$, thus contradicting Axiom 12″. ∎

Figure 233

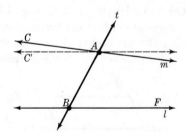

Since one of a pair of alternate interior angles is always vertical to one of a pair of corresponding angles, there is an obvious corollary:

53.2 Corollary: If two parallel lines cut a transversal, each pair of corresponding angles thus formed are congruent.

53.3 Euclidean Exterior-Angle Theorem: The measure of any exterior angle of a triangle equals the sum of the two opposite interior angles.

PROOF: Let $\angle ACD$ be an exterior angle of $\triangle ABC$ and let $\overleftrightarrow{CE}$ be parallel to $\overleftrightarrow{AB}$, with E on the same side of $\overleftrightarrow{BC}$ as A (Fig. 234). Then either $(\overrightarrow{CB}\ \overrightarrow{CE}\ \overrightarrow{CA}\ \overrightarrow{CD})$ or $(\overrightarrow{CB}\ \overrightarrow{CA}\ \overrightarrow{CE}\ \overrightarrow{CD})$. But the former would imply that $\overrightarrow{CE}$ meets $\overline{AB}$ by the crossbar principle, which is impossible. Hence, the latter holds and therefore $(\overrightarrow{CA}\ \overrightarrow{CE}\ \overrightarrow{CD})$. Apply Theorem 53.1 and Corollary 53.2:

$$\angle 1 = \angle A \qquad \text{and} \qquad \angle 2 = \angle B.$$

Hence,

$$\angle ACD = \angle 1 + \angle 2 = \angle A + \angle B. \quad \blacksquare$$

Figure 234

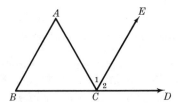

Two easy consequences may be stated, their proofs being left to the reader:

53.4 Angle-Sum Theorem for Euclidean Geometry: The angle sum of any triangle in the Euclidean plane is the measure of a straight angle.

53.5 Corollary: If two angles of one triangle are congruent, respectively, to two angles of another, the third pair of angles are congruent.

54. The Median of a Trapezoid in Euclidean Geometry

We define a **trapezoid** as a quadrilateral which has at least one pair of opposite sides parallel; the parallel sides are called the **bases,** the remaining two sides the **legs,** and the segment joining the midpoints of the legs is the **median.** A trapezoid with *both* pairs of opposite sides parallel is called a **parallelogram.**

Figure 235

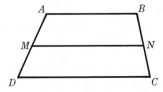

It is a familiar property of Euclidean parallelism that the median $\overleftrightarrow{MN}$ of trapezoid $ABCD$ (Fig. 235) is parallel to the bases $\overleftrightarrow{AB}$ and $\overleftrightarrow{CD}$.[6] This property

---

[6] The transitive property of parallelism is taken for granted here, easily derived from Axiom 12″. (See Exercise 1 below.)

may be derived in an efficient manner without appealing to the concept of the parallelogram and its properties (the usual method in elementary geometry).

54.1   Theorem: The midpoint of the hypotenuse of a right triangle is equidistant from the three vertices.

PROOF: Let M be the midpoint of hypotenuse $\overline{AB}$ of right triangle ABC (Fig. 236). Suppose $MC > MA = MB$. Then by Corollary 31.6

$$\angle A > \angle 1 \qquad \text{and} \qquad \angle B > \angle 2.$$

Therefore,

$$\angle A + \angle B > \angle 1 + \angle 2 = 90,$$

providing a contradiction of the angle-sum theorem. In a similar manner the assumption $MC < MA$ would lead to an absurdity. Therefore,

$$MA = MB = MC. \quad \blacksquare$$

Figure 236

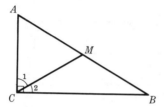

Now consider Fig. 237, where M and N are the midpoints of $\overline{AC}$ and $\overline{AB}$ and $\overset{\leftrightarrow}{AD} \perp \overset{\leftrightarrow}{BC}$. By the preceding theorem M and N are each equidistant from A and D and hence, determine the perpendicular bisector of $\overline{AD}$. Hence, $\overset{\leftrightarrow}{MN} \perp \overset{\leftrightarrow}{AD}$, and since two lines perpendicular to the same line must be parallel, $\overset{\leftrightarrow}{MN} \parallel \overset{\leftrightarrow}{BC}$. Thus:

Figure 237

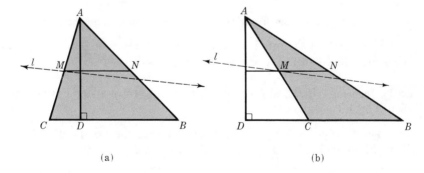

(a)   (b)

54.2   Corollary: The line joining the midpoints of two sides of a triangle is parallel to the third side, and conversely, a line which passes through the midpoint of one side of a triangle and is parallel to a second, bisects the third side.

PROOF: The previous discussion proves the first part. For the converse, if in Fig. 237(a) $l \parallel \overleftrightarrow{BC}$, and M is the midpoint of $\overline{AB}$, $M \in l$, then the uniqueness of the parallel to $\overleftrightarrow{BC}$ through M implies $\overleftrightarrow{MN} = l$. Hence, l passes through N, the midpoint of $\overline{AB}$.

Another corollary follows; the proof is left to the reader.

54.3 Corollary: The segment joining the midpoints of two sides of a triangle has length one half that of the third side.[7]

Corollary 54.2 may be extended to trapezoids, thus proving the property mentioned at the outset.

54.4 Theorem: If a line bisects one leg of a trapezoid and is parallel to each base, it bisects the other leg also and contains the median. Conversely, the median of a trapezoid lies on a line which is parallel to each of the bases.

Figure 238

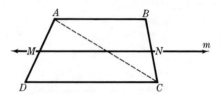

PROOF: Let m pass through the midpoint M of $\overline{AD}$ and suppose $m \parallel \overleftrightarrow{AB} \parallel \overleftrightarrow{CD}$ (Fig. 238). By Corollary 54.2, m bisects $\overline{AC}$ and therefore also $\overline{BC}$. Conversely, if N is the midpoint of $\overline{BC}$, it was just proved that the line m passing through M and parallel to $\overleftrightarrow{AB}$ passes through N. Therefore, $m = \overleftrightarrow{MN}$, and $\overleftrightarrow{MN}$ is parallel to $\overleftrightarrow{AB}$ and $\overleftrightarrow{CD}$. ∎

55. Similar Triangles and the Pythagorean Theorem

The results of the previous section are useful in establishing a basis for similar triangles. The following theorem is the key result needed:

55.1 Theorem: If D and E lie on sides $\overline{AB}$ and $\overline{AC}$ of triangle ABC and $\overleftrightarrow{DE} \parallel \overleftrightarrow{BC}$, then $AD/AB = AE/AC$.

PROOF: For each positive integer n locate the points $A = B_0, B_1, B_2, \cdots$, $B_n = B$, as shown in Fig. 239, such that $B_0B_1 = B_1B_2 = \cdots = B_{n-1}B_n = r_n$.

---

[7] This proposition, Theorem 54.1, and Corollary 54.2 were used along with a few simple properties of parallelograms and rectangles to establish the existence of the nine-point circle in Section 8 (Chapter 1). The reader should at this time study that section in detail if he has not already done so.

The lines through $B_1, B_2, \cdots, B_n$ parallel to $\overleftrightarrow{BC}$ must, by the postulate of Pasch, meet $\overline{AC}$ at corresponding points $C_1, C_2, \cdots, C_n$; moreover, if $(B_iB_jB_k)$ then since $\overleftrightarrow{B_jC_j}$ cannot cut $\overline{B_kC_k}$ it must intersect $\overline{B_iC_i}$, and, in turn, $\overline{C_iC_k}$, proving that $(C_iC_jC_k)$ holds. In the same manner, if (B_mDB_{m+1}) then (C_mEC_{m+1}). By repeated application of Theorem 54.4 and mathematical induction, it follows that

$$C_0C_1 = C_1C_2 = \cdots = C_{n-1}C_n = s_n.$$

Put $u_n = B_mD/r_n$ and $v_n = C_mE/s_n$. Thus,

$$\begin{aligned} AD = mr_n + u_nr_n, && AE = ms_n + v_ns_n, \\ AB = nr_n, && AC = ns_n. \end{aligned} \tag{55.2}$$

By algebra,

$$\frac{AD}{AB} = \frac{m}{n} + \frac{u_n}{n}, \qquad \frac{AE}{AC} = \frac{m}{n} + \frac{v_n}{n}, \tag{55.3}$$

or

$$\frac{AD}{AB} - \frac{AE}{AC} = \frac{u_n}{n} - \frac{v_n}{n}. \tag{55.4}$$

Since $0 \leq u_n \leq 1$ and $0 \leq v_n \leq 1$, u_n/n and v_n/n tend to zero as n becomes infinite. Thus, by taking the limit in (55.4),

$$\frac{AD}{AB} - \frac{AE}{AC} = 0. \quad \blacksquare$$

Figure 239

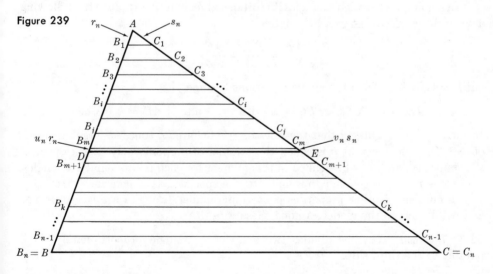

In Fig. 240 it may be observed that if $\overleftrightarrow{DE} \parallel \overleftrightarrow{BC}$, then in $\triangle ADE$ and $\triangle ABC$ $\angle A = \angle A$, $\angle ADE = \angle B$, $\angle AED = \angle C$, and, by the theorem just proved, $AD/AB = AE/AC$. If it could be further proved that $DE/BC = AE/AC$, then the existence of noncongruent similar triangles would be certain. But this is easily accomplished by taking F on $\overline{AC}$ such that $FC = AE$, and G on $\overline{BC}$ such that $\overleftrightarrow{FG} \parallel \overleftrightarrow{AB}$. Then $\angle GFC = \angle A$, $\angle C = \angle AED$ so by ASA $\triangle ADE \cong \triangle FGC$. By Theorem 55.1

$$\frac{AE}{AC} = \frac{FC}{AC} = \frac{GC}{BC} = \frac{DE}{BC}.$$

This proves the following two corollaries of Theorem 55.1 (let the reader finish the remaining details):

55.5 Corollary: In Euclidean geometry there exist similar, noncongruent triangles.

Figure 240

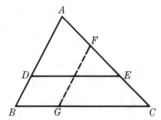

55.6 Corollary: If two triangles have the three angles of one congruent, respectively, to those of the other, the triangles are similar.

In dealing with similar triangles it is often convenient to introduce the following device: Since $\triangle ABC \cong \triangle XYZ$ implies

$$\frac{AB}{XY} = \frac{BC}{YZ} = \frac{AC}{XZ},$$

put $k = AB/XY$. Then the above equations become

$$AB = k \cdot XY, \qquad BC = k \cdot YZ, \qquad \text{and} \qquad AC = k \cdot XZ.$$

Intuitively one might say that the sides of $\triangle ABC$ are "k times as large" as those of $\triangle XYZ$. The constant k is called the **constant of proportionality** with respect to the similarity, and may be defined in this manner for each pair of similar triangles. The value $k = 1$ of course corresponds to the special case of congruent triangles.

A number of other properties concerning similar triangles are easily proved and will be left to the reader. Among these we state:

55.7 Theorem: If a line divides two sides of a triangle proportionately, it is parallel to the third side.

55.8 Corollary: If under some correspondence between the vertices of two triangles two sides of one are proportional to the corresponding sides of the other and the included angles are congruent, the triangles are similar.

The theory resulting from the Euclidean parallel axiom culminates in the

55.9 *Pythagorean Theorem:* In any right triangle the square of the length of the hypotenuse equals the sum of the squares of the lengths of the two legs.

PROOF: As in Fig. 241 let $\triangle ABC$ be the given triangle, with the right angle at C. Since the angles at A and B are acute, the foot D of the perpendicular from C to $\overleftrightarrow{AB}$ will fall interior to $\overline{AB}$ and hence $(\overrightarrow{CA}\ \overrightarrow{CD}\ \overrightarrow{CB})$. Observe the three right triangles thus formed. Since $\angle A = \angle A$ and $\angle ADC = \angle ACB$ it follows that $\angle ACD = \angle B$ and $\angle BCD = \angle A$. By Corollary 55.6, $\triangle ABC \sim \triangle CBD \sim \triangle ACD$, with proportionality constants $x = b/h$ and $y = b/c_1$. Therefore,

$$a = xc_2 = yh, \tag{55.10}$$

$$b = xh = yc_1, \tag{55.11}$$

$$c = xa = yb. \tag{55.12}$$

Multiply both sides of the first equation in (55.10) by a and use (55.12):

$$a^2 = axc_2 = cc_2. \tag{55.13}$$

Multiply both sides of the second equation in (55.11) by b and again use (55.12):

$$b^2 = byc_1 = cc_1. \tag{55.14}$$

Add (55.13) and (55.14) to obtain

$$a^2 + b^2 = cc_1 + cc_2 = c(c_1 + c_2) = c^2. \ \blacksquare$$

Figure 241

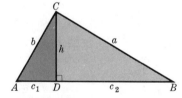 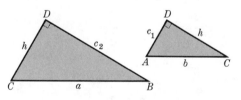

The above equations may be used to obtain at the same time the proposition: *The length of the altitude to the hypotenuse of a right triangle is the geometric mean of the lengths of the segments it determines on that hypotenuse.* For, "cross-multiplying" in (55.10) and (55.11) yields

$$xyh^2 = xyc_1c_2,$$

or

$$h^2 = c_1c_2. \tag{55.15}$$

EXERCISES

1. A line is commonly regarded as being *parallel to itself*. With this convention, prove that Euclidean parallelism is transitive. That is, if $l_1 \parallel l_2$ and $l_2 \parallel l_3$, then $l_1 \parallel l_3$.
2. Parallel lines intercept proportional segments on any two transversals.
3. Prove Theorem 55.7.
4. Prove Corollary 55.8.

5. If the three sides of one triangle are proportional to the three sides of another the triangles are similar. Prove.

6. In Fig. 242 the angles at B and D are right angles. Prove that A, B, C, and D are concyclic.

7. The perpendicular bisector of the segment joining the feet of two altitudes of a triangle passes through the midpoint of the remaining side. *Hint:* There is a very simple proof.

8. Prove that a parallelogram is a convex quadrilateral.

9. **Properties of the parallelogram.** Establish the following simple facts about parallelograms and, without proving it, point out where the convexity property for quadrilaterals is used in each argument:

    (a) The opposite sides of a parallelogram are congruent.
    (b) If a quadrilateral has its opposite sides congruent, it is a parallelogram.
    (c) If a quadrilateral has a pair of opposite sides both congruent and parallel, it is a parallelogram.
    (d) The diagonals of a parallelogram bisect each other.
    (e) If the diagonals of a quadrilateral bisect each other, it is a parallelogram.

10. Prove that in Euclidean geometry the only convex set which contains two intersecting lines is the entire plane.

11. State and prove the converse of the Pythagorean theorem.

★12. Prove that a trapezoid is a convex quadrilateral.

13. Using the convexity property of the trapezoid (Exercise 12) prove that if $\overset{\leftrightarrow}{EF} \parallel \overset{\leftrightarrow}{AB} \parallel \overset{\leftrightarrow}{CD}$, $p = ED/AD$, $q = AE/AD$, and the remaining notation is as indicated (Fig. 243), then

$$\boxed{m = pa + qb.}$$

(55.16)

Figure 242

Figure 243

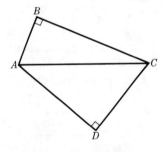

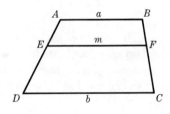

14. Solve Euclid's compass straightedge construction of the regular pentagon (Exercise 27, Chapter 1) at this time if you have not already done so.

15. Solve Exercise 11, Chapter 1 if you have not already done so.

16. If a point is equidistant from the vertices of a right triangle, then it is the midpoint of the hypotenuse. Prove.

17. A **rhombus** is a parallelogram having adjacent sides congruent. Prove that the diagonals of a rhombus are perpendicular.

18. Making use of the formula (55.15) find a short geometric proof of the following well-

known property of the real numbers: *The geometric mean of two positive numbers is less than or equal to their arithmetic mean, with equality only when the two numbers are equal. Hint:* See Fig. 244.

19. The lines joining the respective trisection points of the consecutive sides of a quadrilateral in cyclic order form a parallelogram.

20. Prove the following theorem, the converse of Theorem 54.1:

Figure 244

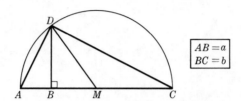

$AB = a$
$BC = b$

55.17 Theorem: If the midpoint of one of the sides of a triangle is equidistant from the three vertices, the triangle is a right triangle, the right angle being opposite the given midpoint.

PART IV

DEVELOPMENT
OF
GEOMETRY
FROM
MODELS

7

Some Modern Geometry
of the Triangle

The development up to this point has brought the reader to an awareness of the nature of rigor in geometry. It is now important that he expand his knowledge of the subject to include a variety of important topics. This is made possible not by the slow, plodding effort characteristic of the previous development, but rather by a more informal and intuitive approach. Indeed, a purely synthetic study of both Euclidean and hyperbolic geometry would be largely redundant.

A license for the less formal approach may be had: Since there exists an easily visualized, but accurate model of axiomatic Euclidean geometry, one can then agree to study the model rather than the logical structure. That model is the *analytic plane*, with all its familiar methods and useful applications.

Accordingly, the development from this point on should be regarded as taking place *in the model*—that is, in the analytic plane. Coordinates and equations will therefore be used whenever convenient. But since the previous axioms for Euclidean geometry are all evidently realized in the analytic plane, the theorems established from them may be used, and the "synthetic" method commonly associated with axiomatic geometry can be employed. There is an obvious advantage in being able to choose either method.

The first several sections will introduce the *tools* which will be needed. The opening section in particular will provide the reader with an informal review of analytic geometry.

56. Analytic Geometry

The representation of points by ordered pairs of real numbers forms the concept of the "analytic plane" or the "analytic method" which is indispensable in the study of calculus. If the pair (x, y) represents the point P, then x and y are called the **coordinates** of P, x the **abscissa**, and y the **ordinate**. In accordance with earlier conventions, we shall write

$$P[x, y].$$

To obtain a model for Euclidean geometry (instead of a *representation* of it) one agrees to let "point" designate any ordered pair of real numbers (x, y) and to let "line" be any locus of "points" (x, y) whose coordinates satisfy an equation of the form

$$ax + by + c = 0, \qquad a^2 + b^2 \neq 0. \tag{56.1}$$

Two distinct lines are therefore represented by two distinct linear equations, such as, for example, the equation (56.1), and an equation of the form $a'x + b'y + c' = 0$, where a', b', c' are not proportional to a, b, c. A pair (x, y) which satisfies both equations clearly corresponds to the point of intersection, and the possibility of solving for such a pair is certain if

$$ab' \neq a'b, \tag{56.2}$$

or, when b and b' are nonzero, if

$$\frac{a}{b} \neq \frac{a'}{b'}. \tag{56.3}$$

Since the numbers $-a/b$ and $-a'/b'$ are the **slopes** of the lines, it follows that two lines intersect whenever they have *different slopes*. Alternatively, *two lines are parallel if and only if they are either vertical* (have equations of the form $x = $ constant) *or have the same slope*. It may be shown that two lines whose slopes exist are *perpendicular if and only if the product of their slopes is* -1.

The *distance* between the points $P[x_1, y_1]$ and $Q[x_2, y_2]$ is given by the formula

$$\sqrt{(x_1 - x_2)^2 + (y_1 - y_2)^2}. \tag{56.4}$$

It immediately follows that a *circle* centered at (h, k) and having radius r is the locus of points $P[x, y]$ such that

$$(x - h)^2 + (y - k)^2 = r^2. \tag{56.5}$$

It may also be shown directly from (56.4) that the *midpoint* of $\overline{PQ}$ is the point

$$M\left[\frac{x_1 + x_2}{2}, \frac{y_1 + y_2}{2}\right]. \tag{56.6}$$

56.7 Example: If the equation $a'x + b'y + c' = 0$ represents a line, one may divide through by $\sqrt{a'^2 + b'^2}$ to arrive at an equation of the form (56.1) with $a^2 + b^2 = 1$. Such an equation is said to be written in **normal form**. Show that if $F(x, y) = ax + by + c = 0$ is the equation of any line l written in normal form, then $|F(x_0, y_0)|$ is the distance from the point $P[x_0, y_0]$ to line l.

Solution: The slope of any line perpendicular to l is b/a. Hence, the equation of the perpendicular through P is

$$y - y_0 = \frac{b}{a}(x - x_0).$$

This equation and the equation of l are solved simultaneously to give the coordinates of the foot of P on l:

$$P'[b^2x_0 - aby_0 - ac, -abx_0 + a^2y_0 - bc].$$

By formula (56.4),

$$PP' = \sqrt{[(1 - b^2)x_0 + aby_0 + ac]^2 + [abx_0 + (1 - a^2)y_0 + bc]^2}.$$

The relation $a^2 + b^2 = 1$ yields

$$\begin{aligned}
PP' &= \sqrt{(a^2x_0 + aby_0 + ac)^2 + (abx_0 + b^2y_0 + bc)^2} \\
&= \sqrt{a^2(ax_0 + by_0 + c)^2 + b^2(ax_0 + by_0 + c)^2} \\
&= \sqrt{(ax_0 + by_0 + c)^2}\sqrt{a^2 + b^2} \\
&= |ax_0 + by_0 + c|. \quad\blacksquare
\end{aligned}$$

To make our review more complete, we show how to derive the trigonometric functions. From now on we shall agree to measure all angles in radians. This may be accomplished formally by simply replacing the value $\beta = 180$, which was originally postulated as the measure of a straight angle, by the value $\beta = \pi = 3.14159\cdots$. Observe that at the origin $O[0, 0]$ (as well as at any other point) the rays may be assigned the coordinates as shown in Fig. 245.

Figure 245

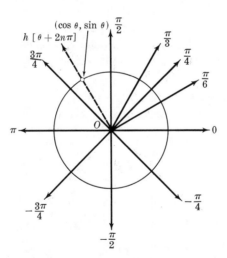

Now let θ be any real number. Since θ will lie on one of the intervals, $\cdots$ $(-3\pi, -\pi], (-\pi, \pi], (\pi, 3\pi], (3\pi, 5\pi], \cdots$ (Fig. 246), there is a unique integer n such that

$$-\pi < \theta + 2n\pi \leq \pi.$$

Figure 246

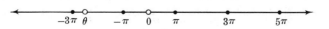

Thus $\phi = \theta + 2n\pi$ will be the coordinate of a unique ray h from O, and this ray will intersect the unit circle

$$x^2 + y^2 = 1 \tag{56.8}$$

at a unique point $P[x, y]$. The sine and cosine functions may then be defined as

$$\cos \theta = x \quad \text{and} \quad \sin \theta = y, \tag{56.9}$$

(sometimes called the *circular functions*). The remaining functions $\tan \theta$, $\cot \theta$, $\sec \theta$, and $\csc \theta$ are defined—with certain restrictions on x, y, and therefore θ necessary—

$$\tan \theta = \frac{y}{x}, \quad \cot \theta = \frac{x}{y}, \quad \sec \theta = \frac{1}{x}, \quad \text{and} \quad \csc \theta = \frac{1}{y}, \tag{56.10}$$

respectively. The fundamental identities readily follow by definition and by (56.8). The ones encountered most frequently are

$$\boxed{\tan \theta = \frac{\sin \theta}{\cos \theta}} \quad \text{and} \quad \boxed{\sin^2 \theta + \cos^2 \theta = 1.} \tag{56.11}$$

Polar coordinates may now be introduced. Let (r, θ) be any pair of real numbers, and determine the integer n such that $-\pi < \theta + 2n\pi \leq \pi$. Consider the ray $h[\phi]$ where $\phi = \theta + 2n\pi$, and the opposite ray h', both having O as origin. Then determine the point P such that $OP = |r|$ with $P \in h$ if $r \geq 0$, and $P \in h'$ if $r < 0$ (Fig. 247). It is clear that conversely, for any point P given in the xy plane one may determine a pair of real coordinates (r, θ) such that the above procedure will yield P. (Of course many pairs of coordinates are possible for each point.) Such coordinates associated with P are called **polar coordinates**. To distinguish from the *rectangular* representation $P[x, y]$ we shall write

$$P(r, \theta).$$

It is then easy to prove the connecting relations

$$\boxed{x = r \cos \theta, \quad y = r \sin \theta.} \tag{56.12}$$

Figure 247

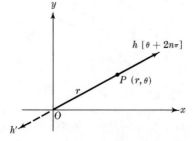

An interesting application of (56.12) is the following derivation of the law of cosines. Assuming a representation in polar coordinates with A as origin, let $A(0, 0)$, $B(c, 0)$, and $C(b, A)$ be the vertices of triangle ABC, where for brevity a, b, and c

denote the lengths of the sides in standard notation and A stands for the measure of $\overline{\angle A}$. It follows from (56.12) that the vertices in rectangular coordinates are $A[0, 0]$, $B[c, 0]$, and $C[b \cos A, b \sin A]$. (See Fig. 248.) Thus by (56.4)

$$
\begin{aligned}
a^2 &= (b \cos A - c)^2 + (b \sin A - 0)^2 \\
&= b^2 \cos^2 A - 2bc \cos A + c^2 + b^2 \sin^2 A \\
&= b^2 + c^2 - 2bc \cos A.
\end{aligned}
$$

Figure 248

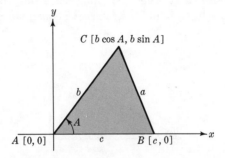

This proves the **law of cosines for triangle** ABC:

$$
\boxed{
\begin{aligned}
a^2 &= b^2 + c^2 - 2bc \cos A, \\
b^2 &= a^2 + c^2 - 2ac \cos B, \\
c^2 &= a^2 + b^2 - 2bc \cos C.
\end{aligned}
}
\qquad \textbf{(56.13)}
$$

One may use the same method to derive the law of sines: In the same triangle ABC, take B as the origin; the rectangular coordinates of the vertices may be taken as $A[-c, 0]$, $B[0, 0]$, and $C[-a \cos B, a \sin B]$. (See Fig. 249.) Since the y coordinate of each of the two representations of C in Figs. 248 and 249 must be the same,

$$
b \sin A = a \sin B.
$$

Figure 249

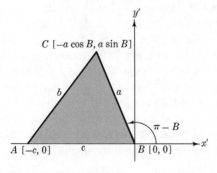

This proves the **law of sines for triangle** ABC:

$$
\boxed{
\frac{a}{\sin A} = \frac{b}{\sin B} = \frac{c}{\sin C}.
}
\qquad \textbf{(56.14)}
$$

We conclude the review by obtaining the trigonometry of the right triangle. If $C = \pi/2$, the third equation in (56.13) reduces to $c^2 = a^2 + b^2$; this is substituted into the first equation to obtain

$$a^2 = 2b^2 + a^2 - 2bc \cos A,$$

which simplifies to

$$b = c \cos A.$$

The remaining relations may then be obtained from the identity $\cos A = \sin(\pi/2 - A)$. Thus we have found the **formulas relating the parts of a right triangle** (Fig. 250):

Figure 250

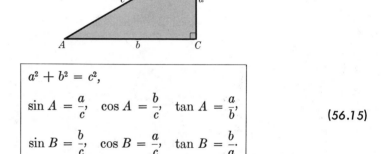

$$a^2 + b^2 = c^2,$$

$$\sin A = \frac{a}{c}, \quad \cos A = \frac{b}{c}, \quad \tan A = \frac{a}{b},$$

$$\sin B = \frac{b}{c}, \quad \cos B = \frac{a}{c}, \quad \tan B = \frac{b}{a}.$$

(56.15)

57. The Analytic versus the Synthetic Method

A question arises each time one faces a problem in geometry which he has never seen before: Should the problem be attacked analytically or synthetically? The beginner often gathers the impression that the analytic method is always the most potent, even if he is mystified by the formidable algebra which sometimes accompanies the proofs. Analytic geometry does afford a method of *discovering* relationships which sometimes lead to important theorems, but quite often the most elegant proofs of those same theorems are synthetic, and on those occasions one would be led to believe that the synthetic method is the more powerful.

Without trying to build a case for either point of view we take the position that the issue is to be decided in each particular situation, and that it is better to allow the two methods to *enhance one another* rather than to adopt one method at the exclusion of the other. We illustrate in the proofs of two familiar theorems.

57.1   Theorem: The medians of a triangle meet at a point which is two thirds the distance from each vertex to the midpoint of the opposite side.

PROOF: (1) *By analytic geometry.* Let the vertices of the triangle be represented as $A[0, 0]$, $B[a, 0]$, and $C[b, c]$ for convenience, and write down the coordinates of the midpoints L, M, and N of the sides opposite A, B, and C (Fig. 251):

$$L\left[\frac{a+b}{2}, \frac{c}{2}\right], \qquad M\left[\frac{b}{2}, \frac{c}{2}\right], \qquad \text{and} \qquad N\left[\frac{a}{2}, 0\right].$$

The equations for $\overleftrightarrow{AL}$ and $\overleftrightarrow{BM}$ are

$$\overleftrightarrow{AL}: \quad y = \frac{cx}{a+b}$$

$$\overleftrightarrow{BM}: \quad y = \frac{c}{b-2a}(x-a).$$

We find the coordinates of G by solving these two equations simultaneously:

$$\frac{cx}{a+b} = \frac{c(x-a)}{b-2a}$$

$$bx - 2ax = ax + bx - a^2 - ab$$

$$3ax = a^2 + ab$$

$$x = \frac{a+b}{3}.$$

From the equation for $\overleftrightarrow{AL}$, $y = \frac{c}{3}$. Thus G is represented by $\left[\dfrac{a+b}{3}, \dfrac{c}{3}\right]$. By the distance formula,

$$AG = \frac{\sqrt{(a+b)^2 + c^2}}{3}, \qquad AL = \frac{\sqrt{(a+b)^2 + c^2}}{2}.$$

Figure 251

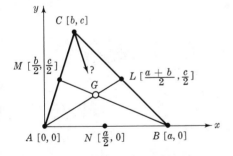

Figure 252

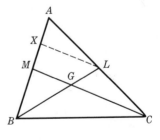

Therefore, $AG/AL = \frac{2}{3}$. Since this result is independent of the coordinate system, it shows that two medians of any triangle intersect at a point which is two thirds the distance from one of the vertices to the opposite midpoint. But this is clearly enough to establish the proposition. ∎

(2) *By the synthetic method.* In Fig. 252 let medians $\overline{BL}$ and $\overline{CM}$ be determined and take $\overleftrightarrow{LX} \parallel \overleftrightarrow{CM}$. $\overline{BL}$ and $\overline{CM}$ will intersect at some point G by the crossbar principle. Since L is the midpoint of $\overline{AC}$, X must also be the midpoint of $\overline{AM}$ (observe $\triangle AMC$). Hence,

$$BM = MA = 2MX$$

and hence,

$$BM = \tfrac{2}{3}BX.$$

But in $\triangle BLX$ we have $\overleftrightarrow{MG} \parallel \overleftrightarrow{XL}$. Therefore,

$$BG = \tfrac{2}{3}BL.$$

The remainder follows as in (1). ∎

Many arguments have been offered for the above theorem, but it would be difficult to improve upon the one just given. Thus here it appears that the synthetic proof is the most concise. It should be pointed out, however, that in attempting an analytic proof in terms of general coordinates one *might* have discovered the interesting formulas

$$x_0 = \frac{a_1 + b_1 + c_1}{3}, \qquad y_0 = \frac{a_2 + b_2 + c_2}{3}, \tag{57.2}$$

which give the coordinates of the centroid of a triangle in terms of the coordinates of the vertices (a_1, a_2), (b_1, b_2), and (c_1, c_2). It is doubtful whether this discovery would have been made by the synthetic approach to the problem.

57.3 Theorem: The altitudes of a triangle are concurrent. If the angles of the triangle are acute, the point of concurrency is the incenter of the orthic triangle.[1]

PROOF: (1) *By analytic geometry.* For convenience take the y axis as one of the altitudes (as in Fig. 253) and determine the coordinates of the vertices. The slopes of $\overleftrightarrow{BE}$ and $\overleftrightarrow{CF}$ must be c/a and b/a, so the equations of the three altitudes are

$$\overleftrightarrow{AD}: \quad x = 0,$$

$$\overleftrightarrow{BE}: \quad y = \frac{c}{a}(x - b),$$

$$\overleftrightarrow{CF}: \quad y = \frac{b}{a}(x - c).$$

Figure 253

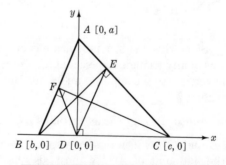

---

[1] Defined in Section 8 as the triangle whose vertices are the feet of the three altitudes.

Hence, the point of concurrency is $(0, -bc/a)$. For the second part, the equations of $\overleftrightarrow{AC}$ and $\overleftrightarrow{AB}$ are

$$\overleftrightarrow{AC}: \quad y = -\frac{a}{c}(x - c) \qquad \text{and} \qquad \overleftrightarrow{AB}: \quad y = -\frac{a}{b}(x - b).$$

Thus the points

$$E\left[\frac{c(a^2 + bc)}{a^2 + c^2}, \frac{ac(c - b)}{a^2 + c^2}\right] \qquad \text{and} \qquad F\left[\frac{b(a^2 + bc)}{a^2 + b^2}, -\frac{ab(c - b)}{a^2 + b^2}\right]$$

may be found. Hence, the slopes of $\overleftrightarrow{DE}$ and $\overleftrightarrow{DF}$ are

$$\frac{ac(c - b)}{c(a^2 + bc)} = \frac{a(c - b)}{a^2 + bc} \qquad \text{and} \qquad -\frac{ab(c - b)}{b(a^2 + bc)} = -\frac{a(c - b)}{a^2 + bc}$$

respectively, proving that $\overleftrightarrow{DE}$ and $\overleftrightarrow{DF}$ are equally inclined and that, therefore, $\overrightarrow{DA}$ bisects $\angle EDF$. Since $\overline{AD}$ was any one of the three altitudes to begin with, the argument applies to any other altitude and the theorem is established. $\blacksquare$

(2) *By the synthetic method.* Let $\overleftrightarrow{YZ}$, $\overleftrightarrow{ZX}$, and $\overleftrightarrow{XY}$ be lines through A, B, and C parallel to the opposite sides $\overleftrightarrow{BC}$, $\overleftrightarrow{CA}$, and $\overleftrightarrow{AB}$ in that order (Fig. 254). The lines will intersect because of Axiom 12″. Hence $ABCY$ and $ZBCA$ are parallelograms, and consequently,

$$ZA = BC = AY.$$

Figure 254

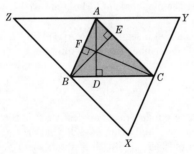

Therefore, A is the midpoint of $\overline{ZY}$. Similarly, B and C are the midpoints of $\overline{XZ}$ and $\overline{XY}$. Therefore, the lines containing the altitudes $\overline{AD}$, $\overline{BE}$, and $\overline{CF}$ are the *perpendicular bisectors* of the sides of triangle ABC, and are concurrent by Theorem 34.5.

For the second part, observe that since $\angle A$ is common to right triangles BAE and ACF, Fig. 255(a),

$$\triangle ABE \sim \triangle ACF.$$

Hence,

$$\frac{AF}{AC} = \frac{AE}{AB}.$$

By Corollary 55.8,

$$\triangle AEF \sim \triangle ABC,$$

and therefore,

$$\angle AEF = \angle B, \qquad \angle AFE = \angle C.$$

The same argument proves that in Fig. 255(b) $\angle BFD = \angle C$, $\angle BDF = \angle A$, $\angle CDE = \angle A$, and $\angle CED = \angle B$. The conclusion is then obvious. ∎

Figure 255

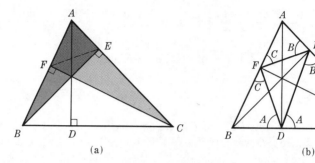

(a)   (b)

EXERCISES

1. Prove by analytic geometry that the diagonals of a parallelogram bisect each other.
2. Prove by analytic geometry that the segment joining the midpoints of two sides of a triangle determines a line parallel to the third side, and has length one half that of the third side.
3. Prove by analytic means the theorem that if a line passes through an interior point of a circle it intersects the circle at two distinct points.
4. Prove analytically that the diagonals of a rhombus are perpendicular. (See Exercise 17, Section 55.)
5. Show that if the endpoints of a segment are $A[r, s]$ and $B[u, v]$, the coordinates of the points C and C' which determine the two equilateral triangles with $\overline{AB}$ as base are given by

$$\left(\frac{1}{2}(u+r) \pm \frac{\sqrt{3}}{2}(v-s), \mp \frac{\sqrt{3}}{2}(u-r) + \frac{1}{2}(v+s) \right).$$

6. Prove analytically that if equilateral triangles $A'BC$, $AB'C$, and ABC' be constructed on the sides of triangle ABC and lying outside that triangle, then $AA' = BB' = CC'$. *Hint:* Use the formula of Exercise 5 with a suitable choice of signs.
7. Prove the result of Exercise 6 synthetically.
8. Squares $ABEF$, $BCGH$, $CDJK$, and $DALM$ are placed externally on the sides of quadrilateral $ABCD$, with X, Y, Z, and W the respective centers of those squares. Prove that $XZ = YW$ and $\overleftrightarrow{XZ} \perp \overleftrightarrow{YW}$. (H. S. M. Coxeter, *Introduction to Geometry* [4, p. 23].)
9. As in Exercise 8, but with $ABCD$ a parallelogram, prove in addition that $XYZW$ is a square. (*Hungarian Problem Book II*, Random House: New York, 1963, p. 9; see p. 23 for an attractive solution. For a different synthetic solution, reflect the figure through O, the point of intersection of the diagonals of $ABCD$, and show that $XYZW$ is a parallelogram. The result of Exercise 8 then shows that $XYZW$ is a square.)
10. The centroids of three equilateral triangles laid externally on the sides of any triangle form another equilateral triangle. *Hint:* Use the formula of Exercise 5 and (57.2).

11. Define the two-variable function

$$F(x, y) = (x - h)^2 + (y - k)^2 - r^2. \tag{57.4}$$

The equation $F(x, y) = 0$ will be recognized as that of a circle having center (h, k) and radius r. If the point $P[x_0, y_0]$ lies outside the circle $F(x, y) = 0$, what geometric interpretation can be given to the number $F(x_0, y_0)$?

12. With F defined as in Exercise 11, interpret $F(x_0, y_0)$ geometrically when $P[x_0, y_0]$ lies *inside* the circle $F(x, y) = 0$. In particular, prove in this case that $F(x_0, y_0)$ is *the negative of the product of the lengths of the segments formed on any chord passing through that point.* Hint: First interpret the number $F(x_0, y_0)$ in terms of the length of the chord through P perpendicular to the line joining P with the center of the circle. The rest may be proved analytically by assuming a coordinate system such that $x_0 = y_0 = h = 0$. The following easily derived fact will be found useful: If $A[x_1, y_1]$ and $B[x_2, y_2]$ are any two points on the line $y = mx$, then $AB = |x_1 - x_2|\sqrt{1 + m^2}$.

13. Using the same choice of coordinates for the vertices of triangle ABC as in the proof of Theorem 57.3 find the coordinates of the orthocenter H, centroid G, and circumcenter O, then prove analytically that H, G, and O are collinear. The line of collinearity is the **Euler line** of the triangle. Find its equation. (See Section 8, Chapter 1.)

14. Find the coordinates of each of the nine points of the nine-point circle of triangle ABC assuming the same coordinate system as in the proof of Theorem 57.3 (refer to Section 8, Chapter 1). Show that the equation of the nine-point circle is $x^2 + y^2 + px + qy = 0$, where

$$p = -\frac{b + c}{2} \quad \text{and} \quad q = \frac{bc - a^2}{2a}.$$

Thus show that the nine points are indeed concyclic. Compare this with the argument given in Section 8.

15. The sum of the distances from any point in the interior of an equilateral triangle is a constant. Prove by any method you wish.

58. Directed Distance and Cross Ratio

It is often useful to have a distance function which can take on negative as well as positive values. By the coordinatization axiom (Section 18) each line l has a coordinate system with any chosen point as origin and another chosen point having positive coordinate. Then for any two points $P_1[x_1]$ and $P_2[x_2]$ on l, it was postulated that $P_1P_2 = |x_1 - x_2|$. Distances on l become *directed* if instead one assumes

$$\boxed{P_1P_2 = x_2 - x_1.} \tag{58.1}$$

We say that (58.1) gives the **directed distance from** P_1 **to** P_2. In order to emphasize the fact that directed distances are being used in the statement of any theorem which follows, the terminology "in magnitude and in sign" will often be employed.

Whenever possible, it will always be assumed that the directed distance from one point to another is *positive*. In view of Axiom 5', for any two points A and B

it is always possible to choose a coordinate system on line $\overleftrightarrow{AB}$ so that $AB > 0$, but having made this choice, the sign for all other pairs (C, D) on $\overleftrightarrow{AB}$ is determined. Thus, unless it is known that the points A, B, C, and D are noncollinear, it cannot be asserted that *both* AB and CD are positive. In that case if we want to denote the *length of the segment* $\overline{CD}$ (which would be the *undirected* distance from C to D), then absolute values must be employed.

The fundamental properties of directed distance are collected in the following statement:

58.2 Theorem: If A, B, C, and X are any four points on a line l, then, in magnitude and in sign,

(a) $AB + BA = 0$,
(b) $AB + BC + CA = 0$,
(c) $AX = AB$ if and only if $X = B$,

and

(d) $XA = AB$ and $A \neq B$ if and only if A is the midpoint of the segment $\overline{XB}$.

PROOF: Let $A[a]$, $B[b]$, $C[c]$, and $X[x]$ be the points considered with their coordinates. Then (a), (b), and (c) follow immediately by simple algebra. For example, in (b) write

$$AB + BC + CA = (b - a) + (c - b) + (a - c) = 0.$$

To prove (d), suppose A is the midpoint of $\overline{XB}$. Then $|XA| = |AB|$ and (XAB). Hence $\pm XA = AB$. If $-XA = AB$, then (a) and (c) together imply that $X = B$, denying (XAB). Therefore, $XA = AB$. Conversely, suppose $XA = AB$ and that A' is the midpoint of $\overline{XB}$. Then by the preceding case, $XA' = A'B$, and by (a), $A'X = BA'$. Hence, $A'X + XA = BA' + AB$, or by (c) $A'A = AA'$. That is, $2\,A'A = 0$, and therefore, $A' = A$. ∎

NOTE: It is sometimes useful to recognize equation (b) of Theorem 58.2 in the equivalent form

$$AB + BC = AC.$$

58.3 Corollary: If points $A[a]$ and $B[b]$ be located on line l and $M[m] \in l$, M is the midpoint of segment $\overline{AB}$ if and only if

$$m = \tfrac{1}{2}(a + b).$$

PROOF: Left to the reader [use part (d) of the previous theorem].

Two classical theorems further illustrate the behavior of directed distances.

58.4 Euler's Theorem: Given any four collinear points A, B, C, and D, then in magnitude and in sign,

$$AB \cdot CD + AC \cdot DB + AD \cdot BC = 0.$$

PROOF: Let C be "inserted" between the pairs A, B and A, D. The left member of the desired equation becomes

$$(AC + CB)CD + AC \cdot DB + (AC + CD)BC$$
$$= AC \cdot CD + AC \cdot DB + AC \cdot BC$$
$$= AC(CD + DB + BC)$$
$$= 0. \blacksquare$$

58.5 Stewart's Theorem: Given three collinear points A, B, and C, and P any other point in the plane, then, in magnitude and in sign,

$$PA^2 \cdot BC + PB^2 \cdot CA + PC^2 \cdot AB + BC \cdot CA \cdot AB = 0.$$

PROOF: The equation is first established for the case when P is collinear with A, B, and C. Insert A between both P, B and P, C in the left member above. Thus

$$PA^2 \cdot BC + (PA^2 + 2PA \cdot AB + AB^2)CA$$
$$+ (PA^2 + 2PA \cdot AC + AC^2)AB + BC \cdot CA \cdot AB$$
$$= PA^2(BC + CA + AB) + 2PA \cdot AB(CA + AC)$$
$$\overline{]} + AB^2 \cdot CA + AC^2 \cdot AB + BC \cdot CA \cdot AB$$
$$= \overline{AB \cdot CA}(AB + CA + BC)$$
$$= 0.$$

For $P \not\in \overleftrightarrow{AB}$, let P' be the foot of P on $\overleftrightarrow{AB}$ (Fig. 256). By the Pythagorean theorem, $PA^2 = PP'^2 + P'A^2$, $PB^2 = PP'^2 + P'B^2$, and $PC^2 = PP'^2 + P'C^2$. If these are substituted in the left member of the desired equation, it will be found that this case reduces to the previous one. The details will be left to the reader. $\blacksquare$

Figure 256

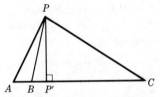

The concept of directed distance is also involved in the definition of the classical "ratio of ratios" known as *cross ratio*, which will be considered next.

58.6 Definition: To each ordered quadruple of distinct collinear points A, B, C, D is associated a number $[AB, CD]$, called its **cross ratio**, and defined by the equation

$$[AB, CD] = \frac{AC \cdot BD}{AD \cdot BC}.$$

Occasionally the definition of cross ratio is used in either of the equivalent forms

$$[AB, CD] = \frac{AC}{CB} \bigg/ \frac{AD}{DB} = \frac{AC}{AD} \bigg/ \frac{BC}{BD},$$

which are perhaps more suggestive of the term *cross ratio*.

It appears from the definition that the value of the cross ratio of four points depends not only on their position, but also on the *order* in which they occur on the line of collinearity. For example, even though *collectively* the points $A[3]$, $B[13]$, $C[8]$, and $D[11]$ are identical with $A'[13]$, $B'[3]$, $C'[8]$, and $D'[11]$ (thus $A' = B$, $B' = A$, $C' = C$, $D' = D$), the resulting cross ratios taken in these orders are not the same (see Fig. 257):

$$[AB, CD] = \frac{5(-2)}{8(-5)} = \frac{1}{4}, \qquad [A'B', C'D'] = \frac{(-5)8}{(-2)5} = 4.$$

Since there are 24 such permutations possible, one might expect to have as many as 24 distinct cross ratios. But it can be shown that there are *never more than 6*. The 6 possible cross ratios among four collinear points may *always* be represented in the form

$$\lambda, \quad \frac{1}{\lambda}, \quad 1 - \lambda, \quad \frac{1}{1 - \lambda}, \quad \frac{\lambda - 1}{\lambda}, \quad \text{and} \quad \frac{\lambda}{\lambda - 1}, \qquad \text{(58.7)}$$

where λ is the value of any one of those cross ratios. (See Exercise 3 below.)

Figure 257

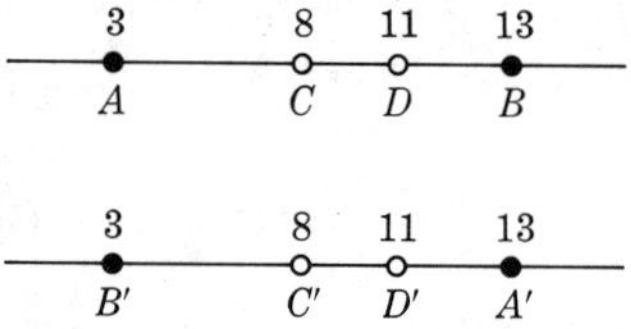

The question arises whether there are *ever* 6 cross ratios. It appears from the forms in (58.7) that, generally speaking, the values are distinct. But it is of significance to find out how arbitrary the cross ratio can be.

Let four points $A[a]$, $B[b]$, $C[c]$, and $X[x]$ be located on some line l (say the x axis), and for convenience assume $c = 0$, $0 < a < b$. By definition,

$$[AB, CX] = \frac{(c - a)(x - b)}{(x - a)(c - b)} = \frac{a}{b} \cdot \frac{x - b}{x - a}.$$

As $X[x]$ varies, the above cross ratio becomes the following function of x:

$$y = f(x) = \frac{a}{b} \cdot \frac{x - b}{x - a}, \qquad x \neq 0, b. \qquad \text{(58.8)}$$

Its graph is the rectangular hyperbola having asymptotes $x = a$ and $y = a/b$, shown in Fig. 258.

It is readily seen from the graph that, except for the values 0, 1, and a/b (by definition $X = B$ and $X = C$ are not allowed) the cross ratio can be made to assume any given real number. But by varying the choice of a/b one can assert: *For each real number* $r \neq 0, 1$ *there exists a quadruple of collinear points whose cross ratio is* r, *having specified two of those points*. It is also seen from the graph that

$$[AB, CD] = [AB, CX] \text{ implies } X = D. \qquad \text{(58.9)}$$

(This fact may also be proved directly from the definition.)

Figure 258

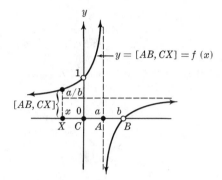

It then follows that, as a general rule, the cross ratio of four points has 6 distinct values. For example, take $\lambda = 3$ in (58.7). There can also be fewer than 6: If $\lambda = -1$, one obtains only the values -1, $\frac{1}{2}$, and 2. It so happens that *this is the only case when there are less than 6 values* for the cross ratio (see Exercise 4 below). This case is important geometrically, as will be seen later.

EXERCISES

1. Show that for any four collinear points A, B, C, and D,
$$AB + BC + CD + DA = 0.$$

2. Finish the proof of Stewart's theorem.

3. Prove that if $[AB, CD] - \lambda$, then: $[AB, DC] = 1/\lambda$; $[AC, BD] = 1 - \lambda$ (*Hint:* Use Euler's theorem); $[AC, DB] = 1/(1 = \lambda)$; $[AD, BC] = (\lambda - 1)/\lambda$; $[AD, CB] = \lambda/(\lambda - 1)$. Make a table for the 24 cross ratios which shows that any one of the 24 will be equal to one of the above 6.

4. Prove that if among the 6 values listed in (58.7) any two are equal, then λ is either $= 1$, $\frac{1}{2}$, or 2, and that in any case the values of the six expressions in (58.7) are $= 1$, $\frac{1}{2}$, and 2.

★5. Show that for four distinct collinear points A, B, C, and D,
$$[AC, BD] = -1 \quad \text{if and only if} \quad \frac{1}{AB} - \frac{1}{BC} + \frac{1}{CD} - \frac{1}{DA} = 0.$$

6. Apply the law of sines to triangles APB and CPB in Fig. 259 to derive the formula
$$\frac{AB}{CB} = \frac{PA \sin \angle APB}{PC \sin \angle CPB}, \tag{58.10}$$
assuming the betweenness relation (ACB).

Figure 259

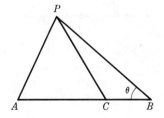

7. **Casey's Theorem.** Let A, B, C, and D be four points on a line such that the order relation $(ACBD)$ holds. Then the six cross ratios of those points are expressible in the terms $\sin^2 \theta$, $\cos^2 \theta$, $-\tan^2 \theta$, $\csc^2 \theta$, $\sec^2 \theta$, and $-\cot^2 \theta$, where 2θ measures the angle between the tangents to the circles O and O' constructed on $\overline{AB}$ and $\overline{CD}$ as diameters (Fig. 260). *Hint:* It is easy to see that $\angle CPB = \theta$ and $\angle APD = \pi - \theta$. Show that $[AC, BD] = \csc^2 \theta$ by making use of (58.10). Recall Theorem 55.17.

Figure 260

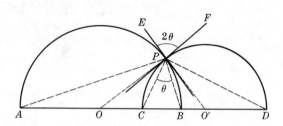

59. A Formula for Cross Ratio

A certain elementary formula for cross ratio has important consequences. First, it is necessary to introduce *directed angle measure.* As in analytic geometry, we orient the rays from any point O by considering, say, the "counterclockwise" direction. This may be made precise by the following analysis: The orientation afforded by the positive x and y axes leads to an orientation at any other point $O[a, b]$ by means of the translation $x' = x - a$, $y' = y - b$. Take the positive x' axis as the origin of a coordinate system for the rays from O with the positive y' axis having positive coordinate $\pi/2$ (Fig. 261). Then, given any two *nonopposing* rays $h[u]$, $k[v]$, we say that h *precedes* k, and write

$$h < k,$$

if and only if either $u < v$ *and* $v - u < \pi$, *or* $u > v$ *and* $u - v > \pi$. (Note that we make no attempt to order h and k when they are opposite rays.) The reader may now check a number of cases to see if this definition corresponds to his idea of "counterclockwise rotation."

Figure 261

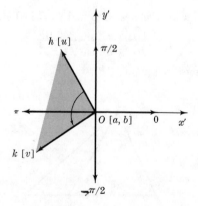

It is now a simple matter to orient angle measure. Let $\overline{hk}$ represent any non-degenerate, nonstraight angle at O. The value hk becomes directed if it is assumed that

$$hk > 0 \quad \text{if} \quad h < k$$

and

$$hk < 0 \quad \text{if} \quad h > k,$$

where $|hk|$ is the original (undirected) measure of the angle as provided by Axiom 6. Note that as a consequence, we have the familiar property

$$hk = -kh$$

of directed measure.[2] From the identity $\sin(-\theta) = -\sin\theta$ for all real θ it immediately follows that

$$\sin hk = -\sin kh,$$

a fact to be used later.

To prove the formula mentioned at the outset, first consider any four points $O, P, Q,$ and R, where $P, Q,$ and R are collinear but $O \notin \overleftrightarrow{PQ}$ (Fig. 262). Let $p, q,$ and r denote the respective rays $\overrightarrow{OP}$, $\overrightarrow{OQ}$ and $\overrightarrow{OR}$, and $\theta = |\angle PQO|$. By the law of sines, $|PQ|/|OP| = \sin|pq|/\sin\theta$ and $|QR|/|OR| = \sin|qr|/\sin\theta$, where one makes use of the identity $\sin(\pi - \theta) = \sin$, if necessary. Divide these two equations to obtain

$$\frac{|PQ|}{|QR|} = \frac{OP \sin|pq|}{OR \sin|qr|},^{3} \tag{59.1}$$

which holds regardless of the order of the points $P, Q,$ and R. Observe that PQ and QR have like signs if and only if (PQR), if and only if (pqr), if and only if pq

Figure 262

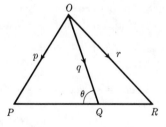

and qr have like signs. Therefore, $PQ/QR > 0$ if and only if $\sin pq/\sin qr > 0$. Thus, the absolute values in (59.1) may be removed and the general formula

---

[2] Other properties may be proved. For example, if concurrent rays $h_1, h_2,$ and h_3 lie in the same half plane, then, regardless of order, $h_1h_2 + h_2h_3 = h_1h_3$.

[3] The absolute values on OP and OR are not needed by the former convention of taking directed distance *positive* whenever possible; it is possible here since $\overleftrightarrow{OP}$ and $\overleftrightarrow{OR}$ are distinct lines. The formula (59.1) is actually identical with (58.10) given in the preceding set of exercises.

$$\frac{PQ}{QR} = \frac{OP \sin pq}{OR \sin qr}$$ (59.2)

is proved, in magnitude and in sign. ▮

This formula may be used to compute the cross ratio of any four collinear points A, B, C, and D, relative to a point $O \not\subset \overleftrightarrow{AB}$. Use a, b, c, and d to denote the rays $\overrightarrow{OA}$, $\overrightarrow{OB}$, $\overrightarrow{OC}$, and $\overrightarrow{OD}$ (Fig. 263). From (59.2) one obtains

$$\frac{AC}{AD} \cdot \frac{BD}{BC} = \frac{OC \sin ac}{OD \sin ad} \cdot \frac{OD \sin bd}{OC \sin bc}$$

and the desired formula

$$[AB, CD] = \frac{\sin ac \sin bd}{\sin ad \sin bc}.$$ (59.3)

A surprising fact is that the right member of (59.3) seems to depend only upon the *lines* containing a, b, c, and d, and not on the particular locations of the points A, B, C, and D on those lines. Indeed, consider the effect of interchanging one or

Figure 263

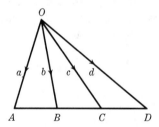

more of the rays a, b, c, and d with their corresponding opposite rays a', b', c', and d'. If we start with a and a', and take h *any other* ray from O (Fig. 264), then $|ah| + |ha'| = \pi$. Since $a > h$ if and only if $h > a'$, we have

$$ah + ha' = \pm\pi,$$

in magnitude and in sign. Hence,

$$\sin ah = \sin ha' = -\sin a'h.$$

Figure 264

Therefore, replacing a by a' in the right member of (59.3) has the following effect:

$$\frac{\sin ac \sin bd}{\sin ad \sin bc} = \frac{-\sin a'c \sin bd}{-\sin a'd \sin bc} = \frac{\sin a'c \sin bd}{\sin a'd \sin bc}.$$

Since this argument obviously applies to any such replacement, the assertion is proved.

EXERCISE

In each of the two figures 265(a) and 265(b) there appear two sets of four collinear points, related in a certain way. Show that in either case $[AB, CD] = [A'B', C'D']$.

Figure 265

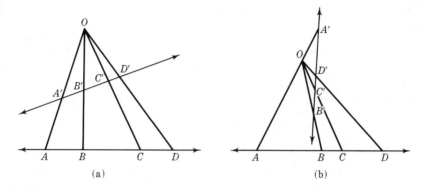

(a)   (b)

60. The Cevian Formula and the Fundamental Relations in a Triangle

This section will contain a unified method for deriving many important formulas involving the lengths of the sides of a triangle.

60.1 Definition: A **cevian** of a triangle is the segment joining any vertex to a point which lies on the opposite side.

Important cevians have already been observed: the *medians, angle bisectors,* and *altitudes* were defined in Section 34, Chapter 4. Of considerable use is the formula for a general cevian in terms of the side lengths of the triangle being considered. Stewart's theorem proved in Section 58 provides the perfect tool.

For notational purposes, let the length of cevian $\overline{AD}$ in triangle ABC (Fig. 266) be denoted d_a to signify the fact that the cevian of which this term represents the length, joins vertex A to a point on the opposite side. Similarly, d_b and d_c denote the lengths of cevians from B and C, respectively (note that the subscripts "a," "b," and "c" as used here do not represent the lengths of the sides of triangle ABC). Further, define $p = BD/BC$ and $q = DC/BC$. The remaining notation is standard.

Figure 266

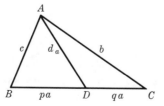

Apply Stewart's theorem to the four points A, B, C, and D:
$$AB^2 \cdot CD + AC^2 \cdot DB + AD^2 \cdot BC + CD \cdot DB \cdot BC = 0.$$

With the above notation, this becomes

$$c^2 \cdot CD + b^2 \cdot DB + d_a^2 \cdot BC + CD \cdot DB \cdot BC = 0.$$

Solve for d_a^2 to obtain

$$d_a^2 = \frac{BD}{BC} \cdot b^2 + \frac{DC}{BC} \cdot c^2 - BD \cdot DC.$$

This may be put into the following form—the **cevian formula for triangle** ABC:

$$\boxed{d_a^2 = pb^2 + qc^2 - pqa^2, \qquad p = \frac{BD}{BC} \text{ and } q = \frac{DC}{BC}.} \qquad (60.2)$$

60.3 Example: In an equilateral triangle having unit side, find the lengths of the cevians which cut the sides at interior points which are at a distance $1/n$ from the endpoints of those sides, n a positive integer.

Solution: Here $p = \dfrac{1}{n}$ and $q = \dfrac{n-1}{n}$. Hence,

$$d^2 = \frac{1}{n} \cdot 1^2 + \frac{n-1}{n} \cdot 1^2 - \frac{1}{n} \cdot \frac{n-1}{n} \cdot 1^2 = 1 - \frac{n-1}{n^2}.$$

Therefore, $d = n^{-1}\sqrt{n^2 - n + 1}$. (In the special case $n = 2$, we get $d = \frac{1}{2}\sqrt{3}$, the correct value for the altitude of an equilateral triangle having unit side.) ∎

60.4 Example: In a triangle whose sides are of length 3, 5, and 6 find the position of the cevian whose length is the arithmetic mean of 3 and 5 and which joins a point on the side whose length is 6.

Solution: In Fig. 267, since $AD = \frac{1}{2}(3 + 5) = 4$, the cevian formula implies

$$4^2 = p \cdot 5^2 + q \cdot 3^2 - pq \cdot 6^2$$
$$16 = 25p + 9(1 - p) - 36p(1 - p)$$
$$36p^2 - 20p - 7 = 0$$
$$\therefore \ p = \frac{5 \pm 2\sqrt{22}}{18}. \ \blacksquare$$

Figure 267

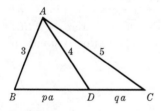

Figure 268

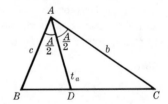

The reader will find the consideration of these questions important later: What is the significance of *two* answers in the solution of Example 60.4, and what must be concluded if there had been only one answer?

Formula 60.2 may be specialized to obtain the formulas for the cevians mentioned previously. For the medians, set $p = q = \frac{1}{2}$. Thus the **formulas for the medians of a triangle in terms of its sides** are

$$
\boxed{
\begin{aligned}
m_a &= \sqrt{\tfrac{1}{2}b^2 + \tfrac{1}{2}c^2 - \tfrac{1}{4}a^2}, \\
m_b &= \sqrt{\tfrac{1}{2}a^2 + \tfrac{1}{2}c^2 - \tfrac{1}{4}b^2}, \\
m_c &= \sqrt{\tfrac{1}{2}a^2 + \tfrac{1}{2}b^2 - \tfrac{1}{4}c^2}.
\end{aligned}
}
\tag{60.5}
$$

For the angle bisectors, a preliminary relation will be needed. Consider in Fig. 268 the bisector $\overrightarrow{AD}$ of $\angle A$. By (59.2),

$$
\frac{BD}{DC} = \frac{AB \sin \tfrac{1}{2} A}{AC \sin \tfrac{1}{2} A},
$$

since (BDC) implies $BD/DC > 0$. Therefore,

$$
\frac{BD}{DC} = \frac{c}{b}.
\tag{60.6}
$$

In terms of p and q this becomes $p/q = c/b$, which, together with $p + q = 1$, yields

$$
p = \frac{c}{b+c}, \qquad q = \frac{b}{b+c}.
$$

Hence if t_a denotes the length of the angle bisector $\overline{AD}$,

$$
\begin{aligned}
t_a{}^2 &= \frac{cb^2}{b+c} + \frac{bc^2}{b+c} - \frac{bca^2}{(b+c)^2} \\
&= bc \cdot \frac{b(b+c) + c(b+c) - a^2}{(b+c)^2} \\
&= \frac{bc}{(b+c)^2} \left[(b+c)^2 - a^2 \right] \\
&= \frac{bc}{(b+c)^2} \left[(b+c+a)(b+c-a) \right] \\
&= \frac{4bcs(s-a)}{(b+c)^2}, \qquad \text{where} \qquad s = \tfrac{1}{2}(a+b+c).
\end{aligned}
$$

Therefore, the **formulas for the angle bisectors of a triangle in terms of its sides** are

$$
\boxed{
\begin{aligned}
t_a &= 2(b+c)^{-1}\sqrt{bcs(s-a)}, \\
t_b &= 2(a+c)^{-1}\sqrt{acs(s-b)}, \\
t_c &= 2(a+b)^{-1}\sqrt{abs(s-c)}.
\end{aligned}
}
\tag{60.7}
$$

To illustrate the power of the "cevian method," we shall demonstrate the famous Steiner–Lehmus proposition that a triangle is isosceles if two of its angle bisectors are congruent. Suppose that $t_a = t_b$. Then we have the following:

$$(a + c)\sqrt{b(s - a)} = (b + c)\sqrt{a(s - b)}$$
$$b(a + c)^2(s - a) = a(b + c)^2(s - b)$$
$$a^2bs + 2abcs + bc^2s - a^3b - 2a^2bc - abc^2$$
$$= ab^2s + 2abcs + ac^2s - ab^3 - 2ab^2c - abc^2$$
$$-abs(a - b) + c^2s(a - b) + ab(a^2 - b^2) + 2abc(a - b) = 0$$
$$[-abs + c^2s + ab(a + b) + 2abc](a - b) = 0$$
$$[-abs + c^2s + ab(2s) + abc](a - b) = 0$$
$$(abs + c^2s + abc)(a - b) = 0.$$

Since all the terms in the factor $(abs + c^2s + abc)$ are positive, that factor cannot vanish. This forces the vanishing of the other factor $(a - b)$, which therefore proves that $a = b$. ∎

Finally, to derive formulas for the altitudes, let h_a, h_b, and h_c denote their lengths and write the equation (Fig. 269)

$$h_a{}^2 = pb^2 + qc^2 - pqa^2.$$

Figure 269

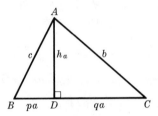

With $q = 1 - p$ this becomes

$$h_a{}^2 = pb^2 + (1 - p)c^2 - p(1 - p)a^2.$$

Regarding this as an equation in the unknown p, we collect like powers of p:

$$a^2p^2 - (a^2 - b^2 + c^2)p + (c^2 - h_a{}^2) = 0.$$

But this is a quadratic equation, and there can be only one solution. (Why?) Hence its discriminant must vanish, and therefore,

$$(a^2 - b^2 + c^2)^2 - 4a^2(c^2 - h_a{}^2) = 0.$$

This gives the desired relation for h_a in terms of a, b, and c. Let us obtain a more elegant form:

$$h_a{}^2 = -\frac{(a^2 - b^2 + c^2)^2 - 4a^2c^2}{4a^2}$$

$$= -\frac{[(a^2 - b^2 + c^2) + 2ac][(a^2 - b^2 + c^2) - 2ac]}{4a^2}$$

$$= -\frac{[(a + c)^2 - b^2][(a - c)^2 - b^2]}{4a^2}$$

$$= -\frac{(a + c + b)(a + c - b)(a - c + b)(a - c - b)}{4a^2}$$

$$= \frac{16s(s - b)(s - c)(s - a)}{4a^2}.$$

Thus, the formulas for the altitudes of a triangle in terms of its sides are

$$
\begin{aligned}
h_a &= \frac{2}{a}\sqrt{s(s-a)(s-b)(s-c)}, \\
h_b &= \frac{2}{b}\sqrt{s(s-a)(s-b)(s-c)}, \\
h_c &= \frac{2}{c}\sqrt{s(s-a)(s-b)(s-c)}.
\end{aligned}
$$

(60.8)

We have hit upon a familiar expression in (60.8) since the radical expression is the substance of Heron's formula. This should not be too surprising in view of the base-altitude formula for the area of triangle ABC:

$$
K = \tfrac{1}{2}ah_a = \tfrac{1}{2}bh_b = \tfrac{1}{2}ch_c.
$$

(60.9)

Heron's formula has thus been deduced by a different method (compare with the discussion of Section 3, Chapter 1)—the formula for the area of a triangle in terms of its sides:

$$
K = \sqrt{s(s-a)(s-b)(s-c)}.
$$

(60.10)

We mention two further useful formulas for area, the derivation of which will be left to the reader (r denotes the inradius of triangle ABC, introduced in Exercise 5, Section 42):

$$
K = \tfrac{1}{2}bc \sin A = \tfrac{1}{2}ac \sin B = \tfrac{1}{2}ab \sin C,
$$

(60.11)

$$
K = rs.
$$

(60.12)

Formulas (60.10) and (60.12) combine to yield the formula for the inradius of a triangle in terms of its sides:

$$
r = \sqrt{\frac{(s-a)(s-b)(s-c)}{s}}.
$$

(60.13)

Example: Prove the following formula relating the radii of the four tritangent circles of triangle ABC (Section 8), with r as the inradius and r_a, r_b, and r_c as the exradii—the radii of the tritangent circles outside the triangle and opposite A, B, and C, respectively (these three circles are called the excircles):

$$
\frac{1}{r} = \frac{1}{r_a} + \frac{1}{r_b} + \frac{1}{r_c}.
$$

(60.14)

Solution: In Fig. 270 observe the similar right triangles IRA and $I_aR'A$. Thus,

$$\frac{r}{r_a} = \frac{AR}{AR'}.$$

It is a simple matter to derive the relations

$$AR = s - a, \qquad AR' = s$$

(recall Exercise 5, Section 42). Therefore,

$$\frac{r}{r_a} = \frac{s - a}{s}.$$

From this follows the **formulas for the exradii of a triangle in terms of its sides:**

$$
\boxed{
\begin{aligned}
r_a &= \frac{rs}{s - a} = \sqrt{\frac{s(s - b)(s - c)}{s - a}}, \\[2mm]
r_b &= \frac{rs}{s - b} = \sqrt{\frac{s(s - b)(s - c)}{s - b}}, \\[2mm]
r_c &= \frac{rs}{s - c} = \sqrt{\frac{s(s - a)(s - b)}{s - c}}.
\end{aligned}
}
$$

(60.15)

Also,

$$\frac{1}{r_a} + \frac{1}{r_b} + \frac{1}{r_c} = \frac{s - a + s - b + s - c}{rs} = \frac{3s - 2s}{rs} = \frac{1}{r}. \quad \blacksquare$$

Figure 270

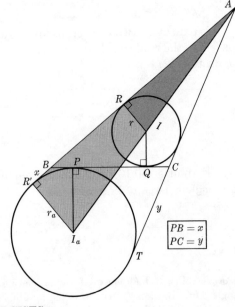

$$PB = x$$
$$PC = y$$

EXERCISES

1. Show that the sum of the squares of the distances of the vertex of the right angle of a right triangle from the two points of trisection of the hypotenuse is equal to $\frac{5}{9}$ the square of the hypotenuse. (N. A. Court, *College Geometry* [6, p. 153].)

2. A cevian of an equilateral triangle of side lengths 1 has length 13/15. Find its possible positions relative to the endpoints of that base.

3. Prove the relation

$$m_a{}^2 + m_b{}^2 + m_c{}^2 = \tfrac{3}{4}(a^2 + b^2 + c^2).$$

(60.16)

4. In triangle ABC with $a = 2$, $b = 3$, $c = 4$ find the medians, angle bisectors, altitudes, and the radii of the four tritangent circles. Verify (60.14) for this case, and the additional formula

$$\frac{1}{r} = \frac{1}{h_a} + \frac{1}{h_b} + \frac{1}{h_c}.$$

5. The midpoint of the hypotenuse of a right triangle is equidistant from the three vertices. Use the "cevian method" to prove this.

6. A triangle XYZ is constructed having side lengths equal to the medians of triangle ABC. Show that the medians of triangle XYZ are $\tfrac{3}{4}$ the sides, respectively, of triangle ABC.

7. Prove:

$$K^2 = rr_a r_b r_c.$$

(60.17)

8. The medians of a triangle are 2, 2, and 3, respectively. Find the sides.

9. If $\overline{AD}$ is the angle bisector from A of triangle ABC, and I and I_a are the centers of the two tritangent circles touching $\overleftrightarrow{AB}$ and $\overleftrightarrow{AC}$, show that

$$[AD, II_a] = -1.$$

Hint: Prove that $AI/AI_a = (s - a)/s$. Drop perpendiculars from I and I_a to $\overleftrightarrow{BC}$ as indicated in Fig. 270.

10. Since for an isosceles triangle the median to the base, the bisector of the vertex angle, and the altitude to the base all coincide, the cevian formulas in this case should reduce to the same expression. Supply the algebra necessary to verify this.

11. The sides of a triangle are $\dfrac{72}{\sqrt{455}}$, $\dfrac{96}{\sqrt{455}}$, and $\dfrac{144}{\sqrt{455}}$, respectively. Find the altitudes.

Ans: 2, 3, and 4.

12. Prove that three positive real numbers x, y, and z are the lengths of the altitudes of some triangle if and only if their reciprocals satisfy the triangle inequality (the sum of any two is greater than the third).

13. A triangle has altitudes $\tfrac{1}{3}$, $\tfrac{1}{4}$, and $\tfrac{1}{5}$. (a) Show that it is a right triangle. (See Exercise 11, Section 55.) (b) Find the sides.

14. A triangle is a right triangle with hypotenuse c if and only if either of the following relations holds:

$$\frac{1}{h_c{}^2} = \frac{1}{h_a{}^2} + \frac{1}{h_b{}^2}.$$

(60.18)

$$5m_c{}^2 = m_a{}^2 + m_b{}^2.$$

(60.19)

15. Establish the formula

$$\frac{1}{r} = \frac{1}{h_a} + \frac{1}{h_b} + \frac{1}{h_c}.$$

(60.20)

16. Find the sides of a triangle (a) whose altitudes are $\frac{1}{6}$, $\frac{1}{8}$, $\frac{1}{12}$, and (b) whose medians are 4, 6, and 8. *Ans:* $3/\sqrt{455}$, $4/\sqrt{455}$, $6/\sqrt{455}$, *and* $\frac{4}{3}\sqrt{46}$, $\frac{4}{3}\sqrt{31}$, $\frac{4}{3}\sqrt{10}$.

17. Establish the following formulas for the *sides* of a triangle in terms of its *medians:*

$$
\begin{aligned}
a &= \tfrac{4}{3}\sqrt{\tfrac{1}{2}m_b{}^2 + \tfrac{1}{2}m_c{}^2 - \tfrac{1}{4}m_a{}^2}, \\
b &= \tfrac{4}{3}\sqrt{\tfrac{1}{2}m_a{}^2 + \tfrac{1}{2}m_c{}^2 - \tfrac{1}{4}m_b{}^2}, \\
c &= \tfrac{4}{3}\sqrt{\tfrac{1}{2}m_a{}^2 + \tfrac{1}{2}m_b{}^2 - \tfrac{1}{4}m_c{}^2}.
\end{aligned}
\tag{60.21}
$$

⋆18. Prove the following formula for the area of a triangle in terms of (a) its three *medians*, and (b) its three *altitudes:*

$$
K = \tfrac{4}{3}\sqrt{m_s(m_s - m_a)(m_s - m_b)(m_s - m_c)},
\tag{60.22}
$$
$$
\text{where } m_s = \tfrac{1}{2}(m_a + m_b + m_c).
$$

$$
\frac{1}{K} = 4\sqrt{\frac{1}{h_s}\left(\frac{1}{h_s} - \frac{1}{h_a}\right)\left(\frac{1}{h_s} - \frac{1}{h_b}\right)\left(\frac{1}{h_s} - \frac{1}{h_c}\right)},
\tag{60.23}
$$
$$
\text{where } \frac{1}{h_s} = \frac{1}{2}\left(\frac{1}{h_a} + \frac{1}{h_b} + \frac{1}{h_c}\right).
$$

19. As in Exercise 6 above, triangle XYZ is constructed having side-lengths equal to the medians of triangle ABC. Prove from (60.22) that the area of triangle XYZ is $\frac{3}{4}$ that of triangle ABC.

20. Establish the following formulas for the *sides* of a triangle in terms of its *altitudes:*

$$
\frac{1}{a} = 2h_a\sqrt{\frac{1}{h_s}\left(\frac{1}{h_s} - \frac{1}{h_a}\right)\left(\frac{1}{h_s} - \frac{1}{h_b}\right)\left(\frac{1}{h_s} - \frac{1}{h_c}\right)}
$$
$$
\frac{1}{b} = 2h_b\sqrt{\frac{1}{h_s}\left(\frac{1}{h_s} - \frac{1}{h_a}\right)\left(\frac{1}{h_s} - \frac{1}{h_b}\right)\left(\frac{1}{h_s} - \frac{1}{h_c}\right)}
\tag{60.24}
$$
$$
\frac{1}{c} = 2h_c\sqrt{\frac{1}{h_s}\left(\frac{1}{h_s} - \frac{1}{h_a}\right)\left(\frac{1}{h_s} - \frac{1}{h_b}\right)\left(\frac{1}{h_s} - \frac{1}{h_c}\right)},
$$
$$
\text{where } \frac{1}{h_s} = \frac{1}{2}\left(\frac{1}{h_a} + \frac{1}{h_b} + \frac{1}{h_c}\right).
$$

21. Complete the details in the following outline of the derivation of a useful formula: Let X be any point in the plane, with L, M, and N the respective midpoints of the sides of triangle ABC opposite A, B, and C, and G the centroid. Then

$$
\begin{aligned}
XG^2 &= \tfrac{1}{3}XA^2 + \tfrac{2}{3}XL^2 - \tfrac{2}{9}m_a{}^2 \\
&= \tfrac{1}{3}XA^2 + \tfrac{2}{3}(\tfrac{1}{2}XB^2 + \tfrac{1}{2}XC^2 - \tfrac{1}{4}a^2) - \tfrac{2}{9}m_a{}^2 \\
&= \tfrac{1}{3}XA^2 + \tfrac{1}{3}XB^2 + \tfrac{1}{3}XC^2 - \tfrac{1}{6}a^2 - \tfrac{2}{9}m_a{}^2.
\end{aligned}
$$

Add the two analogous equations involving (b, m_b) and (c, m_c) to obtain

$$
XA^2 + XB^2 + XC^2 - 3XG^2 = \tfrac{1}{3}(a^2 + b^2 + c^2).
\tag{60.25}
$$

22. With R denoting the circumradius of triangle ABC, use (60.25) to derive the formula for the distance between the centroid and circumcenter of a triangle:

$$OG^2 = R^2 - \tfrac{1}{9}(a^2 + b^2 + c^2).$$ (60.26)

*61. The Area of a Cevian Triangle

A classical problem in elementary geometry is the calculation of the area of a cevian triangle[4] of triangle ABC—the triangle formed by cevians from each of the vertices A, B, and C (such as triangle PQR in Fig. 271).

Figure 271

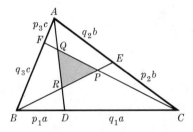

An interesting formula relates the area of the cevian triangle PQR to the area K of triangle ABC and the ratios of the segments which the cevians determine on the sides. Define $p_1 = BD/BC$, $q_1 = DC/BC$, $p_2 = CE/CA$, $q_2 = EA/CA$, $p_3 = AF/AB$, and $q_3 = FB/AB$, and denote the area of triangle PQR by $\triangle$.
  The derivation[5] begins with Fig. 272, where it is observed that

$$\triangle AFE = \tfrac{1}{2}|AF|\cdot|AE| \sin A,$$
$$\triangle ABC = \tfrac{1}{2}|AB|\cdot|AC| \sin A.$$

Figure 272

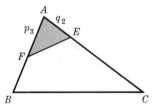

Therefore,

$$\frac{\triangle AFE}{\triangle ABC} = \frac{AF}{AB} \cdot \frac{AE}{AC} = p_3 q_2,$$

which proves

$$\triangle AFE = p_3 q_2 K.$$ (61.1)

This formula is used repeatedly in what follows.
  To calculate the ratio $x = BP/BE$ in terms of $u = FB/AB$ and $v = CE/CA$ (Fig. 273) write

---

[4] Also sometimes called an *aliquot* triangle.
[5] I am indebted to my colleague Professor A. Bernhart who suggested this method of proof.

$$\triangle BFP = xu \, \triangle ABE = xu(1 - v)K,$$
$$\triangle PEC = (1 - x) \, \triangle BEC = (1 - x)vK,$$
$$\triangle BFP + \triangle AFC = \triangle ABE + \triangle PEC,$$
$$xu(1 - v)K + (1 - u)K = (1 - v)K + (1 - x)vK.$$

Thus,

$$xu - xuv + 1 - u = 1 - v + v - xv,$$

and therefore,

$$x = \frac{u}{u + v - uv}. \tag{61.2}$$

Now observe from Fig. 274

$$\triangle = \frac{QR}{QD} \cdot \frac{QP}{QC} \, \triangle QCD$$

$$= \frac{QR}{QD} \cdot \frac{QP}{QC} \cdot \frac{QD}{AD} \, \triangle ACD \tag{61.3}$$

$$= \frac{QR}{AD} \cdot \frac{QP}{QC} \cdot q_1 K.$$

Figure 273

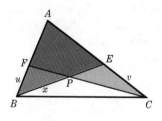

Figure 274

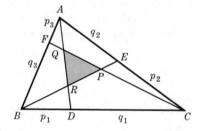

It remains to calculate the ratios QR/AD and QP/QC. Note that (61.2) applied to Figs. 275(a, b) yields

$$\frac{AQ}{AD} = \frac{p_3}{p_3 + q_1 - p_3q_1}, \qquad \frac{AR}{AD} = \frac{q_2}{p_1 + q_2 - p_1q_2}.$$

Thus,

$$\frac{QR}{AD} = \frac{AR - AQ}{AD} = \frac{q_2(p_3 + q_1 - p_3q_1) - p_3(p_1 + q_2 - p_1q_2)}{(p_1 + q_2 - p_1q_2)(p_3 + q_1 - p_3q_1)}.$$

This simplifies to

$$\frac{QR}{AD} = \frac{q_1q_2 - p_3q_1q_2 - p_1p_3 + p_1p_3q_2}{[p_1 + q_2(1 - p_1)][p_3 + q_1(1 - p_3)]}$$

$$= \frac{q_1q_2(1 - p_3) - p_1p_3(1 - q_2)}{(p_1 + q_1q_2)(p_3 + q_1q_3)} \tag{61.4}$$

$$= \frac{q_1q_2q_3 - p_1p_2p_3}{(p_1 + q_1q_2)(p_3 + q_1q_3)}.$$

Similarly, from Figs. 275(c, d) it is observed that

$$\frac{PC}{FC} = \frac{p_2}{p_2 + q_3 - p_2 q_3}, \qquad \frac{QC}{FC} = \frac{q_1}{p_3 + q_1 - p_3 q_1}.$$

Thus,

$$\frac{PC}{QC} = \frac{p_2(p_3 + q_1 - p_3 q_1)}{q_1(p_2 + q_3 - p_2 q_3)}$$

and again by simple algebra,

$$\frac{QP}{QC} = 1 - \frac{PC}{QC} = \frac{q_1 q_2 q_3 - p_1 p_2 p_3}{q_1(p_2 + q_2 q_3)}. \tag{61.5}$$

Figure 275

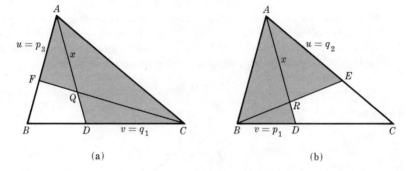

(a)   (b)

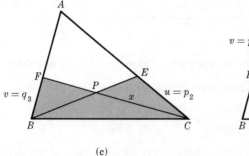 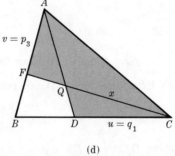

(c)   (d)

Substitute the results of (61.4) and (61.5) in (61.3) and the desired formula is obtained:

$$\triangle = \frac{(p_1 p_2 p_3 - q_1 q_2 q_3)^2 K}{(p_1 + q_1 q_2)(p_2 + q_2 q_3)(p_3 + q_3 q_1)}. \tag{61.6}$$

One application of the above formula is to derive a relationship between the ratios p_k and q_k ($k = 1, 2, 3$) when the cevian lines $\overleftrightarrow{AD}$, $\overleftrightarrow{BE}$, and $\overleftrightarrow{CF}$ are concurrent. This evidently involves the case when the cevian triangle PQR is degenerate, or when $\triangle = 0$. Thus, from (61.6)

$$p_1 p_2 p_3 - q_1 q_2 q_3 = 0,$$

or

$$\frac{p_1 p_2 p_3}{q_1 q_2 q_3} = 1.$$

It then appears that the expression $p_1 p_2 p_3 / q_1 q_2 q_3$ can be used to determine whether the cevian lines are concurrent.

EXERCISES

1. Calculate the ratio $\triangle/K$ when (a) p_1, p_2, and p_3 are $\frac{3}{4}$, $\frac{7}{8}$, and $\frac{1}{2}$, respectively, (b) $p_1 = \frac{1}{3}$ and $p_2 = p_3 = \frac{1}{2}$, and (c) $p_1 = p_2 = \frac{3}{13}$ and $p_3 = \frac{11}{12}$. Find $p_1 p_2 p_3 / q_1 q_2 q_3$ in each case.
   Ans: (a) $\frac{16}{75}$ *and* 21, (b) $\frac{1}{60}$ *and* $\frac{1}{2}$, (c) $1/46 \cdot 139 \cdot 153$ *and* $\frac{99}{100}$.
2. Prove that the area of the triangle formed by constructing cevians to corresponding trisection points on the sides of any triangle is *one seventh* that of the given triangle.
3. With D, E, F as before, triangle DEF is called a **menelaus triangle** associated with triangle ABC (Fig. 276). If $\triangle$ denotes its area prove the formula

$$\triangle = (p_1 p_2 p_3 + q_1 q_2 q_3)K.$$

Figure 276

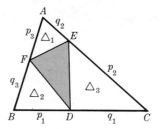

62. Linearity Number: The Theorems of Menelaus, Ceva, and Desargues

Considerations of the preceding section have focused attention on the expression

$$\frac{p_1 p_2 p_3}{q_1 q_2 q_3} = \frac{AF}{FB} \cdot \frac{BD}{DC} \cdot \frac{CE}{EA},$$

where D, E, and F are interior points on the respective sides of triangle ABC. This expression remains meaningful even if the points D, E, and F do not lie on the interior sides of the triangle, but are chosen *arbitrarily* on the *lines* containing those sides and distinct from each of the vertices (Fig. 277).

62.1 Definition: The points D, E, and F described in the preceding paragraph are called **menelaus points** with respect to triangle ABC. In magnitude and in sign, the value

$$\frac{AF}{FB} \cdot \frac{BD}{DC} \cdot \frac{CE}{EA} = \left[\frac{ABC}{DEF}\right]$$

is termed the **linearity number** of the menelaus points D, E, and F with respect to triangle ABC.

Figure 277

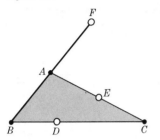

Figure 278

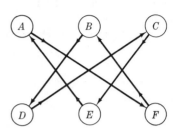

A mnemonic device for the defining relation of the linearity number is shown in Fig. 278.

The reader may easily prove the following fundamental properties for the linearity number in terms of the bracket symbol:

(a) Permutations of the *columns* in the symbol $\begin{bmatrix} ABC \\ DEF \end{bmatrix}$ yield the identities

$$\lambda = \begin{bmatrix} ABC \\ DEF \end{bmatrix} = \begin{bmatrix} BCA \\ EFD \end{bmatrix} = \begin{bmatrix} CAB \\ FDE \end{bmatrix},$$

$$\frac{1}{\lambda} = \begin{bmatrix} ACB \\ DFE \end{bmatrix} = \begin{bmatrix} BAC \\ EDF \end{bmatrix} = \begin{bmatrix} CBA \\ FED \end{bmatrix},$$

which correspond to the even and odd permutations of the letters A, B, and C.

(b) If $\begin{bmatrix} ABC \\ DEF \end{bmatrix} = \begin{bmatrix} ABC \\ XEF \end{bmatrix}$, then $X = D$. It follows from (a) that $\begin{bmatrix} ABC \\ DEF \end{bmatrix}$ $= \begin{bmatrix} ABC \\ DXF \end{bmatrix}$ implies $X = E$ and $\begin{bmatrix} ABC \\ DEF \end{bmatrix} = \begin{bmatrix} ABC \\ DEX \end{bmatrix}$ implies $X = F$.

The results of the preceding section provide a proof of a special case of

62.2   Ceva's Theorem: If D, E, and F are menelaus points with respect to the non-collinear points A, B, and C, then the lines $\overleftrightarrow{AD}$, $\overleftrightarrow{BE}$, and $\overleftrightarrow{CF}$ are concurrent or parallel if and only if $\begin{bmatrix} ABC \\ DEF \end{bmatrix} = 1$.

A more general proof will be indicated in the exercises (see Exercise 1 below). Another property of the linearity number is found in:

62.3   Menelaus' Theorem: If D, E, and F are menelaus points with respect to the noncollinear points A, B, and C, then D, E, and F are collinear if and only if

$$\begin{bmatrix} ABC \\ DEF \end{bmatrix} = -1.$$

PROOF: (1) First assume that D, E, and F lie on a line l. Since the ratio AF/FB is positive if and only if (AFB), and similarly for the other ratios, one positive ratio in the linearity number implies *exactly* one more by the postulate of Pasch (Theorem 24.10). Thus there are always either no positive ratios (that is, three negative ratios), or two positive ratios (that is, one negative ratio) in the linearity number, which proves

$$\begin{bmatrix} ABC \\ DEF \end{bmatrix} < 0.$$

Figure 279

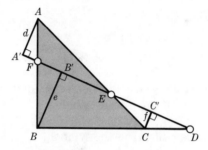

For the algebraic part of the proof, let A', B', and C' be the feet of the perpendiculars from A, B, and C on l (Fig. 279). Then from similar triangles

$$\left| \frac{AF}{FB} \right| = \frac{d}{e}, \quad \left| \frac{BD}{DC} \right| = \frac{e}{f}, \quad \text{and} \quad \left| \frac{CE}{EA} \right| = \frac{f}{d}.$$

Hence,

$$\left| \begin{bmatrix} ABC \\ DEF \end{bmatrix} \right| = \frac{d}{e} \cdot \frac{e}{f} \cdot \frac{f}{d} = 1.$$

Therefore,

$$\begin{bmatrix} ABC \\ DEF \end{bmatrix} = -1.$$

(2) For the converse, suppose $\begin{bmatrix} ABC \\ DEF \end{bmatrix} = -1$ and that $\overleftrightarrow{DE} \parallel \overleftrightarrow{AB}$ [Fig. 280(a)].

Figure 280

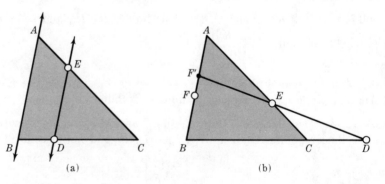

(a)   (b)

Since (BDC) if and only if (AEC), BD/DC is positive if and only if CE/EA is, and hence, these two ratios have like signs. Since $|BD|/|DC| = |EA|/|CE|$, we have

$$\frac{BD}{DC} \cdot \frac{CE}{EA} = 1.$$

The hypothesis then implies that

$$\frac{AF}{FB} = -1,$$

or

$$AF = BF.$$

By Theorem 58.2(c), $A = B$, providing a contradiction. Hence, $\overleftrightarrow{DE}$ meets $\overleftrightarrow{AB}$ at some point F' [Fig. 280(b)], and by (1)

$$\begin{bmatrix} ABC \\ DEF' \end{bmatrix} = \begin{bmatrix} ABC \\ DEF \end{bmatrix} = -1.$$

Therefore, $F = F'$, proving that D, E, and F are collinear. ∎

The theorems of Menelaus and Ceva provide an excellent example of the duality between collinearity and concurrency in elementary geometry. The theorem bearing the name Ceva was discovered by a late seventeenth century Italian mathematician, Giovanni Ceva, who noted that it complemented the theorem proved about 1600 years earlier by the Greek astronomer, Menelaus.

As might be imagined, these two theorems can be used to prove a variety of *incidence* theorems, theorems which primarily involve collinearity and concurrency.

62.4 Example: Prove the theorem of Desargues (Section 4, Chapter 1) by using the linearity number.

Solution: Let the point pairs (A, A'), (B, B'), and (C, C') determine three lines concurrent in P, and suppose $L = \overleftrightarrow{BC} \cap \overleftrightarrow{B'C'}$, $M = \overleftrightarrow{AC} \cap \overleftrightarrow{A'C'}$, and $N = \overleftrightarrow{AB} \cap \overleftrightarrow{A'B'}$ (Fig. 281). We must prove that L, M, and N are collinear. By the theorem of Menelaus,

$$\begin{bmatrix} PAB \\ NB'A' \end{bmatrix} = \begin{bmatrix} PBC \\ LC'B' \end{bmatrix} = \begin{bmatrix} PCA \\ MA'C' \end{bmatrix} = -1.$$

Therefore,

$$\begin{bmatrix} PAB \\ NB'A' \end{bmatrix} \begin{bmatrix} PBC \\ LC'B' \end{bmatrix} \begin{bmatrix} PCA \\ MA'C' \end{bmatrix} = -1,$$

which reduces to

$$\begin{bmatrix} ABC \\ LMN \end{bmatrix} = -1$$

(let the reader show this). Then, again by Menelaus' theorem, L, M, and N are collinear. ∎

Figure 281

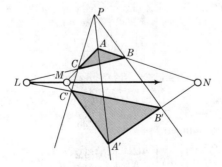

EXERCISES

1. In Fig. 282, $\overleftrightarrow{B'C'}$ is constructed parallel to $\overleftrightarrow{BC}$. From the similar triangles indicated show that

$$\left|\frac{BD}{DC}\right| = \left|\frac{AB'}{AC'}\right|.$$

Also relate the ratios $|CE|/|EA|$ and $|AF|/|FB|$ to the segments on $\overleftrightarrow{B'C'}$ and prove that $\begin{bmatrix} ABC \\ DEF \end{bmatrix} = \pm 1$. Why must $\begin{bmatrix} ABC \\ DEF \end{bmatrix}$ be positive regardless of the position of P? Now complete the proof of Ceva's theorem.

Figure 282

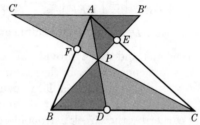

2. Using Ceva's theorem prove (a) the medians, (b) the bisectors, and (c) the altitudes of a triangle are concurrent. *Hint:* For the altitudes note the similarity of triangles ADC and BEC in Fig. 255(a).

3. **The Gergonne point.** The cevians joining the vertices of a triangle with the points of contact of the incircle with the opposite sides are concurrent in a point called the **Gergonne point** of the triangle.

4. **The Nagel point.** Let D, E, F be the points of contact of the sides opposite A, B, and C with the three excircles of triangle ABC. The cevians $\overline{AD}$, $\overline{BE}$, and $\overline{CF}$ are concurrent in the **Nagel point** of the triangle. *Hint:* Show that $|BD| = s - c$, $|DC| = s - b$, $|CE| = s - a$, and so on.

5. **The Fermat point.** Let equilateral triangles $A'BC$, $AB'C$, and ABC' be constructed externally on the respective sides of triangle ABC. Segments $\overline{AA'}$, $\overline{BB'}$, and $\overline{CC'}$ are congruent and concurrent, the point of concurrency being called the **Fermat point** of the triangle. *Hint:* Let D, E, and F be the intersections of $\overline{AA'}$, $\overline{BB'}$, and $\overline{CC'}$ with $\overleftrightarrow{BC}$, $\overleftrightarrow{AC}$, and $\overleftrightarrow{AB}$. Find all the congruent pairs of angles you can, then apply (59.2) and Ceva's theorem.

6. If the sides $\overleftrightarrow{AB}$, $\overleftrightarrow{BC}$, $\overleftrightarrow{CD}$, and $\overleftrightarrow{DA}$ of a quadrilateral $ABCD$ are cut by a transversal in the points X, Y, Z, and W, respectively, show that

$$\frac{AX}{XB} \cdot \frac{BY}{YC} \cdot \frac{CZ}{ZD} \cdot \frac{DW}{WA} = 1.$$

7. **Pappus' theorem.** If (A, B, C) and (A', B', C') be two triples of collinear points (Fig. 283), the pairs of *cross joins* $(\overleftrightarrow{AB'}, \overleftrightarrow{A'B})$, $(\overleftrightarrow{AC'}, \overleftrightarrow{A'C})$, and $(\overleftrightarrow{BC'}, \overleftrightarrow{B'C})$ meet in collinear points (assuming that they meet). Prove this for the special case shown in Fig. 283 by calculating the linearity numbers

$$\begin{bmatrix} PQR \\ CBA \end{bmatrix}, \quad \begin{bmatrix} PRQ \\ A'LB \end{bmatrix}, \quad \begin{bmatrix} RQP \\ B'NC \end{bmatrix}, \quad \begin{bmatrix} QPR \\ C'MA \end{bmatrix}, \quad \text{and} \quad \begin{bmatrix} PQR \\ A'C'B' \end{bmatrix}.$$

Figure 283

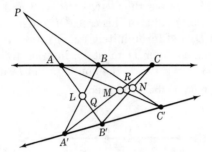

★8. Quadrilaterals $ABCD$ and $A'B'C'D'$ have angle C in common with B' on $\overline{BC}$ and D' on $\overline{CD}$. Prove that if each of the line triples $(\overleftrightarrow{AB}, \overleftrightarrow{A'B'}, \overleftrightarrow{CD})$ and $(\overleftrightarrow{AD}, \overleftrightarrow{A'D'}, \overleftrightarrow{BC})$ are concurrent then the lines $\overleftrightarrow{AA'}$, $\overleftrightarrow{BD'}$, and $\overleftrightarrow{B'D}$ are concurrent. *Hint:* Use the theorem of Pappus.

9. **Simson line.** Let D, E, and F be the feet of the perpendiculars on the sides of a triangle from any point P on the circumcircle of that triangle (Fig. 284). The points D, E, and F are collinear, the line of collinearity being called the **Simson line** of triangle ABC with respect to point P. To prove this using Menelaus' theorem, use a property of circles to be established later and observe the three pairs of congruent angles indicated in Fig. 284. Thus $|BD|/d = \cot \alpha = |EA|/e \cdots$.

Figure 284

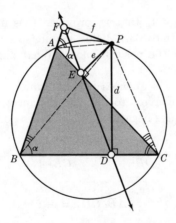

10. A point P is selected at random on the median $\overline{AD}$ of triangle ABC. If E and F are the points of intersection $\overleftrightarrow{BP} \cap \overleftrightarrow{AC}$ and $\overleftrightarrow{CP} \cap \overleftrightarrow{AB}$, show that $\overleftrightarrow{FE} \parallel \overleftrightarrow{BC}$.

63. The Harmonic Sequence

A recurring phenomenon in elementary geometry is the occurrence of four collinear points A, B, C, and D such that

$$\frac{|AC|}{|CB|} = \frac{|AD|}{|DB|}.$$

(63.1)

For example, in Fig. 285 we see that $|AC|/|CB| = 2$ and $|AD|/|DB| = 6/3 = 2$; therefore, the relation (63.1) holds in this case. Of course it can hold for an infinity of points A, B, C, and D, and for infinitely many ratios. It is clear that in general, however, one of the points C or D must be interior to segment $\overline{AB}$, the other exterior to it. When (63.1) is satisfied the points C and D are said to **divide segment** $\overline{AB}$ **internally and externally in the same ratio.**

Figure 285

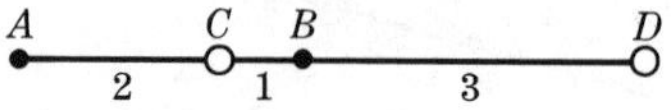

Now suppose that A, B, and C are given points such that (ACB), and it is desired to find the corresponding D which makes (63.1) true. Let the coordinates of A, B, C, and D be, respectively, $a = 0$, $b > 0$, c, and d. In order to determine D so that (63.1) will be satisfied set $r = |AC|/|CB|$, write

$$r = \frac{|AD|}{|DB|} = \frac{AD}{BD} = \frac{d - 0}{d - b},$$

and solve for d. Thus,

$$d = \frac{br}{r - 1}.$$

It follows that provided $r \neq 1$ (that is, provided C is not the midpoint of $\overline{AB}$) point D always exists such that (63.1) holds, having specified C. The reader should study the effect of varying the position of C. (What happens to D if $r < 1$, for example?)

The various properties of sequences of points which satisfy equation (63.1) are more easily studied if we convert to directed distances. Since one of the ratios AC/CB and AD/DB is always negative and the other positive, (63.1) becomes

$$\frac{AC}{CB} = -\frac{AD}{DB} \quad \text{or} \quad \frac{AC \cdot DB}{CB \cdot AD} = -1,$$

which is equivalent to

$$[AB, CD] = -1. \tag{63.2}$$

Note that the condition (63.2) is symmetric in A and B, and in C and D. Furthermore, (63.2) is equivalent to

$$[CD, AB] = -1. \tag{63.3}$$

Thus, if $\overline{AB}$ is divided internally and externally in the same ratio by C and D, then $\overline{CD}$ will be divided internally and externally in that ratio by A and B.

63.4 Definition: Four collinear points A, B, C, and D are said to form a **harmonic sequence,** with D as the **fourth harmonic,** if their cross ratio taken in the order given is -1. Points C and D are also said to **divide segment** $\overline{AB}$ **harmonically,** and C and D are termed **harmonic conjugates** of one another with respect to A and B.

The following elementary property of harmonic sequences will be useful later.

63.5   Theorem: If O is the midpoint of $\overline{AB}$, the points A, B, C, and D form a harmonic sequence of points provided OB is the geometric mean of OC and OD, and conversely.

PROOF: Let O be the midpoint of $\overline{AB}$. Then, in magnitude and in sign,

$$AO = OB.$$

Therefore, from $[AB, CD] = -1$ it follows that

$$AC \cdot BD = AD \cdot CB,$$
$$(AO + OC)(BO + OD) = (AO + OD)(CO + OB),$$
$$(OB + OC)(-OB + OD) = (OB + OD)(-OC + OB),$$
$$-OB^2 + OC \cdot OD = OB^2 - OD \cdot OC,$$
$$OB^2 = OC \cdot OD.$$

Since the steps are reversible the proof is complete. ∎

Figure 286

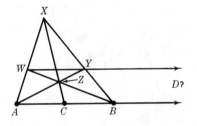

Figure 287

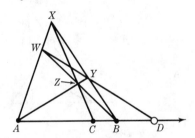

A certain configuration may be associated with the harmonic sequence. Consider first Fig. 286, where C is the midpoint of segment $\overline{AB}$. By a previous exercise (Exercise 10, Section 62) line $\overleftrightarrow{WY}$ is parallel to $\overleftrightarrow{AB}$, regardless of the position of Y on $\overline{BX}$. If we allow C to vary on $\overline{AB}$, a figure like that shown in Fig. 287 will result, where $\overleftrightarrow{WY}$ intersects $\overleftrightarrow{AB}$ at D. (The configuration in Fig. 286 may be regarded as the "limiting case" of that of Fig. 287 as C tends to the midpoint of $\overline{AB}$.) Now by the theorems of Ceva and Menelaus, $\left[\dfrac{XAB}{CYW}\right] = 1$ and

$\left[\dfrac{XAB}{DYW}\right] = -1$. Hence,

$$\left[\frac{XAB}{CYW}\right]\left[\frac{XAB}{DYW}\right]^{-1} = -1,$$

or

$$\frac{XW}{WA} \cdot \frac{AC}{CB} \cdot \frac{BY}{YX} \cdot \frac{WA}{XW} \cdot \frac{DB}{AD} \cdot \frac{YX}{BY} = -1,$$

which reduces to

$$\frac{AC}{CB} \cdot \frac{DB}{AD} = -1.$$

That is,

$$[AB, CD] = -1.$$

This result has an interesting application: Since the fourth harmonic is unique,[6] the configuration of Fig. 287 provides a *straightedge construction of the fourth harmonic* in terms of the first three members of the sequence: Having specified any three collinear points A, B, and C, with C not the midpoint of $\overline{AB}$, take X any point not on $\overleftrightarrow{AB}$, join X with A, B, and C, and choose Y any point on $\overleftrightarrow{BX}$. Determine $Z = \overleftrightarrow{CX} \cap \overleftrightarrow{AY}$ and $W = \overleftrightarrow{AX} \cap \overleftrightarrow{BZ}$. Then, by the property observed above, $D = \overleftrightarrow{WY} \cap \overleftrightarrow{AB}$ is the desired fourth harmonic (Fig. 288). The construction described here will in the future be termed the **harmonic construction**.

Figure 288

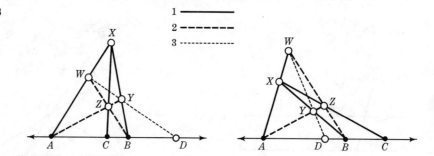

EXERCISES

1. The reader should already be familiar with the *harmonic series* $1 + \frac{1}{2} + \frac{1}{3} + \frac{1}{4} + \cdots$. By definition, a **harmonic sequence** of real numbers is a sequence whose *reciprocals* form an arithmetic sequence $a, a + d, a + 2d, \cdots, a + nd$. The **harmonic mean** of two numbers a and b such that $a \neq -b$ is the unique number h such that a, h, and b form a harmonic sequence. (a) Show that h is the harmonic mean of a and b if and only if

$$\frac{1}{h} = \frac{1}{2}\left(\frac{1}{a} + \frac{1}{b}\right).$$

(b) Prove that the geometric mean of two positive real numbers is the geometric mean of their arithmetic and harmonic means.

---

[6] This may be observed directly from the definition and the property (58.9) of cross ratios: If $[AB, CD] = [AB, CD'] = -1$, then $D = D'$.

2. Prove that A, B, C, and D form a harmonic sequence of points if and only if AB is the harmonic mean of AC and AD, in magnitude and in sign (see Exercise 1). That is, show that $[AB, CD] = -1$ if and only if

$$\frac{1}{AB} = \frac{1}{2}\left(\frac{1}{AC} + \frac{1}{AD}\right).$$

Hint: First prove by coordinates, then try a proof without coordinates.

3. Show that if O is the midpoint of $\overline{AB}$, then $[AB, CD] = -1$ if and only if

$$AD \cdot BD = CD \cdot OD.$$

★4. With O_1 and O_2 the respective midpoints of $\overline{AB}$ and $\overline{CD}$, prove that $[AB, CD] = -1$ if and only if

$$O_1B^2 + O_2C^2 = O_1O_2{}^2.$$

(H. Eves, *A Survey of Geometry* [9, p. 102]; used by permission.)

★5. Show that if O is the midpoint of $\overline{AB}$, then $[AB, CD] = -1$ if and only if

$$\frac{OC}{OD} = \frac{AC^2}{AD^2}.$$

(Eves [9, p. 103]; used by permision.)

6. Prove that the lines $\overleftrightarrow{AB}$, $\overleftrightarrow{YW}$, and $\overleftrightarrow{Y'W'}$ in Fig. 289 are concurrent in D, the fourth harmonic of A, B, and C.

7. Explain why, in Fig. 290, $\overleftrightarrow{Y'W'}$ will pass through point A.

Figure 289   Figure 290

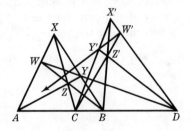

8. The bisector of angle A of triangle ABC intersects the opposite side in D. U and V are the feet of the perpendiculars from B and C upon line $\overleftrightarrow{AD}$. Use similar triangles to deduce $[AD, UV] = -1$.

9. The incenter of a triangle and any excenter divides the angle bisector on whose extension they lie harmonically. (See Exercise 9, Section 60.)

10. The following result provides a converse for the previous exercise: *If E and F divide $\overline{AD}$ harmonically and $\overleftrightarrow{BE} \perp \overleftrightarrow{BF}$, then $\overrightarrow{BE}$ and $\overrightarrow{BF}$ bisect the angles at B.* *Hint:* Let the line parallel to $\overleftrightarrow{BF}$ at E meet $\overleftrightarrow{AB}$ at X and $\overleftrightarrow{BD}$ at Y. From $XE/BF = AE/AF$, $\cdots$, prove that $|XE| = |EY|$.

64. Three Important Transformations in Geometry

For us, a **mapping** or **transformation** T of the elements of a set S_1 to a set S_2 is a one-to-one correspondence between the elements of S_1 and S_2 whereby each element in S_1 may be paired with a unique element in S_2, and conversely. The set S_1 is called the **domain** of the mapping and S_2 its **range**. We write $T(X)$ for the element in S_2 that corresponds to X in S_1.

Sometimes a correspondence is defined formally without regard to its range or domain, and then these sets are deduced from the *definition* of the transformation. For example, suppose S_1' denotes the set of all United States citizens and S_2' the set of all 9-digit numbers. For $A \in S_1'$ and $x \in S_2'$, say that A corresponds to x *if and only if* x *is* A's *Social Security number*. In this case the domain S_1 would not be all of S_1' since not all United States citizens have Social Security numbers, and the range S_2 would not be all of S_2' since not all 9-digit numbers can be Social Security numbers for United States citizens.

Transformations are exceedingly important in geometry. Here we explore those which will be most useful to us in later proofs. Let us begin with a space **perspectivity** (alternately called a **central projection,** or simply **projection**). Consider planes π and π', and point O not lying on either plane (Fig. 291). Let a correspondence T be set up between the points of π and π' by simply intersecting π and π' by lines which pass through O. Thus $T(P) = P'$ is defined by letting P' be the unique point of intersection of the line $\overleftrightarrow{OP}$ with π', if it exists. Evidently the transformation is defined for all those points $P \in \pi$ such that $\overleftrightarrow{OP}$ is not parallel to π' (unless π itself is parallel to π'). These points would consist of line v, the intersection of π with the plane parallel to π' and passing through O. Thus, the domain of this transformation is the set of all points on π except those lying on v. The reader should have no trouble in determining the range of T.

Figure 291

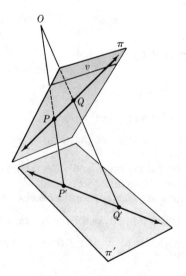

It is clear that a line in π will always map into a line in π' unless that line is v itself. Thus such a transformation is named **linear** ("line preserving"). In general, any property, number, or configuration associated with points or lines in plane π which is left unaltered by the transformation is called an **invariant**. For example, in the case of a *linear* transformation we would say that *collinearity is an invariant*. It is easy to list a few invariants of any perspectivity, as well as some noninvariants:

INVARIANTS OF A PERSPECTIVITY	NONINVARIANTS OF A PERSPECTIVITY
1. Collinearity of three or more points.	1. Perpendicularity of two lines.
2. Concurrency of three or more lines (it is customary to regard three or more *parallel* lines concurrent).	2. Distance between two points.
	3. The measure of an angle.
3. Conic section.	4. Circle.
	5. Ellipse.[7]

Another important invariant for a perspectivity is provided in

64.1   Theorem: Cross ratio is a perspective invariant.

PROOF: Obvious by (59.3) and the remarks immediately following it. (See Fig. 292.) ▌

Figure 292

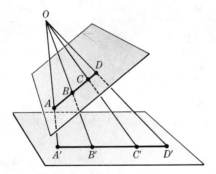

64.2   Corollary: The harmonic division of a segment is a perspective invariant.[8]

In plane geometry one is usually interested primarily in those transformations which map the plane *into itself*, with range and domain both lying in the plane. Obvious examples are the translations and rotations of analytic geometry. We consider here a slightly less restrictive class of mappings.

---

[7] It is recommended that the reader at this time return to Section 4 for a more detailed reading and to examine the image of a circle or ellipse which is tangent to the "horizon" in Fig. 11(b).

[8] Instead of proving this result as a corollary to (64.1), it can be directly inferred from the obvious fact that the harmonic construction itself is invariant.

64.3 *Definition:*   Consider point O and a nonzero real number a. For each point X in the plane, define $T(X) = X'$ as that point on line $\overleftrightarrow{OX}$ such that, in magnitude and sign,

$$OX' = a\, OX.$$

The mapping thus defined is called a **homothetic transformation**, with O as **center** and a as **dilation factor**. A homothetic transformation is said to be **direct** or **opposite** according as its dilation factor is positive or negative.

Since a homothetic transformation T is uniquely determined by its center O and dilation factor a, it will be denoted

$$T = H(O, a).$$

An example in each of the cases $a = \pm 3$ is illustrated in Fig. 293. From these two examples it is apparent that a homothetic transformation maps a triangle into one which is similar to it, with factor of proportionality $|a|$. If this be true, then such mappings will have many other invariants besides collinearity and cross ratio.

Figure 293

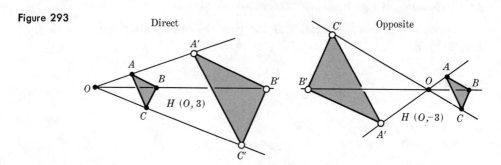

The key to all the properties of homothetic transformations are found in the following result:

64.4 *Theorem:*   A homothetic transformation $H(O, a)$ maps each pair of points (A, B) into a pair (A', B') such that

(a) $\overleftrightarrow{AB} \parallel \overleftrightarrow{A'B'}$, and
(b) $|A'B'| = |a||AB|$[9]

   PROOF: In Fig. 294, $|OA'|/|OA| = |OB'|/|OB| = |a|$. By Corollary 55.8 $\triangle OAB \sim \triangle OA'B'$ and hence, $|A'B'| = |a||AB|$. Since $\angle OAB = \angle OA'B'$, $\overleftrightarrow{AB} \parallel \overleftrightarrow{A'B'}$. ∎

---

[9] Indeed, these two properties together with the fact $T(O) = O$ are characteristic of homothetic transformations. A proof of the converse of Theorem 64.4 may be easily constructed making use of Exercise 6 below (with $a_3 = b_3 = 0$).

Figure 294

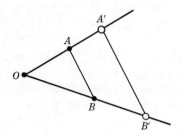

The following forms a list of a few *homothetic invariants*, each being inferred from the preceding theorem:

(a) *Collinearity*: If $\overleftrightarrow{AB} = \overleftrightarrow{BC}$, then $\overleftrightarrow{A'B'}$ and $\overleftrightarrow{B'C'}$ are both parallel to $\overleftrightarrow{AB}$ and both pass through B'. Therefore, $\overleftrightarrow{A'B'} = \overleftrightarrow{B'C'}$.

(b) *Betweenness*: If (ABC), then A, B, and C are collinear and $AB + BC = AC$. By (a), A', B', and C' are collinear and by Theorem 64.4 $|A'B'| + |B'C'| = |a||AB| + |a||BC| = |a||AC| = |A'C'|$. Therefore, $(A'B'C')$.

(c) *Directed ratios*: By Theorem 64.4(b) and the fact that betweenness is preserved, for any three collinear points A, B, and C, we have

$$\frac{AB}{BC} = \frac{A'B'}{B'C'}.$$

(d) *Cross ratio*: A result of (c).

(e) *Angle measure*: Obvious by Theorem 64.4(a) and by (b) above.

(f) *Perpendicularity of lines*: A result of (e).

(g) *Circles*: Let ω be a circle with center at P and radius r. Then $X \in \omega$ if and only if $|PX| = r$. But by Theorem 64.4(b) this is equivalent to $|P'X'| = |a|r$, which holds if and only if X' lies on a circle with center at P' and radius $|a|r$.

Figure 295

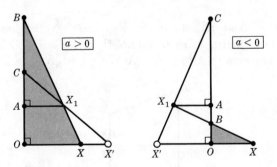

It is an interesting fact that any homothetic mapping in plane π may be obtained from a sequence of two perspectivities. On the perpendicular to π at O locate points A, B, and C such that $[OA, BC] = 1/a$. From the similar triangles in Fig. 295, $OX/AX_1 = OB/AB$ and $AX_1/OX' = AC/OC$. Multiply to obtain

$$\frac{OX}{OX'} = \frac{OB \cdot AC}{AB \cdot OC} = [OA, BC] = \frac{1}{a}.$$

That is,

$$OX' = a\,OX.$$

Hence if π_1 is the plane through A parallel to π (Fig. 296), the homothetic transformation $H(O, a)$ may be realized by centrally projecting π to π_1 from center B, then projecting π_1 back to π from center C. ∎

Figure 296

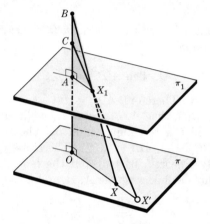

Another class of transformations useful in geometry are the *reflections*. If l is any line, the **axial reflection in** l is defined by associating with each point X in the plane that point X' such that l is the perpendicular bisector of $\overline{XX'}$; the line l is called the **axis** of the reflection. If O is any point a **central reflection in** O may be defined as the homothetic mapping $H(O, -1)$, and O is called the **center**.

All seven invariants mentioned for homothetic mappings are also invariants of the reflections, but for a different reason, providing yet another invariant for reflections:

(h)   *Distance*: For axial reflections, in Fig. 297 quadrilaterals XX_1Y_1Y and $X'X_1Y_1Y'$ are congruent, so it follows that $|XY| = |X'Y'|$. And for central reflections Theorem 64.4(b) yields $|X'Y'| = |-1||XY| = |XY|$.

Figure 297

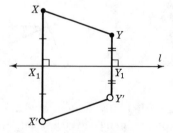

Any such transformation in the plane is called an **isometry** (meaning "samemetric" or -"distance"). The reader should already be familiar with two types of isometries from his knowledge of analytic geometry, since the *rotations* and *transla-tions* are isometries.

An axial reflection may also be obtained as a sequence of perspectivities. Let π_1 be the unique plane perpendicular to π and passing through l (Fig. 298). Choose any two points A and B on either side of π_1 and *equidistant* from π_1, with line $\overset{\leftrightarrow}{AB}$ parallel to π. Then the reflection in l in the plane π is obtained by projecting π to π_1 from center A, and following that by the projection from π_1 back to π from center B. We leave to the reader the problem of expressing a central reflection as a product of space perspectivities. (*Hint:* A central reflection may be expressed as the product of *two axial reflections.*)

Figure 298

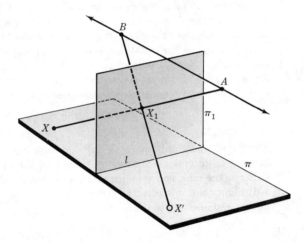

EXERCISES

1. Show from Theorem 64.4 that if $O[0, 0]$ is the center of the homothetic mapping $H(O, a)$, then $P[x, y]$ is related to its image $P'[x', y']$ by the equations

$$\begin{array}{l} x' = ax, \\ y' = ay. \end{array} \qquad (64.5)$$

2. Prove from the transformation equations (64.5) that the circle

$$(x - h)^2 + (y - k)^2 = r^2$$

maps into a circle whose radius is $|a|r$ and whose center is (ah, ak).

3. **Similitudes.** A plane correspondence $X \leftrightarrow X'$ which has the property that each non-collinear point triple (X, Y, Z) is mapped into a corresponding triple (X', Y', Z') such that $\triangle XYZ \sim \triangle X'Y'Z'$ is called a **similitude,** or a **similarity mapping.** (a) Show that such a correspondence is one-to-one and linear. (b) Prove further that there is a dila-tion factor $a > 0$ such that for any pair of points (X, Y), $|X'Y'| = a|XY|$. (This implies that the similitudes have the same seven invariants listed above for homothetic transformations.) (c) Is then every similitude merely a homothetic transformation in disguise?

4. Give an example of a linear transformation in the plane which is not a similitude.

5. Every isometry preserves angle measure, and therefore, perpendicularity.

6. *Most general linear transformation.* It can be shown that any *continuous* linear transformation in the plane has the analytic form

$$\begin{cases} x' = \dfrac{a_1x + b_1y + c_1}{a_3x + b_3y + c_3}, \\[2mm] y' = \dfrac{a_2x + b_2y + c_2}{a_3x + b_3y + c_3}, \end{cases} \quad \begin{vmatrix} a_1 & b_1 & c_1 \\ a_2 & b_2 & c_2 \\ a_3 & b_3 & c_3 \end{vmatrix} \neq 0, \tag{64.6}$$

where a_k, b_k, and c_k ($k = 1, 2, 3$) are constants. (a) Find the range and domain of such a transformation. (b) Making use of the information in (a), show that parallel lines are *not* preserved if a_3 and b_3 are not both zero. (c) For the special case $x' = (x + y)/x$, $y' = (y + 1)/x$, find the images of the parallel lines $y = x$ and $y = x + 1$. Show that for this particular example of the transformation (64.6) the circle $(x - 1)^2 + y^2 = 1$ is tangent to the line $a_3x + b_3y + c_3 = 0$ and maps into a parabola. (d) Find at least one invariant besides collinearity.

7. *Affine mappings.* If the equations in (64.6) be specialized to the case $a_3 = b_3 = 0$ and $c_3 = 1$, the result is an **affine transformation**

$$\begin{cases} x' = a_1x + b_1y + c_1, \\ y' = a_2x + b_2y + c_2, \end{cases} \quad \begin{vmatrix} a_1 & b_1 \\ a_2 & b_2 \end{vmatrix} \neq 0. \tag{64.7}$$

Show that under this mapping (a) parallelism is preserved, (b) distance and angle measure in general are not preserved, and (c) the ratio of the lengths of two segments either on the same line or on parallel lines is preserved (thus cross ratio is an invariant). ★(d) Show that any affine mapping is a product of rotations, translations, and **simple affine transformations** of the form

$$\begin{cases} x' = ax, \\ y' = by, \end{cases} \quad ab \neq 0. \tag{64.7'}$$

★(e) Prove that an affine transformation is *area preserving* if and only if

$$\begin{vmatrix} a_1 & b_1 \\ a_2 & b_2 \end{vmatrix} = 1.$$

★8. *Hjelmslev's theorem.* If T is any isometry in the plane prove that as X varies on a line the midpoint $M(X) = X_1$ of the segment $\overline{XX'}$, where $X' = T(X)$, either varies on a line or remains fixed. Prove an analogous conclusion when X varies on a circle.

NOTE: In this connection it has been pointed out that the generalized Hjelmslev's theorem implies that if two identical road maps be thrown down at random, the midpoint of the segment joining corresponding points will trace a map similar to, but

Figure 299

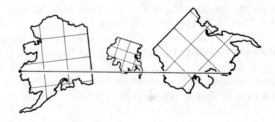

perhaps smaller than, the original map (parts, or all of which may be degenerate). See P. C. Joseph, "A Generalization of Hjelmslev's Theorem," *American Mathematical Monthly*, **74** (1967), pp. 574, 575.

65.   Applications of Plane Transformations: The Euler Line and Fagnano's Theorem

To show how the transformations introduced in the previous section can be used rather effectively in geometry we shall apply them in the proofs of two classical results. The first is the observation that the centroid, orthocenter, and circumcenter of a triangle are collinear. (See Exercise 10, Section 57.)

The standard synthetic proof of the theorem follows this pattern (Fig. 300):

Let $\overleftrightarrow{OG}$ meet altitude $\overleftrightarrow{AD}$ at H_1 (where O is the circumcenter and G the centroid of triangle ABC). If L is the midpoint of side $\overline{BC}$, then $\overleftrightarrow{AD}$ and $\overleftrightarrow{OL}$ are parallel and thus $\triangle GOL \sim \triangle GH_1A$. Since $|AG| = 2|GL|$ the factor of proportionality is 2 and hence, $|H_1G| = 2|GO|$. Since this argument applies to the intersection H_2 of $\overleftrightarrow{OG}$ with any altitude, $|H_2G| = 2|GO|$ and $H_1 = H_2 = H$, where H is the orthocenter. Obviously, certain crucial details are lacking in this argument as, for example, the fact that $\overleftrightarrow{OG}$ meets $\overleftrightarrow{AD}$ was not actually proved.

Figure 300

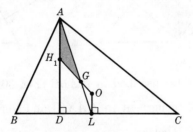

The proof involving the homothetic transformation is at least as elegant as the above and eliminates some of its difficulties. We make use of the observation (easily obtained from the known invariants for such mappings) that the orthocenter, centroid, and circumcenter of a triangle are mapped by any homothetic transformation into, respectively, the orthocenter, centroid, and circumcenter of the image triangle.

65.1   Theorem: The centroid of a triangle lies on the segment which joins the orthocenter and circumcenter of that triangle, assuming a position which is two thirds the distance from the orthocenter to the circumcenter.

PROOF (by use of homothetic mappings):   Consider the homothetic transformation $T = H(G, -\frac{1}{2})$. Since in magnitude and in sign

$$GL = -\tfrac{1}{2}GA, \qquad GM = -\tfrac{1}{2}GB, \qquad \text{and} \qquad GN = -\tfrac{1}{2}GC$$

(where L, M, and N are the midpoints of the sides of $\triangle ABC$—Fig. 300), it follows that

$$T(A) = L, \qquad T(B) = M, \qquad \text{and} \qquad T(C) = N.$$

Thus, T maps $\triangle ABC$ into $\triangle LMN$. But the circumcenter of $\triangle ABC$ is the orthocenter of $\triangle LMN$. Hence, T maps H into O and therefore, $O \in \overleftrightarrow{GH}$ and

$$GO = -\tfrac{1}{2}GH.$$

$HG = \tfrac{2}{3}HO$ immediately follows. ∎

The second application involves a singularly interesting theorem discovered by J. F. Fagnano in 1775. The reflection proof given here is due to L. Fejer and refutes a favorite claim of calculus students to the effect that the only way to attack problems involving maxima and minima is to differentiate.

65.2 *Fagnano's Theorem:*   In acute-angled triangles the inscribed triangle having minimal perimeter is the orthic triangle.

Figure 301

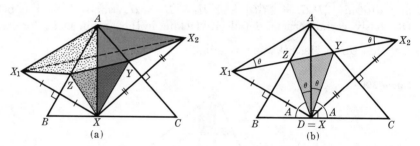

(a)   (b)

PROOF: Let $\triangle XYZ$ be inscribed in $\triangle ABC$ [Fig. 301(a)]. Reflect $\triangle AXZ$ in line $\overleftrightarrow{AB}$ forming $\triangle AX_1Z$ (congruent to $\triangle AXZ$) and similarly, reflect $\triangle AXY$ in line $\overleftrightarrow{AC}$ forming $\triangle AX_2Y$. Then

$$XY + YZ + ZX = X_2Y + YZ + ZX_1 \geqq X_1X_2. \tag{65.3}$$

But from the trigonometry of isosceles triangle AX_1X_2,

$$\tfrac{1}{2}X_1X_2 = AX_1 \sin \tfrac{1}{2}\angle X_1AX_2.$$

$AX_1 = AX$ and $\angle X_1AX_2 = 2\angle A$ reduces this to

$$X_1X_2 = 2AX \sin A,$$

from which may be deduced

$$XY + YZ + ZX \geqq X_1X_2 = 2AX \sin A. \tag{65.4}$$

But if $\overline{AD}$ is the altitude from A, we obtain the general inequality which holds *for all inscribed triangles of* $\triangle$ABC,

$$XY + YZ + ZX \geqq X_1X_2 = 2AX \sin A \geqq 2AD \sin A. \tag{65.5}$$

If equality can occur throughout for some inscribed triangle XYZ, then it will necessarily have minimal perimeter. But it is clear that equality will hold if and only if $\overline{AX}$ is the altitude to side $\overleftrightarrow{BC}$ and X_1, Z, Y, and X_2 are collinear, as shown in Fig. 301(b). It remains to prove that in this case triangle XYZ is the orthic triangle. Since $\triangle AX_1X_2$ is isosceles,

$$2\theta + \angle X_1AX_2 = \pi,$$

or, since $\angle X_1AX_2 = 2A$,

$$\theta = \pi/2 - A.$$

It now easily follows that since $\angle BDA = \angle ADC = \pi/2$,

$$\angle BDZ = \angle YDC = A.$$

By a property of the orthic triangle proved in Theorem 57.3 [see Fig. 255(b) in that connection] it follows that Y and Z are the feet of the altitudes from B and C. ∎

In the course of the above proof we found the **formula for the semiperimeter of the orthic triangle of triangle** ABC, inferred from equality in (65.5):

$$\boxed{s' = h_a \sin A = h_b \sin B = h_c \sin C.} \tag{65.6}$$

EXERCISES

1. A rancher wants to supply water to each of two sprinkler systems located at A and B by laying a pipeline from A and B to some point C on a river bank and installing a pump at C. Assuming the river bank is a straight line, how should C be chosen? (Solve in two different ways: First by calculus, then by the synthetic method by means of a reflection.)

2. Assuming that light always travels by the shortest path when reflected from a mirror, prove mathematically that *the angle of incidence equals the angle of reflection*. (Eves [9, pp. 39–40]; see the previous exercise. Used by permission.)

3. A variable line x passes through a fixed point A and intersects each of two intersecting lines l and m at the respective points X and Y. If Z is determined on x such that $[AX, YZ] = k = $ const., what is the locus of Z?

4. **Scheiner's Pantograph.** An instrument invented in 1630 by Christolph Scheiner called the *pantograph* may be used to increase (or decrease) the size of any figure proportionately. The mechanism is fastened at a point of pivot O with the remaining rods allowed to move in the manner illustrated in Fig. 302(a). The heart of the linkage is parallelogram $AA'BC$, Fig. 302(b). Discuss its operation and how it relates to the homothetic transformation.

5. Establish the following relationship for the distances from the orthocenter H to the vertices of triangle ABC:

$$\boxed{HA^2 + HB^2 + HC^2 = 12R^2 - (a^2 + b^2 + c^2).} \tag{65.7}$$

6. The Euler line of a non-equilateral triangle passes through a vertex of the triangle if and only if it is either isosceles or right.

Figure 302

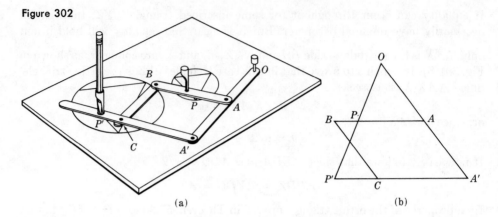

(a)   (b)

7. The line joining the centroid of a triangle to a point P on the circumcircle bisects the line joining the diametric opposite of P to the orthocenter. *Hint:* Locate the centroid of triangle HPP', where H is the orthocenter and P' is the point diametrically opposite P. (Court [6, p. 103]; used by permission.)

8
Circles

The wheel is considered man's most revolutionary invention. But by mastering its geometric counterpart—the *circle*—man has been able to exploit it to the present high level of refinement, now a foremost factor in his advancing civilization. The circle ever remains the inspiration of poet and artist, craftsman and mathematician alike. Entire treatises have been devoted to the subject, and it appears that the relationships between the properties of the triangle and circle are inexhaustible.

The present chapter will be devoted to a study of the more fundamental properties of circles. To indicate some of the fascinating interplay between circles and triangles, an optional section on Steiner's hypocycloid has been included.

66. Elementary Properties: The Chord-Secant Theorem [1]

Among the early experiences one has with the study of circles is observing that three noncollinear points always determine a unique circle. Unless its proof be examined closely, this deceptively simple fact seems not to depend on the strictly Euclidean properties of the plane and is apparently an elementary proposition of absolute geometry.

However, consider the following argument: Let A, B, and C be any three points not lying on a line, and suppose l_1 and l_2 are the respective perpendicular

---

[1] This section will take for granted all the properties of circles established for absolute geometry (Sections 40–42, Chapter 5).

bisectors of $\overline{AB}$ and $\overline{BC}$ (Fig. 303). If l_1 *does not* meet l_2, then $\overleftrightarrow{AB}$, being perpendicular to one of the two parallel lines l_1 and l_2, is perpendicular to the other also. (Recall that this is the essence of Theorem 53.1, a characteristically Euclidean property.) Thus $\overleftrightarrow{AB}$ and $\overleftrightarrow{BC}$ are both perpendicular to l_2 and therefore must coincide, denying the assumption that A, B, and C were not collinear. Then it must be concluded that l_1 and l_2 intersect at some point O, and that point is equidistant from A, B, and C. Hence O is the center of the desired circle.

Figure 303

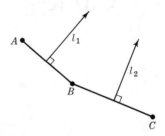

Of course this property is already built into the model, for in analytic geometry the equation of the circle to be determined may be written

$$x^2 + y^2 + ux + vy + w = 0,$$

where u, v, and w are regarded as unknowns, and this circle passes through the given points $A[a_1, a_2]$, $B[b_1, b_2]$, and $C[c_1, c_2]$ if and only if

$$\begin{cases} a_1 u + a_2 v + w = -a_1{}^2 - a_2{}^2, \\ b_1 u + b_2 v + w = -b_1{}^2 - b_2{}^2, \\ c_1 u + c_2 v + w = -c_1{}^2 - c_2{}^2. \end{cases}$$

The system has a unique solution for u, v, and w if and only if

$$\begin{vmatrix} a_1 & a_2 & 1 \\ b_1 & b_2 & 1 \\ c_1 & c_2 & 1 \end{vmatrix} \neq 0,$$

a condition which is clearly equivalent to the noncollinearity of A, B, C (one may choose a coordinate system which makes A the origin), and thus the desired circle exists.

Since each three noncollinear points A, B, and C determines a unique circle, we shall refer to the *circle* $[A, B, C]$. Similarly, it will be convenient to speak of the circle $[O, r]$, being the unique circle centered at O having radius r.

We have not yet introduced the properties of circles needed to prove the classical formula which gives the radius R of the circumcircle of triangle ABC in terms of its sides. The particular property needed is the fact that *inscribed angles of a circle intercepting the same arc are congruent*. In elementary treatments this result is often allowed to depend on the rather delicate concept of arc measure. It is interesting to reverse the procedure and make arc measure depend on the above result. The initial step is the following important property of circles (the "cevian" method of proof is once again employed):

66.1 Chord-Secant Theorem: Let P be any point collinear with two points A and B lying on a circle $[O, r]$. Then, in magnitude and in sign

$$PA \cdot PB = PO^2 - r^2.$$

PROOF: (See in this connection Exercises 11, 12, Section 57.) In Fig. 304, apply the cevian formula to $\triangle AOB$ (which may be degenerate) with $\overline{OP}$ as cevian:

$$PO^2 = \frac{AP}{AB} \cdot r^2 + \frac{PB}{AB} \cdot r^2 - \frac{AP}{AB} \cdot \frac{PB}{AB} \cdot AB^2$$

$$= r^2 + PA \cdot PB. \ \blacksquare$$

Figure 304

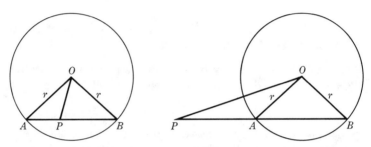

The equation in Theorem 66.1 proves a remarkable property: The product $PA \cdot PB$ is *constant* relative to a variation of the line $\overleftrightarrow{AB}$ through P and the corresponding variation of points A and B. This observation leads to three important corollaries.

66.2 Corollary: The products of the lengths of the segments on each of two intersecting chords of a circle determined by the point of intersection are equal.

PROOF: By Theorem 66.1 [see Fig. 305(a)],

$$|AP| \cdot |PB| = -PA \cdot PB = r^2 - PO^2 = -PC \cdot PD = |CP| \cdot |PD|. \ \blacksquare$$

66.3 Corollary: The products of the lengths of the segments on each of two intersecting secants of a circle, each pair of segments having the point of intersection as one endpoint and the respective intersections of the secants with the circle as the other, are equal.

PROOF: By Theorem 66.1 [see Fig. 305(b)],

$$|PA| \cdot |PB| = PA \cdot PB = PO^2 - r^2 = PC \cdot PD = |PC| \cdot |PD|. \ \blacksquare$$

66.4 Corollary: If a secant and tangent of a circle intersect, the product of the length of the segments from the point of intersection on the secant is the square of the distance from the point of intersection to the point of contact.

PROOF: By Theorem 66.1 [see Fig. 305(c)] and the Pythagorean theorem,

$$|PA| \cdot |PB| = PA \cdot PB = PO^2 - r^2 = |PT|^2. \ \blacksquare$$

NOTE: The result in Corollary 66.4 may be regarded as the limiting case of the formula $PA \cdot PB = PC \cdot PD$ as C and D each approach a point of tangency T, in Figs. 305(b, c).

Figure 305

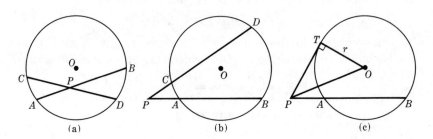

(a)   (b)   (c)

The expression $PO^2 - r^2$ involved in the proof of the chord-secant theorem turns out to be of significance by itself.

66.5   Definition: For any point P in the plane and any circle $[O, r]$, the number $PO^2 - r^2$ is termed the **power** of P with respect to the given circle.

It is immediately observed that the power of a point with respect to a circle is either *positive, zero,* or *negative* according as the point is *exterior to, lies on,* or is *interior to* the circle; the chord-secant theorem shows that the power of a point P may be found by intersecting the circle with any line through P at, say, A and B as in Figs. 305(a, b), and then calculating the product $PA \cdot PB$. Further discussion of this concept will take place later.

EXERCISES

1.   Why must three collinear points never lie on a circle?
2.   Prove that a circle is uniquely determined by specifying its radius and two of its points, given that its center is to lie on a specified side of the line determined by those two points.
3.   Prove that a circle is uniquely determined by specifying two non-parallel tangents, the point of contact on one of those tangents, and a given side of that tangent.
★4.   Let A and B be two points lying on one side of line l. Find a synthetic construction for one of the two circles which is tangent to l and passes through A and B.
5.   Two chords intersect at right angles such that the segments on one chord are of length 2 and 6 and those on the other, 3 and x. Find x and the radius of the circle.
6.   Two chords intersect at right angles such that the segments on one of the chords are of length 3 and 4, and the distance from the point of intersection to the center is 2. Find the length of the other chord.
7.   Two chords intersect at an angle $\pi/4$ such that the segments on those chords are, respectively, of length $3\sqrt{2}$ and $9\sqrt{2}$ on one, and $3\sqrt{7} - 3$ and $3\sqrt{7} + 3$ on the other. Find the radius of the circle.

8. Two chords intersect at an angle $\pi/6$ such that the segments on those chords are, respectively, of length $\sqrt{3}$ and $9\sqrt{3}$ on one, and $\sqrt{91} - 8$ and $\sqrt{91} + 8$ on the other. Find the radius of the circle.

9. Two perpendiculars passing through an external point P of a circle intersect that circle, one at A and B and the other at C and D, such that $|PA| = 2, |AB| = 10$, and $|CD| = 5$. Find the radius of the circle. Also, find the distance from P to the center of the circle by the Pythagorean theorem, then check your answer by the use of the formula in Theorem 66.1.

10. Find the equation of the circle passing through the points $A[2, 0]$, $B[2, 8]$, and $C[9, 1]$.

11. Prove that the power of each vertex of triangle ABC with respect to the circle determined by the centroid G and the other two vertices is $\frac{1}{3}(a^2 + b^2 + c^2)$.

12. Solve Exercise 11 above by the analytic method. (See Exercise 11, Section 57.)

13. Prove that if a point lies on the line containing the common chord of two circles its powers with respect to those two circles are equal.

67. Antiparallelism

The intriguing concept of "antiparallelism" is quite useful in geometry. This term does not refer to lines which are *not* parallel, but rather to lines which *would be* parallel if one of them were reflected in a certain manner.

It is clear that if $\overleftrightarrow{AC}$ is parallel to BD in Fig. 306, then, in magnitude and in sign,

$$\frac{PA}{PB} = \frac{PC}{PD}.$$

Indeed, this equation is characteristic of parallelism. Suppose that instead of the *ratios* one considers the relation involving the *products*

$$PA \cdot PB = PC \cdot PD. \tag{67.1}$$

In this case, ruling out certain special cases (see Exercise 1 below) the lines $\overleftrightarrow{AC}$ and $\overleftrightarrow{BD}$ will *not* be parallel.

Figure 306

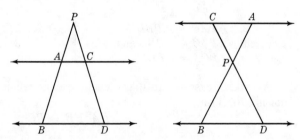

67.2 **Definition:** If two transversals intersect at P and cut each of two lines in the respective point pairs (A, B) and (C, D) such that

$$PA \cdot PB = PC \cdot PD,$$

then the first pair is said to be **antiparallel** with respect to the second pair.

This terminology underscores the fact that in Fig. 307 if relation (67.1) holds, then but for a reflection, *line* $\overleftrightarrow{AC}$ *would be parallel to line* $\overleftrightarrow{BD}$. To be more specific, suppose $\overleftrightarrow{AC}$ is reflected in the *bisector* of $\angle P$, with the rest of the figure being left unchanged. Then if A' and C' are the reflected images of A and C, respectively, the relation (67.1) can be written in the form

$$\frac{PC'}{PB} = \frac{PA'}{PD},$$

and it follows that $\overleftrightarrow{A'C'} \parallel \overleftrightarrow{BD}$.

Figure 307

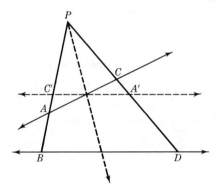

Having made this observation it is easy to compose a list of the contrasts between parallelism and antiparallelism. Here we make use of the notation in Fig. 307.

$\overleftrightarrow{AC}$ AND $\overleftrightarrow{BD}$ PARALLEL	$\overleftrightarrow{AC}$ AND $\overleftrightarrow{BD}$ ANTIPARALLEL
1. $\dfrac{PA}{PB} = \dfrac{PC}{PD}$	1. $PA \cdot PB = PC \cdot PD$
2. $\triangle PAC \sim \triangle PBD$	2. $\triangle PAC \sim \triangle PDB$
3. $\angle PAC = \angle PBD$ and $\angle PCA = \angle PDB$	3. $\angle PAC = \angle PDB$ and $\angle PCA = \angle PBD$

A special case of antiparallelism occurs when C and D coincide, reducing equation (67.1) to

$$PA \cdot PB = PC^2.$$

Thus it follows that in Fig. 305(c) lines $\overleftrightarrow{AT}$ and $\overleftrightarrow{BT}$ are antiparallel with respect to $\overleftrightarrow{PA}$ and $\overleftrightarrow{PT}$. The reader may easily point to numerous other instances of antiparallelism already encountered. The situations depicted in Figs. 305(a, b) provide examples ($\overleftrightarrow{AC}$ is antiparallel to $\overleftrightarrow{BD}$), and the sides of any triangle are antiparallel to the corresponding sides of its orthic triangle [in Fig. 255(b), for example, $\overset{\leftrightarrow}{FE}$ is antiparallel to $\overset{\leftrightarrow}{BC}$ with respect to $\overset{\leftrightarrow}{AB}$ and $\overset{\leftrightarrow}{AC}$].

68.   The Inscribed-Angle Theorems

The theorem introduced here is of fundamental importance to the study of circles. The proof shows how one may make use of the concept of antiparallelism discussed in the preceding section.

Figure 308

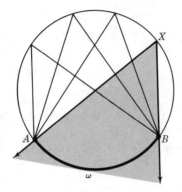

68.1   Theorem: In any circle, inscribed angles intercepting the same arc are congruent.

PROOF: If the given intercepted arc is ω and A and B are its endpoints, our method will be to show that as X varies on the circle with $\angle AXB$ intercepting ω, $\angle AXB$ remains constant (Fig. 308). Let t be the tangent at A and take C any point on t lying on the same side of $\overset{\leftrightarrow}{AB}$ as ω. We show that in all cases, $\angle X = \angle BAC$. If $\overset{\leftrightarrow}{BX} \parallel t$ [Fig. 309(a)], then the radius at A, being perpendicular

Figure 309

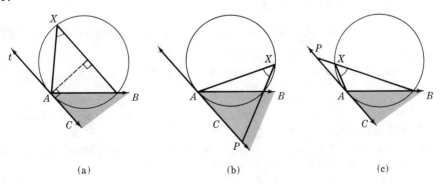

(a)   (b)   (c)

to t, must also be perpendicular to $\overleftrightarrow{BX}$, and hence bisects $\overline{BX}$. Since $\triangle ABX$ would then be isosceles,

$$\angle X = \angle ABX = \angle BAC.$$

Otherwise, $\overleftrightarrow{BX}$ meets t in some point P, Figs. 309(b, c). Since $PB \cdot PX = PA^2$, lines $\overleftrightarrow{AB}$ and $\overleftrightarrow{AX}$ are antiparallel. Therefore,

$$\triangle PAX \sim \triangle PBA,$$

and hence,

$$\angle AXP = \angle BAP.$$

In the two possible cases illustrated in Figs. 309(b, c), it follows that either

$$\angle X = \angle AXP = \angle BAP = \angle BAC,$$

or

$$\angle X = \pi - \angle AXP = \pi - \angle BAP = \angle BAC. \ \blacksquare$$

Note that the above argument also proves

68.2   Corollary: An angle formed by a tangent and chord of a circle (with the vertex at the point of contact) is congruent to any inscribed angle which intercepts the same arc the given angle does. (In Fig. 310, $\angle ABC = \angle ACD$.)

Figure 310

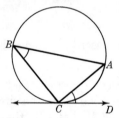

The following corollary is a very useful special case:

68.3   Corollary: Any angle inscribed in a semicircle is a right angle.

PROOF: In Fig. 310 let $\overline{AC}$ be a diameter. Then $\overleftrightarrow{AC} \perp \overleftrightarrow{CD}$ and it follows that $\angle B = \angle ACD = \pi/2. \ \blacksquare$

68.4   Example: Referring to Fig. 311, show that $\overleftrightarrow{AC}$ bisects $\overline{\angle BAD}$ if and only if $BC = CD$. (Assume from the figure that C lies interior to $\angle BAD$.)

Solution: First let us assume that $\overleftrightarrow{AC}$ bisects $\overline{\angle BAD}$. Let $\overline{CE}$ be the diameter passing through C. Then by Theorem 68.1, $\angle BAC = \angle BEC$ and $\angle CAD = \angle CED$ so we have $\angle BEC = \angle CED$. But also $\angle EBC = \angle EDC = \pi/2$ (Corollary 68.3) and hence, $\triangle EBC \cong \triangle EDC$. Therefore, $BC = CD$. Conversely, if $BC = CD$, then again $\triangle EBC \cong \triangle EDC$ and $\angle BEC = \angle CEC$, which implies that $\angle BAC = \angle CAD. \ \blacksquare$

68.5   Example: If points A and B lie on opposite sides of chord $\overline{CD}$, show that the inscribed angles $\overline{\angle CAD}$ and $\overline{\angle CBD}$ are supplementary (Fig. 312).

Figure 311

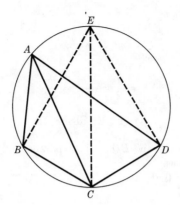

Figure 312   Figure 313

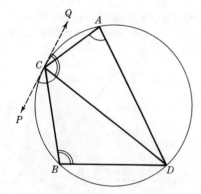

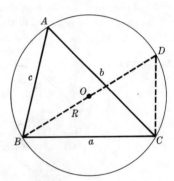

Solution: Construct the tangent $\overleftrightarrow{PQ}$ at C with P on the opposite side of $\overleftrightarrow{CD}$ as A, and Q on the opposite side as B. Then

$$\angle A + \angle B = \angle PCD + \angle QCD = \pi. \ \blacksquare$$

The final example shows how to derive the formula for the circumradius of a triangle, mentioned earlier.

68.6 *Example:* Prove that the area K and circumradius R of the triangle ABC are related by the formula

$$K = \frac{abc}{4R}. \tag{68.7}$$

Solution: Begin with $K = \frac{1}{2}bc \sin A$. In Fig. 313, let $\overline{BD}$ be a diameter. Then $\angle D$ is either congruent or supplementary to $\angle A$, depending on which side of $\overleftrightarrow{BC}$ point D lies. But in either case,

$$\sin D = \sin A.$$

But in right triangle BCD we have

$$\sin D = \frac{a}{2R}.$$

Therefore,

$$K = \frac{1}{2} bc \cdot \frac{a}{2R} = \frac{abc}{4R}. \blacksquare$$

EXERCISES

1. Prove that if two lines $\overleftrightarrow{AC}$ and $\overleftrightarrow{BD}$ are both parallel and antiparallel (Fig. 307), then triangle PAC is isosceles.

2. In Figs. 314(a,b) show that

   (a) $\angle P = \angle E + \angle F$
   (b) $\angle P = \angle E - \angle F$.

   Hint: E and F can assume various positions on the circle without changing the measures of the angles involved. Use the Euclidean exterior angle theorem.

3. Prove that if $\overleftrightarrow{AB} \parallel \overleftrightarrow{CD}$, then $\angle E = \angle F$, and conversely (Fig. 315). Show that the first part may be regarded as the limiting case of Exercise 2(b), Fig. 314(b), as AP becomes infinite.

Figure 314

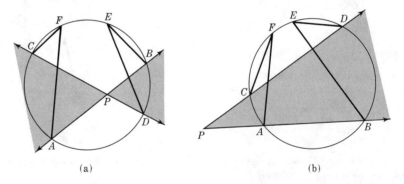

(a)   (b)

Figure 315

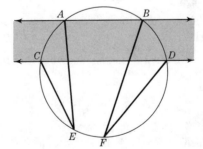

4. If two circles have a common tangent at a point A and one circle passes through the center of the other, then the smaller circle bisects every chord of the larger which passes through A.

5. *The measure of arcs.* Define the **measure of circular arc** ω, denoted $m(\omega)$, to be *twice* that of any inscribed angle of the circle determined by ω which intercepts ω. If ω is the entire circle then define $m(\omega) = 2\pi$. (a) Show that this arc measure is *well defined* (does not depend on the particular inscribed angle used) and is *additive: if ω_1 and ω_2 are any two circular arcs such that $\omega_1 \cup \omega_2$ is an arc and $\omega_1 \cap \omega_2$ consists of at most two points, then $m(\omega_1 \cup \omega_2) = m(\omega_1) + m(\omega_2)$.* (b) Show that ω is a semicircle if and only if $m(\omega) = \pi$.

6. Prove the following relationships between angles and the measures of their intercepted arcs (see Exercise 5 above).

   (a) The measure of a central angle is equal to that of its intercepted arc.

   (b) The measure of any inscribed angle is one half that of its intercepted arc.

   (c) The measure of an angle formed by a chord and tangent is one half that of its intercepted arc.

   (d) The measure of an angle formed by two chords intersecting inside the circle is one half the sum of the measures of the arcs intercepted by it and its vertical angle.

   (e) The measure of an angle formed by two secants of a circle intersecting outside the circle is one half the difference between the measures of the two intercepted arcs.

7. Arcs having equal measures are called *congruent* (see Exercise 5 above). Prove that congruent chords subtend congruent arcs and conversely.

8. Two secants of a circle are parallel if and only if they cut off congruent arcs on that circle.

9. Prove the **formula for the circumradius of a triangle in terms of its sides:**

$$R = \frac{abc}{\sqrt{(a + b + c)(a + b - c)(a - b + c)(-a + b + c)}}. \tag{68.8}$$

★10. With the usual notation prove (a) $t_a{}^2 + t_b{}^2 + t_c{}^2 \leq s^2$, (b) $t_a{}^2 t_b{}^2 + t_b{}^2 t_c{}^2 + t_c{}^2 t_a{}^2 \leq rs^2(4R + r)$, (c) $(t_a t_b)^{-2} + (t_b t_c)^{-2} + (t_c t_a)^{-2} \geq (rs)^{-2}$, and (d) $t_a t_b t_c \leq rs^2$, where equality holds in each case if and only if the triangle is equilateral. *Hint:* Establish the formula $ab + bc + ca = s^2 + 4rR + r^2$. In the formula for t_a, use the fact that $\sqrt{bc} \leq \frac{1}{2}(b + c)$, with equality only when $b = c$.

11. Prove that in a right triangle,

$$s = 2R + r.$$

   NOTE: The converse is also true, but somewhat harder to prove. For an interesting generalization which includes this theorem as a special case, see W. J. Blundon, "Generalization of a Relation Involving Right Triangles," *American Mathematical Monthly*, **74** (1967), pp. 566–568.

12. Find a compass, straightedge construction of the tangents to a circle from a given external point.

13. An interesting and elementary device to solve the problem of constructing the trisectors of an angle is the so-called *tomahawk*. The inventor is unknown but the device appeared in a book in 1835. To construct a tomahawk, start with a line segment $\overline{RU}$ trisected at S and T (as shown in Fig. 316). Draw a semicircle on $\overline{SU}$ as diameter and draw $\overleftrightarrow{SV}$ perpendicular to $\overleftrightarrow{RU}$. Then draw in the remaining lines (inessential to its

function as a trisector). To trisect an angle ABC with the tomahawk, place the instrument on the angle so that R falls on $\overrightarrow{BA}$, $\overleftrightarrow{SV}$ passes through B, and the semicircle touches $\overrightarrow{BC}$ at some point D. Then $\overrightarrow{BS}$ and $\overrightarrow{BT}$ will trisect the given angle. (Explain why.) Is this a legitimate straightedge and compass construction? (Eves [9, p. 215]; used by permission of the publisher.)

Figure 316

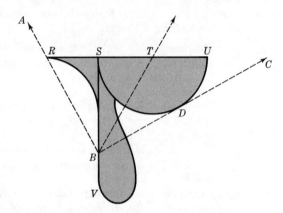

14. In triangle ABC the relation

$$\sin A \cos B = \sin C \cos C$$

implies that the circumdiameter through A, the median through B, and the angle bisector through C are concurrent. Prove, using properties of circles and Ceva's theorem. (Problem E1574, *American Mathematical Monthly*, **73** (1964), p. 531.)

69. Concyclic Points

Points which lie on the same circle are called **concyclic**. Certain simple conditions imposed on four points make them concyclic, forming the basis for reversing the statements of the preceding section.

69.1 *Theorem*: If two congruent angles subtend a common segment and their vertices lie on the same side of the line determined by that segment, those vertices are concyclic with the endpoints of that segment.

PROOF: In Fig. 317 suppose $\overline{AB}$ is the given segment and $\angle C = \angle D$. If D lies outside the circle $[A, B, C]$, then either $\overline{AD}$ or $\overline{BD}$ cuts the circle in some point E; we may assume that $\overline{AD}$ does so. By Theorem 68.1

$$\angle AEB = \angle C = \angle D.$$

But this contradicts the exterior-angle theorem. A similar argument may be used if D falls inside circle $[A, B, C]$. ∎

Figure 317

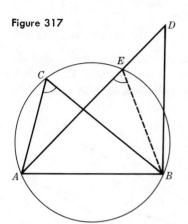

Figure 318

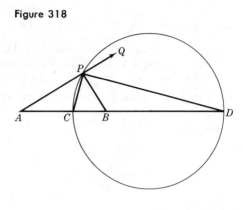

Theorem 69.1 is often stated as a locus property: *The locus of the vertex of an angle of constant measure subtending a fixed segment and lying on one side of the line containing that segment is a circular arc* (less endpoints).

An important corollary to this has already been observed: *The vertex of the right angle of a variable right triangle having a fixed hypotenuse as base varies on a circle having that hypotenuse as a diameter.*

EXERCISES

1. The midpoints of the sides of a convex quadrilateral are concyclic if and only if the quadrilateral is a rhombus. True or false?
2. If the opposite pairs of angles of a convex quadrilateral are supplementary, the quadrilateral is cyclic.
3. One endpoint of a chord of a circle is fixed while the other varies on the circle. Prove that the locus of the midpoint of that chord is a circle.
4. Point X varies on the circumcircle of triangle ABC, on the same side of $\overleftrightarrow{BC}$ as A. Point P is located on $\overrightarrow{XB}$ such that $XP = XC$. Find the locus of P.
5. **Circle of Apollonius.** The locus of a point P, such that the ratio of the (undirected) distances from P to two fixed points A and B is constant ($k \neq 1$), is a circle which has its center on the line of A and B, cutting that line at points which divide $\overline{AB}$ harmonically. Complete the details in the following proof (Fig. 318): Locate the unique points C and D on $\overline{AB}$ such that, in magnitude and in sign, $AC/CB = k = AD/BD$. Hence, $[AB, CD] = -1$. If P lies on the circle whose diameter is $\overline{CD}$ then by the result of Exercise 10, Section 63, $\overrightarrow{PC}$ and $\overrightarrow{PD}$ are the bisectors of the angles at P and hence $PA/PB = AC/CB = k$.

*70.   Straightedge Construction of the Tangents
to a Circle from a Given External Point

The reader may find it entertaining to carry out the following experiment (Fig. 319): Draw a circle and locate any exterior point P. Through P draw any three secants, $\overleftrightarrow{PA}$, $\overleftrightarrow{PC}$, and $\overleftrightarrow{PE}$. Determine L and M, the intersections of the cross joins $\overleftrightarrow{AD}$ and $\overleftrightarrow{BC}$, $\overleftrightarrow{CF}$ and $\overleftrightarrow{DE}$. Draw $\overleftrightarrow{LM}$, cutting the circle in R and S. It will be found that R *and* S *are the points of contact of the tangents from* P.

The properties of circles now established are sufficient to prove this. We shall make use of a special result concerning the harmonic sequence.

Figure 319

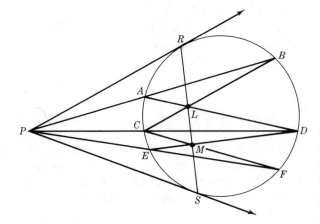

70.1 Lemma: If A, B, C, and D are four distinct collinear points, with O the midpoint of $\overline{AB}$, then $[AB, CD] = -1$ if and only if

$$AD \cdot BD = CD \cdot OD.$$

PROOF: Suppose $[AB, CD] = -1$. Therefore, with $AO = OB$,

$$AC \cdot BD = -AD \cdot BC$$
$$(AD + DC) \cdot BD = -AD \cdot (BD + DC)$$
$$2AD \cdot BD = CD \cdot (AD + BD)$$
$$= CD \cdot (AO + OD + BD)$$
$$= CD \cdot (OD + OD)$$
$$= 2CD \cdot OD.$$

Since the steps are reversible the proof is complete. ∎

The basis for the straightedge construction mentioned above is provided in the following:

70.2 Theorem: The line determined by the points of contact of the tangents from any external point P of a circle intersects any secant through P at a point Q which is the harmonic conjugate of P with respect to the two points of intersection of the secant and the circle.

PROOF: Let $\overleftrightarrow{PR}$ and $\overleftrightarrow{PS}$ be the tangents and $\overleftrightarrow{AB}$ any secant through P (Fig. 320). Let O be the midpoint of $\overline{AB}$ and let the secant through P which passes through the center O' intersect the circle at A' and B', and take $Q' = \overline{RS} \cap \overline{A'B'}$. Since $\angle PQ'R$ is a right angle the circle which has $\overline{PR}$ as diameter will pass through Q'. Hence, by the chord-secant theorem

$$O'A'^2 = O'R^2 = O'Q' \cdot O'P,$$

and it follows that since O' is the midpoint of $\overline{A'B'}$, $[A'B', Q'P] = -1$ (Theorem 63.5).

By the lemma,

$$A'P \cdot B'P = Q'P \cdot O'P. \tag{70.3}$$

But, again by the chord-secant theorem, $A'P \cdot B'P = AP \cdot BP$. By similar triangles, $Q'P \cdot O'P = QP \cdot OP$. Hence (70.3) becomes

$$AP \cdot BP = QP \cdot OP. \tag{70.4}$$

By the lemma $[AB, QP] = -1$, the desired result. ▮

Figure 320

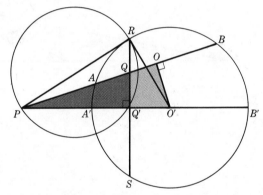

To see how this theorem implies the straightedge construction of the tangents from P, let $\overleftrightarrow{AB}$ and $\overleftrightarrow{CD}$ be any two secants (Fig. 321) and let the cross joins $\overleftrightarrow{AD}$ and $\overleftrightarrow{BC}$ meet at L, and lines $\overleftrightarrow{AC}$ and $\overleftrightarrow{BD}$ at W. If $\overleftrightarrow{WL}$ meets $\overleftrightarrow{AB}$ at Q and $\overleftrightarrow{CD}$ at Q',

Figure 321

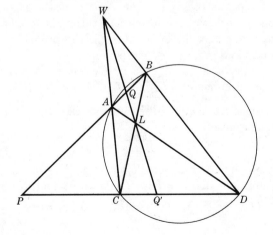

we can recognize the harmonic construction which shows that $[AB, PQ] = [CD,$ $PQ'] = -1$. By the previous theorem $\overleftrightarrow{RS}$ then passes through Q and Q' and hence L lies on $\overleftrightarrow{RS}$. In exactly the same manner the point of intersection M of another pair of cross joins (as in Fig. 319) would also lie on $\overleftrightarrow{RS}$ and therefore, $\overleftrightarrow{LM} = \overleftrightarrow{RS}$. ∎

EXERCISE

Prove that the straightedge construction shown in Fig. 322 will also yield the external tangents from P.

Figure 322

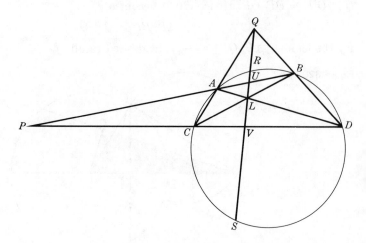

71. Two Classical Theorems of Geometry

Both theorems presented in this section go back to Euler. The first concerns the existence of the nine-point circle of a triangle (see Section 8); the second provides a relationship between the circumradius, the inradius, and the distance between the circumcenter and incenter of any triangle. A proof of the existence of the nine-point circle based on homothetic transformations turns out to be not only elegant, but quite informative; one may derive at once all the properties (a)—(e) of the nine-point circle as previously discussed in Section 8.

With the same notation as introduced in Section 8, observe Fig. 323, which shows, in addition, the circumcircle of triangle ABC and the points of intersection D', L', M', $\cdots$ of the circumcircle with the rays $\overrightarrow{HD}$, $\overrightarrow{HL}$, $\overrightarrow{HM}$, $\cdots$. Consider the homothetic transformation

$$T = H(H, \tfrac{1}{2}),$$

which maps the circumcircle O into some circle ω, with O being mapped into the center of ω. By definition $HX = \tfrac{1}{2}HA$, $HY = \tfrac{1}{2}HB$, and $HZ = \tfrac{1}{2}HC$ so that

$$T(A) = X, \qquad T(B) = Y, \qquad T(C) = Z.$$

Hence X, Y, and Z are points of ω. It remains to prove, therefore, that D, L, M, $\cdots$ are the respective midpoints of $\overline{HD'}$, $\overline{HL'}$, $\overline{HM'}$, $\cdots$. Since the proofs are analogous, it suffices to prove this for just the first two points mentioned.

Figure 323

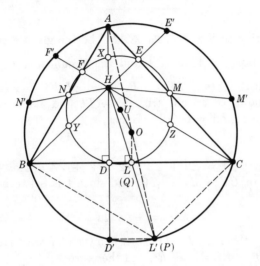

Let P be the intersection of $\overleftrightarrow{AO}$ with the circumcircle and take Q as the intersection of $\overleftrightarrow{HP}$ and $\overleftrightarrow{BC}$. Now $\angle ACP = \angle ABP = \pi/2$ so it follows that $BPCH$ is a parallelogram. Hence, Q is the midpoint of $\overline{HP}$ and $\overline{BC}$, and therefore, $Q = L$, $P = L'$. This proves that L is the midpoint of $\overline{HL'}$, as desired. Now observe also that $\angle AD'L' = \angle AD'P = \pi/2$ and therefore, $\overleftrightarrow{DL} \parallel \overleftrightarrow{D'L'}$. Thus since L is the midpoint of $\overline{HL'}$, D is the midpoint of $\overline{HD'}$. It follows that

$$T(D') = D, \qquad T(L') = L, \qquad T(M') = M, \qquad T(E') = E,$$
$$T(F') = F, \qquad \text{and} \qquad T(N') = N,$$

which shows that the original nine points *lie on the homothetic image of the circumcircle* O. Hence they lie on a *circle*, termed the **nine-point circle** of the triangle.

We have thus proved:

71.1 Theorem: The nine-point circle of a triangle is homothetic to the circumcircle of the triangle under the homothetic transformation which has the orthocenter of that triangle as center and $\frac{1}{2}$ as dilation factor.

The remaining properties may be established with very little additional effort. For example, it is immediate that the center U of the nine-point circle (called the **nine-point center**) lies on the Euler line, since O maps into U and hence $U \in \overleftrightarrow{HO}$. Indeed, one obtains the added information that U *is the midpoint of segment* $\overline{HO}$. Further properties obtainable by this method of proof are indicated in the exercises.

71.2 *Euler's Formula:* If R, r, and d represent the circumradius, inradius, and distance between the circumcenter and incenter of any triangle, then:

$$d^2 = R(R - 2r).$$

PROOF: Referring to Fig. 324, let $\triangle ABC$ be the given triangle, with V the intersection of the bisector $\overrightarrow{AI}$ of $\angle A$ and the circumcircle O, and with $\overline{VW}$ a circumdiameter. Further, take $\overleftrightarrow{IT} \perp \overleftrightarrow{AB}$. Now $\angle BWV = \angle BAV = \angle VAC = \angle VWC$, and it follows that right triangles ATI and WCV are similar. Then $TI/IA = CV/VW$, or $r/IA = CV/2R$. Hence,

$$2rR = IA \cdot CV. \tag{71.3}$$

But $BV = CV$ so that $\angle 1 = \angle 2 + \angle 3 = \angle BCV + \angle 4 = \angle 5 + \angle 4$, and $\triangle BVI$ is isosceles. Therefore, $CV = BV = |IV|$ and since I is between A and V, (71.3) becomes

$$2rR = -IA \cdot IV. \tag{71.4}$$

By the chord-secant theorem, $IA \cdot IV$ is the power of I with respect to circle O. Hence $IA \cdot IV = IO^2 - R^2 = d^2 - R^2$, and therefore

$$2rR = R^2 - d^2. \quad \blacksquare$$

Figure 324

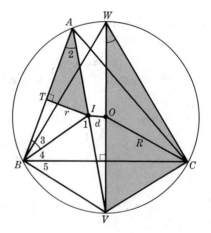

EXERCISES

1. Prove that if H, U, G, and O represent the orthocenter, nine-point center, centroid, and circumcenter of a triangle, then U is the harmonic conjugate of O with respect to H and G. That is, $[HG, UO] = -1$.

Figure 325

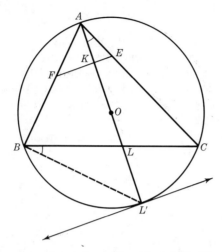

2. The tangents to the nine-point circle at the midpoints L, M, and N of triangle ABC are parallel to the respective sides of the orthic triangle DEF. Prove this by first proving that the line containing circumdiameter $\overline{AL'}$ is perpendicular to $\overleftrightarrow{EF}$ (Fig. 325). The tangent to the circumcircle would then be parallel to $\overleftrightarrow{EF}$. *Hint:* Recall that $\angle AEF = \angle B$.

3. Prove that in Fig. 323 $\triangle XYZ \cong \triangle LMN$.

4. The incircle of an equilateral triangle bisects each of the segments which join the orthocenter with the vertices. Give two independent proofs, one involving the nine-point circle, the other not.

★5. The orthocenter, the midpoint of the base, and the direction of the base of a variable triangle are fixed. Find the locus of the nine-point center. (N. A. Court, *College Geometry* [6, p. 108].)

★6. The base $\overline{BC}$ and the opposite angle A of a variable triangle ABC are fixed. Show that (a) the line $\overleftrightarrow{MY}$ [Fig. 323] has a fixed direction; (b) the nine-point circle is tangent to a fixed circle. (Court [6, p. 119].)

7. Show that the circumcircle of triangle ABC bisects each of the segments $\overline{II_a}$, $\overline{II_b}$, $\overline{II_c}$, $\overline{I_aI_b}$, $\overline{I_bI_c}$, and $\overline{I_aI_c}$, the usual notation understood.

8. Prove the following corollary to Euler's formula: If two circles $[O, R]$ and $[I, r]$ are such that $R > 2r$ and the line of centers has length $\sqrt{R(R - 2r)}$, then an infinity of triangles may be circumscribed about the circle $[I, r]$ which are inscribed in the circle $[O, R]$. *Hint:* Let A be any point on the larger circle and inscribe $\triangle ABC$ in that circle such that $\overleftrightarrow{AB}$ and $\overleftrightarrow{AC}$ are the tangents from A to the smaller circle; reconstruct a portion of the argument used for Euler's formula.

9. Following a method similar to that used to prove Euler's formula, establish the following relations involving the distances d_a, d_b, d_c from the circumcenter O to the excenters opposite A, B, and C, respectively, in terms of the exradii:

$$\begin{array}{|l}
d_a{}^2 = R(R + 2r_a), \\
d_a{}^2 = R(R + 2r_b), \\
d_c{}^2 = R(R + 2r_c).
\end{array}$$ (71.5)

*72.   Steiner's Curve

If P is any point on the circumcircle of triangle ABC and P_a, P_b, P_c are the feet of the perpendiculars from P to the sides opposite A, B, and C, the interesting fact that P_a, P_b, and P_c are collinear may be derived by any one of several elementary arguments. One proof was indicated in Exercise 9, Section 62. The line of collinearity is commonly called the *Simson line corresponding to* P, although it apparently was not the discovery of the person whose name it bears, Robert Simson (1687–1768), but rather that of another eighteenth century mathematician William Wallace (1768–1873).

Figure 326

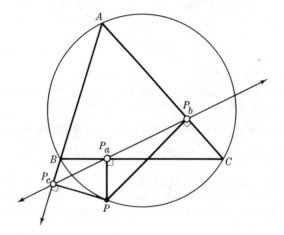

As P varies, the Simson line rotates in a curious manner: It is the variable tangent of a fixed *hypocycloid*. To be more precise, *the family of Simson lines is the family of tangents to a hypocycloid of three cusps*.[2] Indeed, this hypocycloid was originally known as **Steiner's curve** in dual recognition of this property which the Simson line has, and the fact that Jacob Steiner first discovered it.

It is the purpose of this section to establish this intriguing result and to indicate several other attractive properties. We use the methods of E. H. Lockwood.[3]

Begin with a polar-coordinate representation of points on the circumcircle of triangle ABC, where the circumcenter is taken as the origin, and the vertices of the triangle are $A(R, \alpha)$, $B(R, \beta)$, and $C(R, \gamma)$, with $P(R, \theta)$ any (variable) point on circle O (Fig. 327). Take $\overline{P_a'H}$ as the reflection of $\overline{PD'}$ in line $\overleftrightarrow{BC}$ (recall from the previous section that D is the midpoint of $\overline{HD'}$), and define $\overline{P_b'H}$ and $\overline{P_c'H}$ the reflections of $\overline{PE'}$ and $\overline{PF'}$ in $\overleftrightarrow{AC}$ and $\overleftrightarrow{AB}$. We shall derive formulas for the slopes of $\overleftrightarrow{P_a'H}$, $\overleftrightarrow{P_b'H}$, and $\overleftrightarrow{P_c'H}$.

---

[2] In the usual terminology, a hypocycloid of three cusps is the *envelope* of the Simson lines.
[3] E. H. Lockwood, "Simson's Line and Its Envelope," *Mathematical Gazette*, **37** (1953), pp. 124–125.

Figure 327

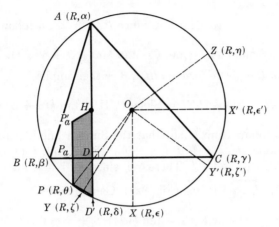

Locate the points X, X', Y, Y', and Z on the circumcircle, with polar coordinates as indicated in Fig. 327, such that X is the midpoint of arc $BD'C$, $\angle XOX' = \pi/2$, Y is the midpoint of arc PD', $\angle YOY' = \pi/2$, and where Z is such that $\angle ZOX' = \angle X'OY'$. Polar coordinates for these points may be easily deduced by the following sequence of steps:

$$\epsilon = \frac{\beta + \gamma}{2} + n_1\pi, \text{ where } n_1 = 0 \text{ or } 1.$$

$$\epsilon' = \epsilon + \frac{\pi}{2} = \frac{\beta + \gamma + n_2\pi}{2}, \text{ where } n_2 = 2n_1 + 1. \tag{72.1}$$

$$\epsilon' = \frac{\alpha + \delta}{2} + n_3\pi, \text{ where } n_3 = 0 \text{ or } 1. \tag{72.2}$$

Therefore, from (72.1) and (72.2) one obtains δ in terms of α, β, γ:

$$\delta = -\alpha + \beta + \gamma + n_4\pi, \text{ where } n_4 = n_2 - 2n_3 = 2(n_1 - n_3) + 1.$$

$$\zeta = \frac{\theta + \delta}{2} + n_5\pi, \text{ where } n_5 = 0 \text{ or } 1.$$

$$\zeta' = \zeta + \frac{\pi}{2} = \frac{\theta + \delta + n_6\pi}{2}, \text{ where } n_6 = 2n_5 + 1.$$

$$\zeta' = \frac{\theta - \alpha + \beta + \gamma + n_7\pi}{2}, \text{ where } n_7 = n_4 + n_6 = 2(n_1 - n_3 + n_5 + 1). \tag{72.3}$$

$$\epsilon' = \frac{\eta + \zeta'}{2} + n_8\pi, \text{ where } n_8 = 0 \text{ or } 1. \tag{72.4}$$

From (72.1) and (72.4), $\beta + \gamma + n_2\pi = \eta + \zeta' + 2n_8\pi$. From (72.3) one obtains the equation

$$\beta + \gamma + n_2\pi = \eta + \tfrac{1}{2}(\theta - \alpha + \beta + \gamma + n_7\pi) + 2n_8\pi.$$

Thus,

$$\eta = \tfrac{1}{2}(\alpha + \beta + \gamma - \theta) + m\pi, \tag{72.5}$$

where $m = n_2 - \frac{1}{2}n_7 - 2n_8 =$ integer. Since $\overleftrightarrow{P_a'H} \parallel \overleftrightarrow{OZ}$, it follows that

$$\text{Slope } \overleftrightarrow{P_a'H} = \tan \eta = \tan \tfrac{1}{2}(\alpha + \beta + \gamma - \theta).$$

The symmetry of the above formula in α, β, γ then implies

$$\text{Slope } \overleftrightarrow{P_a'H} = \text{Slope } \overleftrightarrow{P_b'H} = \text{Slope } \overleftrightarrow{P_c'H} = \tan \tfrac{1}{2}(\alpha + \beta + \gamma - \theta). \quad \textbf{(72.6)}$$

It will then follow from (72.6) that the Simson line is *parallel* to $\overleftrightarrow{P_a'H}$ and, therefore, has slope $\tan \frac{1}{2}(\alpha + \beta + \gamma - \theta)$. First, note that (72.6) implies the collinearity of P_a', P_b', and P_c'. Therefore, since P_a is the midpoint of $\overline{PP_a'}$ and P_b is that of $\overline{PP_b'}$, $\overleftrightarrow{P_aP_b} \parallel \overleftrightarrow{P_a'P_b'}$ (Fig. 328). Consequently,

$$\text{Slope } \overleftrightarrow{P_aP_b} = \tan \tfrac{1}{2}(\alpha + \beta + \gamma - \theta). \quad \textbf{(72.7)}$$

Figure 328

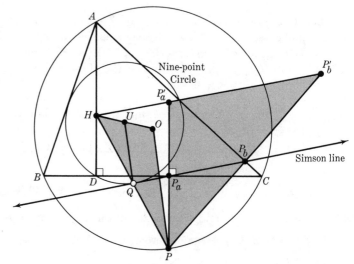

(The reader might notice that this reasoning proves the claim that P_a, P_b, and P_c are collinear.)

Also observe that if Q is the midpoint of $\overline{HP}$, then (from the preceding section) Q lies on the nine-point circle U of triangle ABC. But since P_a is the midpoint of side $\overline{PP_a'}$ in triangle PHP_a', $\overleftrightarrow{P_aQ} \parallel \overleftrightarrow{P_a'H}$, and therefore, the *Simson line passes through* Q. Moreover, $\overleftrightarrow{UQ} \parallel \overleftrightarrow{OP}$ since, in triangle HOP, Q is the midpoint of side $\overline{HP}$ and U is the midpoint of side $\overline{HO}$.

Recall that a hypocycloid of three cusps has parametric equations of the form

$$\begin{cases} x = \dfrac{2a}{3} \cos t + \dfrac{a}{3} \cos 2t, \\[2mm] y = \dfrac{2a}{3} \sin t - \dfrac{a}{3} \sin 2t, \end{cases} \quad \textbf{(72.8)}$$

where $2a$ is the diameter. The point $T[x, y]$ (in rectangular coordinates) will trace the curve exactly once as the parameter t ranges from 0 to 2π (Fig. 329). It can be easily proved from the equations (72.8) and by the use of calculus that the family

Figure 329

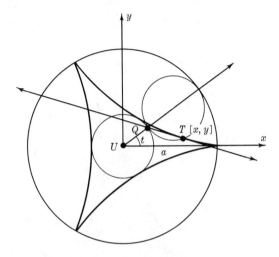

of tangents of such a curve *consists of the lines which pass through* $Q(\tfrac{1}{3}a, t)$ (polar coordinates with center U as origin) *and have slope* $-\tan \tfrac{1}{2}t$, $0 \leq t \leq 2\pi$.

On the circumcircle of triangle ABC, consider the points

$$C_1\left(R, \frac{\alpha + \beta + \gamma}{3}\right), \qquad C_2\left(R, \frac{\alpha + \beta + \gamma + 2\pi}{3}\right), \qquad C_3\left(R, \frac{\alpha + \beta + \gamma + 4\pi}{3}\right),$$

and choose a new coordinate system (x', y') by a rotation through the angle $\tfrac{1}{3}(\alpha + \beta + \gamma)$. (The ray $\overrightarrow{OC_1}$ then becomes the positive x' axis.) The new polar coordinates of A, B, and C will be (R, α'), (R, β'), and (R, γ'), respectively, where

$$\alpha' = \frac{2\alpha - \beta - \gamma}{3}, \qquad \beta' = \frac{-\alpha + 2\beta - \gamma}{3}, \qquad \gamma' = \frac{-\alpha - \beta + 2\gamma}{3}.$$

Note that $\alpha' + \beta' + \gamma' = 0$. Then if $P(R, t)$ be a variable point on circle O relative to the new coordinate system, by (72.7), Slope $\overleftrightarrow{P_a P_b} = \tan \tfrac{1}{2}(\alpha' + \beta' + \gamma' - t) = -\tan \tfrac{1}{2}t$.

If we now translate to the nine-point center U as origin, the slope of the Simson line remains $-\tan \tfrac{1}{2}t$, and by preceding comments it passes through the point Q (since $\overleftrightarrow{UQ} \parallel \overleftrightarrow{OP}$), whose polar coordinates are now $(\tfrac{1}{2}R, t)$. Setting $\tfrac{1}{3}a = \tfrac{1}{2}R$, we have found:

72.9 Theorem: The envelope of the Simson lines of a triangle is a hypocycloid of three cusps whose center is the nine-point center of the triangle and whose diameter is three times the circumradius. Moreover, if the lines $\overleftrightarrow{AO}$, $\overleftrightarrow{BO}$, and $\overleftrightarrow{CO}$

make angles α, β, and γ with the horizontal, then the cusps of the hypocycloid occur at

$$C_1\left(\frac{3R}{2}, \frac{\alpha + \beta + \gamma}{3}\right), \qquad C_2\left(\frac{3R}{2}, \frac{\alpha + \beta + \gamma + 2\pi}{3}\right), \qquad C_3\left(\frac{3R}{2}, \frac{\alpha + \beta + \gamma + 4\pi}{3}\right)$$

indicated by their polar coordinates with the nine-point center as origin (see Fig. 330).

Figure 330

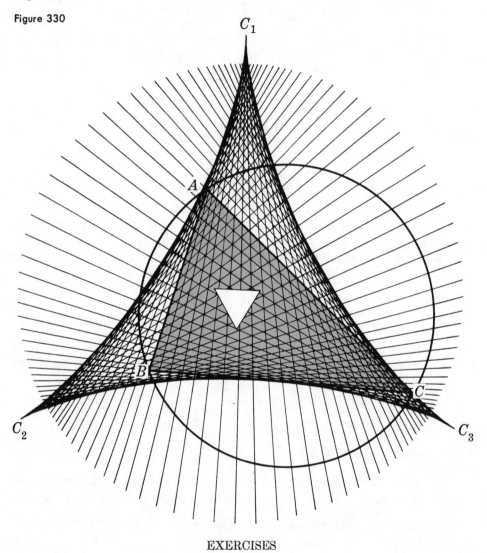

EXERCISES

1. Prove the assertion made above that the tangent to the hypocycloid (72.8) has slope $-\tan \tfrac{1}{2}t$ and passes through the point $Q[\tfrac{1}{3}a \cos t, \tfrac{1}{3}a \sin t] = Q(\tfrac{1}{3}a, t)$. *Hint:* You will need the identity $\tan \tfrac{1}{2}t = (1 - \cos t)/\sin t = \sin t/(1 + \cos t)$.

2. (a) Prove that the hypocycloid (72.8) is the envelope of the lines $\overleftrightarrow{Q_1Q_2}$, where Q_1 and Q_2 are the points whose polar coordinates are $(\frac{1}{3}a, t)$ and $(\frac{1}{3}a, \pi - 2t)$, respectively (Fig. 331). (b) As an experiment, apply the result of (a) by using a straightedge and by drawing the successive lines $\overleftrightarrow{Q_1Q_2}$ for which $t = 5°$, $10°$, $15°$, and so on.

★3. Figure 330 shows the Morley triangle of $\triangle ABC$ (Section 7). Using the method of assigning polar coordinates to points on the circumcircle O (as was done in proving Theorem 72.9), prove that the Morley triangle *is oriented in the same direction as the hypocycloid*. More specifically, prove that in general the altitudes of the Morley triangle are parallel to the tangents of the hypocycloid at the three cusp points C_1, C_2, and C_3. (For additional relationships between the triangle and its associated hypocycloid see E. H. Lockwood, *A Book of Curves* [17, pp. 72–79].)

Figure 331

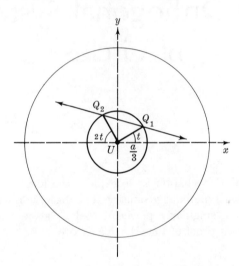

9

Plane Inversion
and Orthogonal Systems
of Circles

The main purpose of this chapter is to explore the ideas needed to understand the Poincaré model for hyperbolic geometry. The discussion will include the inversion transformation—a singularly powerful tool in geometric reasoning—which bears fruit in an elegant proof of Feuerbach's theorem.

73. Orthogonal Circles

The orthogonality of curves is a vital concept in engineering mathematics and plays a key role in the study of geometry. In the case of circles the synthetic method produces a beautiful theory as we shall see.

If two arcs AB and BC meet at B and are oriented away from the common endpoint B, we may term the configuration a **curvilinear angle,** denoted

$$\overset{\frown}{\angle ABC}.$$

The orientation of the *tangents* of those arcs is uniquely determined, and this then leads to a unique **associated Euclidean angle.** Thus, in Fig. 332 angle $A'BC'$ is the angle associated with curvilinear angle ABC. A natural measure for curvilinear angles is provided by the definition

$$\overset{\frown}{\angle ABC} = \text{Measure } \overset{\frown}{\angle ABC} = \angle A'BC',$$

where $\overline{\angle A'BC'}$ is the Euclidean angle associated with $\overset{\frown}{\angle ABC}$, and of course,

$\angle A'BC'$ denotes the measure of that angle as introduced previously. It is then natural to define **congruent curvilinear angles** as curvilinear angles having the same measure, and **right curvilinear angles** as those having measure $\pi/2$. Note that the theory of curvilinear angles includes the usual Euclidean angle theory as a special case.

The following definition introduces the main concept we want to consider:

Figure 332

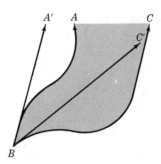

73.1   Definition: Two curves are said to be **orthogonal at a point** if they intersect at that point and if the curvilinear angles thus formed are right angles. If two intersecting curves are orthogonal at every point of intersection, they are said to be **orthogonal.** If curves ω_1 and ω_2 are orthogonal we write

$$\omega_1 \perp \omega_2.$$

It goes without saying that two intersecting curves can be orthogonal at one point of intersection without being orthogonal elsewhere. However, the curves represented by the equations $y = \sqrt{2} \sin x$ and $y = \sqrt{2} \cos x$ in the xy plane provide an example of a pair of curves which are othogonal at all points of intersection, and hence are othogonal. It is easy to see that intersecting *circles* form a pair of congruent angles at the two points of intersection (Fig. 333). Therefore, if two circles are orthogonal at either point of intersection, *they are orthogonal.*

In the case of orthogonal circles a few characteristic properties may be quickly derived. Let circles $[A, a]$ and $[B, b]$ meet orthogonally at P and Q (Fig. 334). Then it is immediately clear that

(a)   The tangent at P (or Q) to either one of the circles passes through the center of the other.

(b)   The radii $\overline{AP}$ and $\overline{BP}$ are perpendicular.

(c) $AB^2 = a^2 + b^2.$

The reader should verify also that any one of the above conditions (a), (b), or (c) is enough to guarantee that circles $[A, a]$ and $[B, b]$ are orthogonal.

A very important theorem now follows.

73.2   Theorem: If a line intersects each of two circles and passes through the center of one of them, then the circles are orthogonal if and only if the points of intersection form a harmonic sequence. Specifically, if line l meets circles $[A, a]$ and $[B, b]$ in the points X, Y and U, V, respectively, and passes through A, then $[A, a] \perp [B, b]$ implies $[XY, UV] = -1$, and conversely.

Figure 333   **Figure 334**

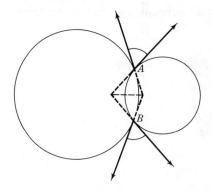   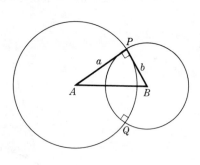

PROOF: First, suppose we know that circle $[A, a]$ is orthogonal to circle $[B, b]$ at P and Q (Fig. 335). Hence by Corollary 66.4,

$$AP^2 = AU \cdot AV.$$

Since $AP = AY$,

$$AY^2 = AU \cdot AV,$$

and therefore $[XY, UV] = -1$ (Theorem 63.5).

Conversely, let the condition $[XY, UV] = -1$ be satisfied for some line intersecting two circles $[A, a]$ and $[B, b]$, with A the midpoint of $\overline{XY}$. We must first prove that the circles intersect. However, this can be observed from the fact that either U or V must fall interior to segment $\overline{XY}$, and thus to circle $[A, a]$. Hence, assume that P lies on both $[A, a]$ and $[B, b]$ as shown in Fig. 336. Then

$$\begin{aligned} a^2 = AY^2 &= AU \cdot AV \\ &= \text{Power } (A) \\ &= AB^2 - b^2. \end{aligned}$$

Therefore, $a^2 + b^2 = AB^2$ and $[A, a] \perp [B, b]$. ∎

Figure 335   **Figure 336**

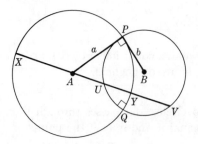   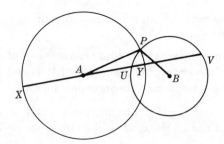

73.3 *Example:* Given a circle and two points not in line with the center, find a construction for the circle which passes through the two given points and orthogonal to the given circle.

Solution: Suppose the given points are P and Q, and A is the center of the given circle. Determine the ends X and Y of the diameter through P, and then locate R on $\overleftrightarrow{XY}$ such that

$$[XY, PR] = -1.$$

(Fig. 337 shows the appropriate straightedge construction.) By the previous theorem, every circle through P and R will be orthogonal to circle A, so in particular, the circle $[P, Q, R]$ will be the desired circle. ∎

Figure 337

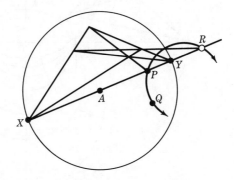

The above example will be extremely important later when we explore the model for hyperbolic geometry. The reader might find it an interesting challenge to solve Example 73.3 by the analytic method.

74. Coaxial Systems

Theorem 73.2 shows that if P, Q, U, and V are four points lying on line l such that $[PQ, UV] = -1$, then *any circle passing through* U *and* V *will be orthogonal to the circle ω whose diameter is* $\overline{PQ}$. Thus we have an entire *family* $\mathfrak{F}$ of circles orthogonal to ω, namely, those having $\overline{UV}$ as common chord (Fig. 338). The centers of all these circles will lie on the perpendicular bisector m of $\overline{UV}$.

Now suppose we locate a different pair of points (P', Q') such that $[P'Q', UV] = -1$. Then the circle ω' having $\overline{P'Q'}$ as diameter will be orthogonal *to any circle passing through* U *and* V, and hence to any member of $\mathfrak{F}$, for the same reason as before (Fig. 339). There is an infinity of pairs (P', Q') for which $[P'Q', UV] = -1$; thus we conclude that the family $\mathfrak{F}$ of circles considered above (with centers on m) possesses a *complementary* family $\mathcal{G}$ (with centers on l) such that *every circle of* $\mathcal{G}$ *is orthogonal to every circle of* $\mathfrak{F}$ (Fig. 340).

It is to be noted that the two families $\mathfrak{F}$ and $\mathcal{G}$ have distinctly different characters: Those of $\mathfrak{F}$ all have the points U and V in common, while no pair of circles of $\mathcal{G}$ have any points in common. While they do have their centers on line l, this property does not characterize the family $\mathcal{G}$, for certainly there are circles whose centers lie on l which are not orthogonal to the members of $\mathfrak{F}$ and therefore do not

Figure 338

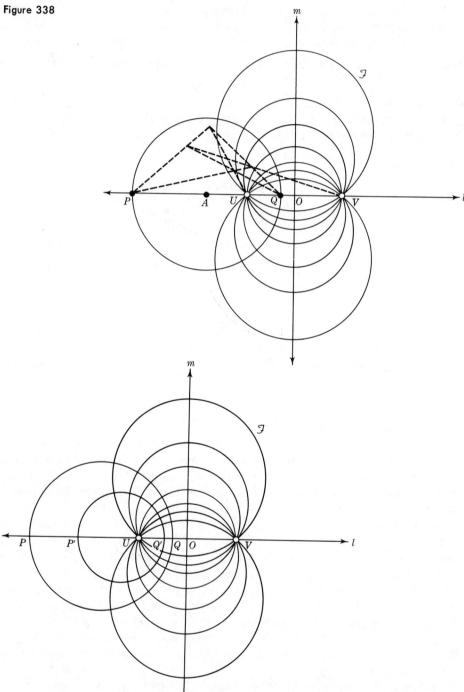

Figure 339

belong to G. The topic of the next section neatly solves the problem of character-izing the families F and G in a unified manner.

NOTE: A degenerate case occurs when the members of F have l as a common tangent (regarded as the limiting case as $U, V \to O$ above). It is clear that in this case the members of G have m as common tangent.

Figure 340

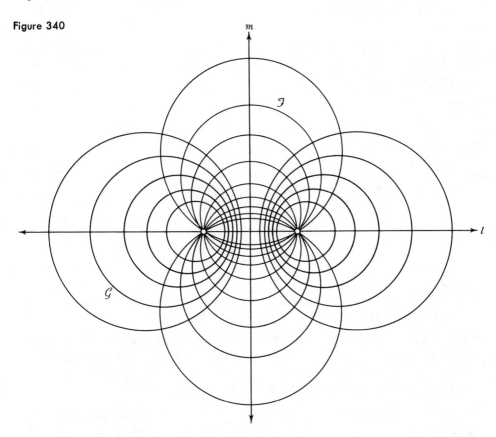

75. Radical Axis

The reader may recall a previous exercise where he was asked to prove that the powers of a point on a line containing the common chord of two circles with respect to those two circles are equal (Exercise 13, Section 66). One might well ask what becomes of this property if the circles assume a nonintersecting position, as suggested by Fig. 341.

The answer emerges from a consideration of the following problem: *To find the locus of points whose powers with respect to two circles are equal.* The proof of the general result (which contains the common-chord property as a special case) depends on the existence of a point C on the line of centers $\overset{\leftrightarrow}{A B}$ of two given circles $[A, a]$

Figure 341

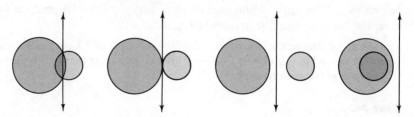

and $[B, b]$ such that the powers of C with respect to the two circles are equal, and the uniqueness of that point. But this is easily answered by converting to coordinates on $\overleftrightarrow{AB}$: Let $A = A[0]$, $B = B[d]$, and $C = C[x]$. The condition to be satisfied is

$$CA^2 - a^2 = CB^2 - b^2. \tag{75.1}$$

Figure 342

Thus, we seek the possible solutions of the equation:

$$x^2 - a^2 = (x - d)^2 - b^2$$
$$x^2 - a^2 = x^2 - 2dx + d^2 - b^2$$
$$2dx = a^2 - b^2 + d^2$$
$$x = \frac{a^2 - b^2 + d^2}{2d}, \qquad d \neq 0.$$

This shows that, provided the circles are not concentric (do not have the same centers), the desired point exists and is unique. ∎

75.2   Theorem: The locus of points whose powers with respect to two nonconcentric circles are equal is a line perpendicular to the line of centers of the two circles. That line is the line containing the common chord of the two circles if they intersect, and the common tangent if they touch.

PROOF: With the same notation as before, locate the point C of the locus on $\overleftrightarrow{AB}$ and let l be the perpendicular to $\overleftrightarrow{AB}$ at C (Fig. 343). If $P \in l$, then by the Pythagorean theorem

$$PA^2 - CA^2 = PC^2 = PB^2 - CB^2$$
$$PA^2 - PB^2 = CA^2 - CB^2.$$

From (75.1),

$$PA^2 - PB^2 = a^2 - b^2$$

or,

$$PA^2 - a^2 = PB^2 - b^2. \tag{75.3}$$

That is, the powers of P with respect to the two circles are equal. Conversely, suppose (75.3) holds; we must show that $P \in l$. Let Q be the foot of the perpendicular from P on $\overleftrightarrow{AB}$ (Fig. 344). Then

$$PA^2 - QA^2 = PQ^2 = PB^2 - QB^2.$$

Figure 343

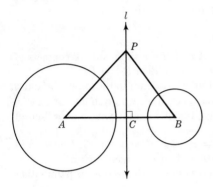

Figure 344

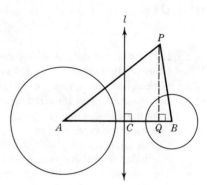

Hence, by (75.3),

$$QA^2 - QB^2 = PA^2 - PB^2 = a^2 - b^2$$

and therefore,

$$QA^2 - a^2 = QB^2 - b^2,$$

and Q is a point of the locus on $\overleftrightarrow{AB}$. Thus $Q = C$ and $P \in l$. This finishes the proof that line l is the desired locus, and the remaining statements in the theorem are obvious. ∎

75.4 Definition: The line whose existence was proved in Theorem 75.2 is called the **radical axis** of the two circles.

An answer to the question (Section 74) of how to characterize the families ℱ and 𝒢 emerges (Fig. 340): ℱ consists of the circles with m as line of centers and having l as radical axis, and 𝒢 *consists of the circles whose line of centers is* l *and having* m *as radical axis*. To see this, consider two circles $[A, a]$ and $[A', a']$ of 𝒢. These circles accordingly meet l at diametrically opposite pairs (P, Q) and (P', Q'), respectively, such that $[PQ, UV] = [P'Q', UV] = -1$ (Fig. 345). By Theorem 63.5 (Chapter 7) and Theorem 66.1 (Chapter 8),

$$OA^2 - a^2 = OP \cdot OQ = OU^2 = OP' \cdot OQ' = OA'^2 - a'^2.$$

Figure 345

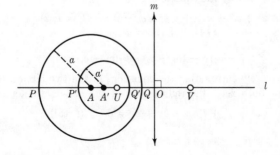

Therefore, the powers of O with respect to the two given circles of $\mathcal{G}$ are equal. Then O lies on the radical axis and line m must be that radical axis.

EXERCISES

1. If P lies on the radical axis of two circles, and A and B are the points of contacts of either of the tangents from P to each of the two circles, then $PA = PB$.
2. Explain how the concept of the radical axis and its relationship to the families $\mathcal{F}$ and $\mathcal{G}$ can be used to find the circle which is orthogonal to each of two given nonconcentric, nonintersecting circles and passing through a given point not lying on the line of centers of the given circles.
3. Circles $[A, a]$, $[B, b]$, and $[C, c]$ are given, $A \neq B$, $A \neq C$, and $B \neq C$. Under what conditions does there exist a circle orthogonal to each of the given circles?
4. The lines containing the common chords of three circles intersecting in pairs either coincide, are parallel, or concurrent.
5. **Orthogonality of circles treated analytically.** (a) Show that the circle

$$x^2 + y^2 + ax + by + c = 0$$

    is orthogonal to the circle $x^2 + y^2 = r^2$ if and only if

$$c = r^2.$$

    Hint: Do not use calculus. (b) Use this to prove Theorem 73.2 analytically. (c) Identify the one-parameter family of circles

$$x^2 + y^2 + tx + 1 = 0 \qquad (t < -2),$$

    and find the point to which the centers of the circles of this family converge as their radii tend to zero. (d) Generalize the condition in (a) to any pair of orthogonal circles. Specifically, prove that the circles

$$x^2 + y^2 + a_k x + b_k y + c_k = 0 \qquad (k = 1, 2)$$

    are orthogonal if and only if

$$a_1 a_2 + b_1 b_2 = 2c_1 + 2c_2.$$

6. **Power and radical axis treated analytically.** (a) Show that the power of the point $P[x_0, y_0]$ with respect to the circle whose equation is

$$x^2 + y^2 + ax + by + c = 0$$

    is given by the formula

$$F(x_0, y_0) = x_0{}^2 + y_0{}^2 + ax_0 + by_0 + c.$$

    (b) Prove Theorem 75.2 by the analytic method. (c) Show that the radical axis of the two circles represented by the equations

$$x^2 + y^2 + a_k x + b_k y + c_k = 0 \qquad (k = 1, 2)$$

    has the equation

$$(a_1 - a_2)x + (b_1 - b_2)y + (c_1 - c_2) = 0$$

    —the result of merely subtracting the equations of the two circles. (d) Show that if a and b are not both zero, the radical axis of the one-parameter family of circles

$$x^2 + y^2 + t(ax + by + c) = 0 \qquad (t \text{ real}),$$

    is the line

$$ax + by + c = 0.$$

Find the line of centers of the circles in this family.

7. Referring to Fig. 340, it is observed that the circles of $\mathcal{G}$ seem to "cluster" around the endpoints of the common chord of the family $\mathcal{F}$. More precisely, as the radii of the circles of $\mathcal{G}$ tend to zero the corresponding centers converge to one of the two endpoints, provided the two sides of m are taken separately. Show that, if $c \neq 0$, the one-parameter family of circles

$$x^2 + y^2 + t(ax + by + c) = 0, \qquad a^2 + b^2 \neq 0 \text{ and } t \text{ real}, \qquad (75.5)$$

is of the same type as the family $\mathcal{G}$ (no two members intersect) and find the two endpoints of the common chord of the family of circles orthogonal to those of (75.5). Show analytically that as the radius becomes small the centers converge to one of those endpoints. What happens if $c = 0$?

8. Using the result of Exercise 7, or by some other method, find the parameter equation of the family of circles orthogonal to the family (75.5). *Ans:* $abx^2 + aby^2 + b(bs + c)x + a(-as + c)y = 0$, s *real, provided* $a \neq 0$ *and* $b \neq 0$; $bx^2 + by^2 + 2bsx + 2cy = 0$ *if* $a = 0$, *and* $ax^2 + ay^2 + 2cx + 2asy = 0$ *if* $b = 0$.

★9. *Orthogonal systems of circles treated analytically.* (a) Show that the one-parameter family of circles

$$x^2 + y^2 + a(t)x + b(t)y + c(t) = 0, \qquad t \text{ real}, \qquad (75.6)$$

can represent a family of the type $\mathcal{F}$ or $\mathcal{G}$ only if $a(t)$, $b(t)$, and $c(t)$ are *linear* functions of t. That is, the above equation must have the special form

$$x^2 + y^2 + (at + a')x + (bt + b')y + (ct + c') = 0, \qquad (75.7)$$

where a, a', b, b', c, and c' are constants, but for a change of parameter. (*Hint:* Use the fact that any two members must have a common line of centers and a common radical axis.) (b) Using (a), find the conditions on the defining constants in order for (75.7) to represent a family of the type $\mathcal{F}$, $\mathcal{G}$, or the degenerate case mentioned in the Note of Section 74. *Ans: Define* $\triangle = (aa' + bb' - 2c)^2 - (a^2 + b^2)(a'^2 + b'^2 - 4c')$. *Then* (75.7) *represents a family of the type* $\mathcal{F}$, *the degenerate case, or* $\mathcal{G}$ *according as* $\triangle < 0$, $\triangle = 0$, *or* $\triangle > 0$.

★10. The reader may find it interesting to pursue the analogous theory of orthogonal families of *conics*. Show that if $\mathcal{F}$ is a family of *confocal* ellipse (ellipses having the same foci), then the family $\mathcal{G}$ of confocal hyperbolas having the same foci as $\mathcal{F}$ is such that each member of $\mathcal{F}$ meets each member of $\mathcal{G}$ orthogonally.

11. *Synthetic construction of the radical axis of two circles.* Let ω_1 and ω_2 be any two non-concentric, nonintersecting circles, and take ω_3 any circle cutting ω_1 at A_1, B_1 and ω_2 at A_2, B_2. The perpendicular to the line of centers of ω_1 and ω_2 from the point of intersection of $\overleftrightarrow{A_1B_1}$ and $\overleftrightarrow{A_2B_2}$ is the desired radical axis of ω_1 and ω_2. Prove. (H. Eves, *A Survey of Geometry* [9, p. 113].)

12. Find the locus of points the difference of whose powers with respect to two given circles is constant.

13. *Casey's power theorem.* The difference of the powers of a point with respect to two circles equals twice the product of the distance from the point to the radical axis and the distance between the centers of the given circles. Prove analytically. (For a synthetic proof, see Court [6, p. 211].)

76. Plane Inversion

To invert a number in arithmetic usually means to "take its reciprocal." A closely related idea in geometry is that of "inverting" a point. Suppose O is the origin and P is any other point. Then point Q may be located on ray $\overrightarrow{OP}$ such that the distance from Q to O is the *reciprocal* of the distance from P to O. This idea will be slightly generalized in a formal definition of the inversion transformation.

76.1 Definition: Let circle $[O, r]$ be given. To each point P in the plane distinct from O corresponds a unique point $P' = T(P)$ on the line OP such that, in magnitude and in sign,

$$OP' = \frac{r^2}{OP}.$$

The correspondence T defined in this manner is called an **inversion** with respect to the given circle. The point O, the number r, and the circle $[O, r]$ are, respectively, the **center of inversion, radius of inversion,** and **circle of inversion** (Fig. 346); P and P' are called **inverse points** with respect to the inversion.

Figure 346

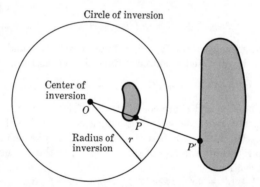

Circle of inversion

Center of inversion

O

Radius of inversion r

P

P'

We then have in T a transformation of the plane into itself whose domain and range is the *punctured plane*—the plane minus the point O. Note that as a point moves closer to the center of inversion its inverse moves farther away from that center, with all points on the inside of the circle of inversion being mapped to points outside, points on the outside mapped to the inside, and the points on the circle itself remaining fixed. As a point moves to the outside of the circle of inversion and its distance to the center becomes large, its inverse approaches the center.

It is to be noted that if P and P' are inverse points with respect to circle $[O, r]$ and $\overleftrightarrow{OP}$ meets that circle at M and N (Fig. 347), then $r^2 = ON^2 = OP \cdot OP'$ implies that $[MN, PP'] = -1$. Thus a *straightedge construction exists for the inverse of a point.* Our above comments are ramifications of the fact that if P and P' divide $\overline{MN}$ harmonically and P tends to the midpoint of $\overline{MN}$, then MP' becomes infinite.

Further properties of the inversion transformation are

(a) If (P, P') and (Q, Q') are inverse pairs, then lines $\overleftrightarrow{PQ}$ and $\overleftrightarrow{P'Q'}$ are antiparallel (Fig. 348), since $OP \cdot OP' = OQ \cdot OQ'$.

(b) By Property (a) the triangles POQ and $Q'OP'$ are similar. Therefore,

$$\angle OPQ = \angle OQ'P' \quad \text{and} \quad \angle OQP = \angle OP'Q'.$$

(c) Lines through O map into lines through O.

Figure 347

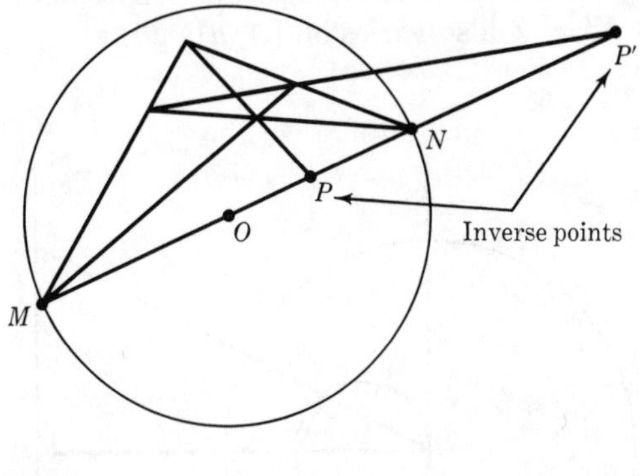

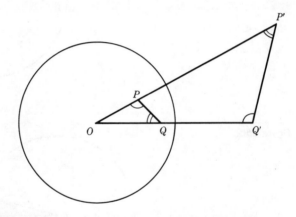

Figure 348

Inverse points

(d) A circle through O maps into a line that is parallel to the tangent to the given circle at O (Fig. 349). This follows from Property (a): As X varies on the circle $\angle X'P'O$ remains equal to $\angle OXP$, but $\overline{ZOXP}$, being inscribed in a semicircle, remains a right angle. Therefore X' varies on the perpendicular to $\overleftrightarrow{OP}$ at P'.

(e) A line not through O maps into a circle through O. This is the "inverse" of Property (d).

(f) A circle not through O maps into a circle not through O. (See Fig. 350 for the method of proof, left to the reader.)

A previously established property of orthogonal circles provides an easy proof of yet another property of the inversion transformation.

76.2 Theorem: A circle which is orthogonal to the circle of inversion is self inverse.

PROOF: In Fig. 351 let circle $[A, a]$ be orthogonal to the circle of inversion

$[O, r]$. Then if X lies on $[A, a]$, $\overrightarrow{OX}$ cuts $[A, a]$ again at Z and $[O, r]$ at Y such that, by Theorem 73.2,

$$OY^2 = OX \cdot OZ.$$

But $OY = r$, so $OX \cdot OZ = r^2$. Therefore, X and Z are inverse points. Hence, as X varies on circle $[A, a]$ the inverse point $X' = Z$ also varies on $[A, a]$. ∎

Figure 349

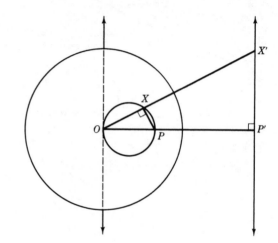

Figure 350

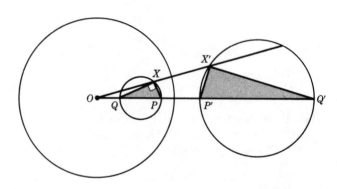

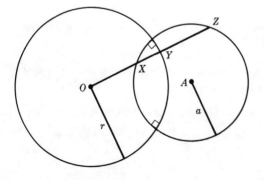

Figure 351

77. Invariants of the Inversion Mapping

In view of the definition of the inversion mapping, we are dealing with a non-linear transformation which certainly does not have the more common types of invariants, such as collinearity and distance. In a certain sense, however, it does preserve cross ratio, and it is **conformal** (angle preserving).

Since the inversion mapping is not linear it would not make sense to examine the invariance of cross ratio as originally defined since it would not have any meaning for noncollinear points. If however one removes the restriction of collinearity in defining cross ratio, an invariant for the inversion transformation emerges.

77.1 Definition: If A, B, C, and D be four distinct points in the plane, collinear or not, the **planar cross ratio** of those four points taken in the order given is defined to be the number

$$\frac{|AC||BD|}{|AD||BC|}.$$

To distinguish from the usual cross ratio this number will be denoted by the symbol $R[AB, CD]$.

From its definition $R[AB, CD] > 0$ and $R[AB, CD] \neq 1$ for any four points A, B, C, and D. Also one observes the property

$$R[AB, CD] = \frac{1}{R[BA, CD]},$$

and the six permutations of B, C, and D as before lead to the six possible planar cross ratios, except here we do not have the specific values λ, $1/\lambda$, $1 - \lambda$, $1/(1 - \lambda)$, $\lambda/(\lambda - 1)$, $1 - 1/\lambda$ (as we did for the "linear" cross ratio defined previously).

77.2 Theorem: Planar cross ratio is an inversive invariant.

PROOF: Suppose $A \to A'$, $B \to B'$, $C \to C'$, and $D \to D'$ indicates the inversion of four points with respect to the circle $[O, r]$ (Fig. 352). Since $\triangle AOC \sim \triangle C'OA'$ and $\triangle AOD \sim \triangle D'OA'$ [Property (a), Section 76] then, in magnitude only[1]

$$\frac{AC}{OA} = \frac{A'C'}{OC'}, \qquad \frac{AD}{OA} = \frac{A'D'}{OD'}.$$

Therefore,

$$\frac{AC}{AD} = \frac{OD'}{OC'} \cdot \frac{A'C'}{A'D'}. \tag{77.3}$$

Similarly, using $\triangle BOC \sim \triangle C'OB'$ and $\triangle BOD \sim \triangle D'OB'$,

$$\frac{BD}{BC} = \frac{OC'}{OD'} \cdot \frac{B'D'}{B'C'}. \tag{77.4}$$

Multiplying equations (77.3) and (77.4) then yields

$$R[AB, CD] = R[A'B', C'D']. \ \blacksquare$$

---

[1] This means that the absolute values of all distances would be understood if any three of the points were collinear.

Figure 352

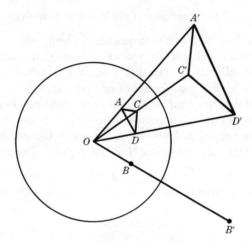

The general conformal property of the inversion transformation can be proved analytically (see Exercise 10, below). The special case of angles whose sides are circles or lines can be handled synthetically, however. Let us first examine an angle one of whose sides is a line passing through O, the center of inversion, and the other an arc of a circle passing through O. Referring to Fig. 353, if curvilinear angle ABC is mapped by inversion to curvilinear angle $A'B'C'$, the sides of the latter will be

Figure 353

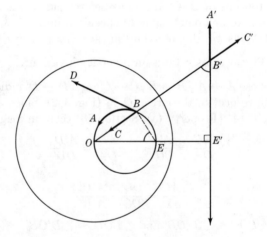

rays. By one of the inscribed angle theorems therefore, with $\overleftrightarrow{BD}$ tangent to circle $[E, B, A]$,

$$\widehat{\angle ABC} = \angle DBC = \angle BEO = \angle E'B'O = \angle A'B'C' = \widehat{\angle A'B'C'},$$

and the assertion is proved for this relatively simple case.

Now if $\triangle ABC$ has a line through O as one of its sides and an arc of a circle not through O as the other, its image, $\overset{\frown}{\angle A'B'C'}$, will be a curvilinear angle also of that type (Fig. 354). With $\overrightarrow{BD}$ and $\overrightarrow{B'E}$ as the appropriate tangents we have

$$\begin{aligned}
\overset{\frown}{\angle ABC} &= \angle DBC \\
&= \angle BFG \\
&= \angle BFO - \angle GFO \\
&= \angle F'B'O - \angle F'G'O \\
&= \angle B'F'G' \\
&= \angle EB'G' \\
&= \overset{\frown}{\angle A'B'C'}.
\end{aligned}$$

Figure 354

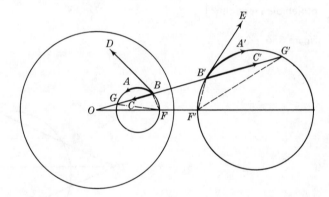

Since the measure of any other curvilinear angle with circles or lines for sides can be expressed as either a sum or difference of two angles of the types considered in the previous two cases (for example, see Fig. 355), the general case follows. Thus:

77.5 Theorem: Curvilinear angle measure for angles whose sides are circles or lines is an inversive invariant.

Figure 355

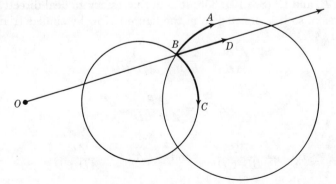

78. Applications of Inversion: Ptolemy, Steiner, and Feuerbach

To illustrate the ease with which many important results may be proved using inversion, we discuss here several classical theorems. The method usually works as follows: A desired property concerning a particular configuration is inverted in such a manner that it may be easily "read off" from a familiar property of the inverted figure.

The first theorem is a generalization of Ptolemy's theorem (discussed in Section 3; the reader might at this time reread that section). The desired property may be stated: *If* A, B, C, *and* D *are any four points in the plane, then in magnitude only,*

$$AB\cdot CD + AC\cdot BD \ge AD\cdot BC, \tag{78.1}$$

where equality occurs if and only if A, B, C, *and* D *are concyclic or collinear in the order* A, B, D, *and* C, *or cyclic rearrangement thereof.* [Inequality (80.1) is called the **ptolemaic inequality.**]

Figure 356

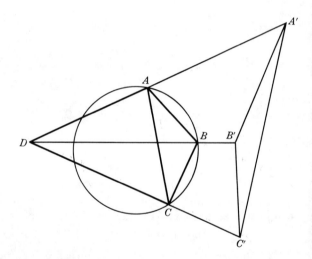

To prove this by the method of inversion take D as center of inversion, with any convenient radius r of inversion, say $r = 1$, and let A', B', and C' denote the inverses of A, B, and C. (See Fig. 356; it is not necessary to deal directly with the circle of inversion, so it is omitted from the figure.) Now by similar triangles, and since $A'D = 1/AD$,

$$\frac{A'B'}{AB} = \frac{A'D}{BD} = \frac{1}{AD\cdot BD}.$$

That is

$$A'B' = \frac{AB}{AD\cdot BD}.$$

Similarly,

$$B'C' = \frac{BC}{BD\cdot CD} \quad \text{and} \quad A'C' = \frac{AC}{AD\cdot CD}.$$

These equations clearly hold even if D is collinear with two or more of the points A, B, and C. But in triangle $A'B'C'$, $A'B' + A'C' \geqq B'C'$. Hence,

$$\frac{AB}{AD \cdot BD} + \frac{AC}{AD \cdot CD} \geqq \frac{BC}{BD \cdot CD},$$

which is easily simplified to the desired inequality (78.1). It is obvious that the equality occurs if and only if $A'B' + A'C' = B'C'$, which in turn occurs if and only if A', B', and C' are collinear and $(B'A'C')$. This happens whenever the circle (or line) passing through A, B, and C passes through the center of inversion D, which then finishes the proof. ∎

The ptolemaic inequality may be used to provide an elegant proof of a result which is one of the more surprising properties of the triangle. It is hard to resist the temptation of including this here, even though it has no connection with the inversion mapping, except through the proof of (78.1) itself. Pierre Fermat (1601–1665) proposed the problem of locating a point P in a given acute-angled triangle ABC whose distances from A, B, and C have the smallest possible sum. Any attempt to solve this by means of calculus would most probably end in considerable frustration.

Consider however the following neat solution: We assume that the angles of the triangle are each less than $2\pi/3$. In Fig. 357, point C' is located on the opposite side of $\overleftrightarrow{AB}$ as C such that $\triangle ABC'$ is equilateral. Then for P any point in the plane we have, by the ptolemaic inequality (in magnitude only)

$$PA \cdot BC' + PB \cdot AC' \geqq PC' \cdot AB.$$

Figure 357

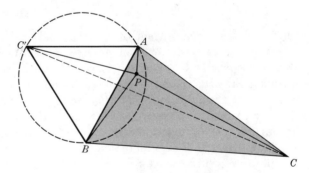

Since $BC' = AC' = AB$ this reduces to

$$PA + PB \geqq PC'. \tag{78.2}$$

Add PC to both sides and apply the triangle inequality:

$$PA + PB + PC \geqq PC' + PC \geqq C'C. \tag{78.3}$$

If P lies on both the circle $[A, B, C']$ and the segment $\overline{C'C}$, equality prevails in (78.3) (since in that case the order of the points on $[A, B, C']$ would be P, A, C', B), and such a point would give us a minimum value for $PA + PB + PC$. If the same argument be applied to the equilateral triangles $AB'C$ and $A'BC$ on the remaining two sides of $\triangle ABC$, we find that the desired minimum occurs *when* P *lies on all three lines* $\overleftrightarrow{AA'}$, $\overleftrightarrow{BB'}$, *and* $\overleftrightarrow{CC'}$, that is, *when* P *is the Fermat point of* $\triangle ABC$ (see Exercise 5, Section 62), and moreover that minimum *is the length of any one of the congruent segments* $\overline{AA'}$, $\overline{BB'}$, *or* $\overline{CC'}$. ∎

NOTE: A slight variation in the above argument would prove that lines $\overleftrightarrow{AA'}$, $\overleftrightarrow{BB'}$, and $\overleftrightarrow{CC'}$ are concurrent.

The next application of inversion involves what is known as *Steiner's porism*. Given two nonintersecting circles ω_1 and ω_2—one inside the other—begin with circle A_1 tangent to the given circles as shown in Fig. 358, and inscribe circles

Figure 358

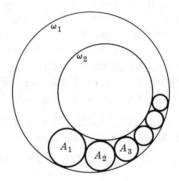

$A_2, A_3, A_4, \cdots$ each tangent to ω_1 and ω_2 and to their predecessors. A **Steiner chain** of circles occurs when, for some n, circle A_n is tangent to A_1, and thus circle A_{n+1} = circle A_1, circle A_{n+2} = circle A_2, and so on (the chain repeats itself). The porism occurs in Steiner's claim that if one such chain exists infinitely many exist, with the existence of the chain *independent of the position of the initial circle* A_1.

Now this is all perfectly self evident in the case when ω_1 and ω_2 are concentric circles. Thus, if an inversion can be found which maps ω_1 and ω_2 into a pair of concentric circles with the center of inversion not lying on any of the circles $A_1, A_2, \cdots$, the problem will be solved. Since it may be assumed that ω_1 and ω_2 are not concentric, as in Fig. 359(a), the radical axis of ω_1 and ω_2 exists and is the line of centers of a family $\mathfrak{F}$ of circles orthogonal to ω_1 and ω_2 (see concluding comments in Section 75). The circles of $\mathfrak{F}$ have a common chord $\overline{UV}$, with, say U, lying inside both ω_1 and ω_2. Now take U as center of inversion and any convenient radius, and invert the configuration to one having the appearance of Fig. 359(b); the images of all the circles of $\mathfrak{F}$ are lines since the members of $\mathfrak{F}$ all pass through the center of inversion. Because orthogonality is an inversive invariant, and since a line is orthogonal to a circle if and only if it passes through its center, ω_1 and ω_2 are mapped into concentric circles ω_1' and ω_2'. Steiner's statement may then be immediately inferred from the resulting configuration. ∎

Figure 359

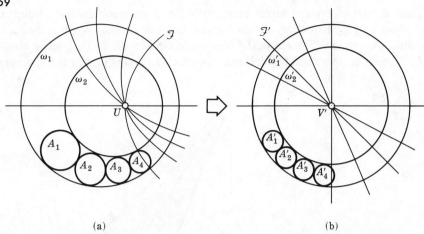

(a)   (b)

The final application will establish *Feuerbach's theorem* (Section 8): *The nine-point circle of any triangle is tangent to the four tritangent circles of the triangle.* Let the incircle I and excircle I_a be determined in triangle ABC (Fig. 360), and let $\overleftrightarrow{RS}$ be the remaining common tangent to circles I and I_a. Take K as the intersection of $\overleftrightarrow{BC}$ and $\overleftrightarrow{RS}$. It is easy to see that $\overline{AK}$ is the angle bisector at A ($\overrightarrow{KI}$ and $\overrightarrow{KI_a}$ bisect the vertical angles at K and are therefore collinear). Hence $[AK, II_a] = -1$ (the result of Exercise 9, Section 60). Since the perpendiculars from A, I, K, and I_a to $\overleftrightarrow{BC}$ meet $\overleftrightarrow{BC}$ at D, P, K, and Q, it follows that

$$[DK, PQ] = -1.$$

Recall that $|BP| = |QC| = s - b$. Therefore, the midpoint L of side $\overline{BC}$ is also the midpoint of $\overline{PQ}$, and by Theorem 63.5

$$LP^2 = LD \cdot LK,$$

in magnitude and in sign. Apply inversion with respect to circle $[L, |LP|]$. Then D maps to K, line $\overleftrightarrow{KD}$ maps to itself, and since the nine-point circle passes through L, it maps into the line l which passes through K and is parallel to the tangent t of the nine-point circle at L. Thus line l is parallel to $\overleftrightarrow{EF}$ (see Exercise 2, Section 71). Recall that $\overleftrightarrow{EF}$ is antiparallel to $\overleftrightarrow{BC}$ with respect to $\overleftrightarrow{AB}$ and $\overleftrightarrow{AC}$. But so is $\overleftrightarrow{RS}$. Therefore,

$$l \,\|\overleftrightarrow{EF}\| \,\overleftrightarrow{RS},$$

and since l and $\overleftrightarrow{RS}$ both pass through K,

$$l = \overleftrightarrow{RS}.$$

That is, the nine-point circle maps into $\overleftrightarrow{RS}$, a common internal tangent to the circles I and I_a. However, circles I and I_a are self-inverse, being orthogonal to the circle of inversion. Therefore, the nine-point circle must be tangent to circles I and I_a. Since it is immaterial which excircle we choose, this proves the theorem. ∎

Figure 360

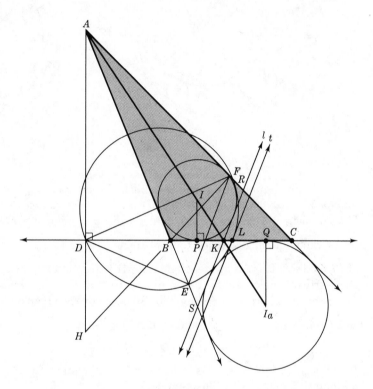

EXERCISES

1. Show that if ω_1 is mapped to line l under inversion with respect to circle ω_2, then l is the radical axis of ω_1 and ω_2.
2. Let l be the radical axis of two circles ω_1 and ω_2, and suppose ω_1, ω_2 are inverted into circles ω_1', ω_2' and l remains fixed under the inversion. Show that l is also the radical axis of the image circles ω_1' and ω_2'.
3. Show that the center of a circle need not map into the center of the inverted circle under an inversion.
4. Find the following analytic form for an inversion with respect to the circle $x^2 + y^2 = r^2$:

$$\begin{cases} x' = \dfrac{r^2 x}{x^2 + y^2}, \\[2mm] y' = \dfrac{r^2 y}{x^2 + y^2}. \end{cases} \tag{78.4}$$

Verify analytically that the circle $x^2 + y^2 + ax = 0$, $a \neq 0$, inverts into a line parallel to the tangent to this circle at $(0, 0)$.

5. By using the result of the preceding exercise and that of Exercise 5, Section 75, give an analytic proof of the assertion that any circle orthogonal to the circle of inversion is self inverse.

6. In the above proof of Feuerbach's theorem the actual locations of the points of contact of the nine-point circle with circles I and I_a are readily apparent. Explain how they may be located.

7. **Peaucellier's cell.** A linkage for drawing a straight line was invented by a French army officer, A. Peaucellier (1832–1913). Depicted in the sketch in Fig. 361(a), it is based on an application of inversion in Fig. 361(b). Point O is fixed and $OA = OB$, with $\Diamond APBP'$ a rhombus. Show that P and P' are inverse with respect to the circle of inversion $[O, r]$, where $r = \sqrt{OA^2 - AP^2}$. Thus if P varies on the circle $[D, DO]$, P' will describe a line segment.

Figure 361

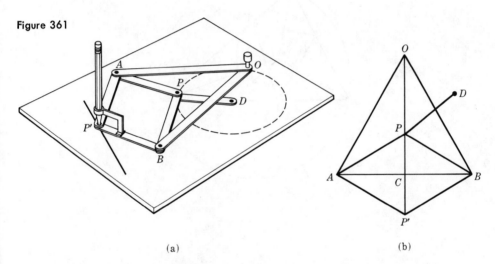

(a)   (b)

8. Let circle $[A, a]$ map into circle $[B, b]$ under inversion with respect to circle $[C, c]$. Show that if $d = AC > 0$ and $a > d$,

$$b = \frac{ac^2}{a^2 - d^2}.$$ **(78.5)**

Hint: Consider this a problem in analytic geometry. If $\overrightarrow{CA}$ is taken as the positive x axis, C the origin, then the point $P[x]$ inverts to the point $P'[c^2/x]$. Locate $D[d + a]$, $E[d - a]$ and find the coordinates of D' and E'. *WARNING:* B is not in general the inverse of A.

⋆9. **The Euler formula via inversion.** (a) Prove that the inverse of the circumcircle O of a triangle ABC with respect to the incircle I, as circle of inversion, is the nine-point circle of the triangle XYZ determined by the points of contact of I with the sides of ABC; (b) prove, by inversion, Euler's formula $d^2 = R^2 - 2Rr$; (c) state and prove by inversion the analogous propositions regarding the excircles [formulas (71.5)].

Hint: Note that $\overleftrightarrow{AA'}$ as shown in Fig. 362 is the perpendicular bisector of $\overline{YZ}$. (Court [6, p. 243]; used by permission.)

★10. Recall a formula in calculus for the (undirected) angle ψ between the radius vector and the tangent vector to a curve whose polar equation is $r = f(\theta)$:

$$\cot \psi = \frac{f'(\theta)}{f(\theta)}, \qquad 0 < \psi < \pi.$$

Using this, derive a proof of the conformal property of inversion. *Hint:* Take the origin as the center of inversion and show that $r = f(\theta)$ is mapped to $r_1 = a^2/f(\theta)$, where a is the radius of the inversion, and that $\psi_1 = \pi - \psi$, where ψ_1 is the angle between the radius vector and the tangent vector to the image curve.

Figure 362

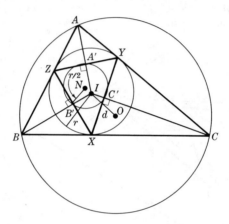

10

A Study
of Elliptic Geometry
from the Sphere

It is possible to develop axiomatically all of elliptic geometry, including the spherical trigonometry to be derived in this chapter.[1] But the same theory can also be obtained, and much more readily, by using the sphere as a *model* for the axioms. Following the precedent set for the previous development of Euclidean geometry, we shall be content to derive the further properties of elliptic geometry, and of hyperbolic geometry in the next chapter, from models.

79. Description of the Model

Let ε be the elliptic plane as previously introduced formally from Axioms 1–12. It is now our purpose to obtain a concrete realization of the plane ε. A model for ε will undoubtedly be a sphere in three-dimensional Euclidean space, since our axioms were originally stated with the geometry of the sphere in mind. However, let us be very explicit at this point.

Let $\mathcal{S}$ be a sphere in three-dimensional Euclidean space, with O as center and $r > 0$ as radius. Since no loss in generality will result, we shall assume that $r = 1$. Now take as the "points" the points lying on $\mathcal{S}$ and as "lines," the great circles.

---

[1] See H. E. Wolfe, *Non-Euclidean Geometry* [30] for a readable axiomatic development of both elliptic and hyperbolic geometry. The methods of projective geometry permit a *unified* treatment of the elliptic, hyperbolic, and Euclidean geometries. See C. E. Springer, *Geometry and Analysis of Projective Spaces* [25, Chapter 9] for a complete description of this method, as well as H. S. M. Coxeter, *Non-Euclidean Geometry* [5, Chapter 8].

Since a great circle is the intersection of S with a plane passing through O, a "line" is uniquely determined by each pair of "points" (A, B) which are not in line with O (are not *antipodal*). However, if $\overleftrightarrow{AB}$ is a diameter, then infinitely many "lines" pass through A and B. (See Figs. 363 and 364.)

Figure 363   **Figure 364**

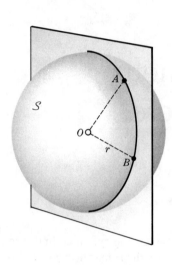

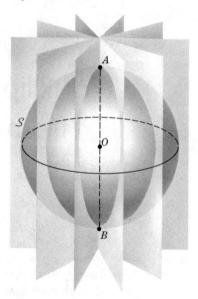

Next we associate the concept of distance in $\mathcal{E}$ with certain properties of sphere S. Let A and B be two points. Taking ω as a great circle passing through A and B, let $\overline{\angle AOB}$ be the unique central angle determined by A and B. We know that the Euclidean length of the minor arc AB subtended by $\overline{\angle AOB}$ is the (radian) measure of that angle times the radius. Since here the radius is of unit length, define the "distance" from A to B to be the number $\angle AOB$. To distinguish this from the ordinary Euclidean distance (which in this case would be the length of the *chord* $\overline{AB}$), we shall in this and the following section use the notation $\widehat{AB}$ for spherical "distance" and continue using the symbol AB for Euclidean distance. Thus, for any two points A and B in S,

$$\widehat{AB} = \angle AOB. \tag{79.1}$$

Immediately we notice that the least upper bound α of all "distances" is the measure of a straight angle. That is, $\alpha = \pi$. Two "points" A and B on S are obviously antipodal if and only if $\widehat{AB} = \pi$. Our previous comments regarding "lines" then show that Axioms 1–4 are thus far realized on S. The coordinatization axiom, Axiom 5, also follows. Hence the terms "segment," "ray," and "angle" have meaning and may be realized as objects of S. As we have seen before, those objects are, respectively, the minor arc of a great circle, a semi-great-circle, and two semi-great-circles meeting at antipodal points.

To define spherical "angle measure" the precedent already established (and fully discussed in Section 23) will be followed: Take the "measure" of "angle" ABC in S to be the measure of the *Euclidean angle* between the *tangent rays* $\overrightarrow{BA'}$ and $\overrightarrow{BC'}$ (Fig. 365). That is, for "angle measure" on S we have

$$\widetilde{\angle ABC} = \angle A'BC'. \tag{79.2}$$

Axioms 6–10 may then be easily verified (with $\beta = \pi$ rather than 180). The verification of Axiom 11, the SAS postulate, is more difficult and will be postponed until some spherical trigonometry is developed. Finally, Axiom 12 is clear from the fact that any two planes through O meet in a line through O, which in turn meets the sphere in a pair of antipodal "points." Hence any two "lines" in S meet in a pair of antipodal "points," and therefore parallel "lines" do not exist.

Figure 365

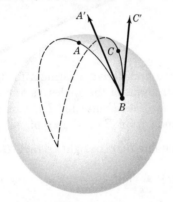

80. Elliptic Trigonometry of the Triangle

A simple derivation of the basic formula relating the parts of a spherical triangle is afforded by the methods of analytic geometry in three dimensions. Recall that points in space may be represented by ordered triples of reals (x, y, z) in such a manner that the Euclidean distance between the points $P_1[x_1, y_1, z_1]$ and $P_2[x_2, y_2, z_2]$ is given by the formula

$$P_1P_2 = \sqrt{(x_1 - x_2)^2 + (y_1 - y_2)^2 + (z_1 - z_2)^2}. \tag{80.1}$$

Now consider spherical "triangle" ABC on S, with the standard notation

$$a = \widetilde{BC}, \qquad b = \widetilde{AC}, \qquad c = \widetilde{AB}$$

$$A = \widetilde{\angle BAC}, \qquad B = \widetilde{\angle ABC}, \qquad \text{and} \qquad C = \widetilde{\angle ACB}.$$

Assume a three-dimensional coordinate system with origin A, such that the xy plane is tangent to sphere O and the x axis is tangent to "line" AB (Fig. 366).

Figure 366

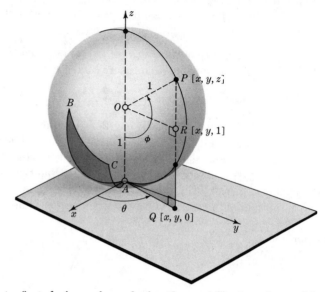

It will be convenient to first derive a formula for the coordinates of an arbitrary point $P[x, y, z]$ on S in terms of the angles θ and ϕ shown in the figure (which may be regarded as *spherical coordinates* for P). Consider the points $Q[x, y, 0]$ (the foot of the perpendicular from P on the xy plane) and $R[x, y, 1]$ (the foot of the perpendicular from O on line $\overleftrightarrow{PQ}$). In right triangle OPR we have, in magnitude and in sign,

$$RP = \sin (\phi - \pi/2) = -\cos \phi.$$

But $z = QP = QR + RP = 1 + RP$. Therefore,

$$z = 1 - \cos \phi.$$

To determine x and y note that the pair (AQ, θ) constitute polar coordinates for Q and hence, since $AQ = OR = \cos (\phi - \pi/2) = \sin \phi = r$,

$$x = \sin \phi \cos \theta$$
$$y = \sin \phi \sin \theta.$$

Thus P has coordinates

$$(\sin \phi \cos \theta, \sin \phi \sin \theta, 1 - \cos \phi). \tag{80.2}$$

Applying (80.2) to the two points B and C one obtains

$$B[\sin c, 0, 1 - \cos c] \qquad \text{and} \qquad C[\sin b \cos A, \sin b \sin A, 1 - \cos b].$$

From (80.1),

$$BC^2 = (\sin c - \sin b \cos A)^2 + (\sin b \sin A)^2 + (-\cos c + \cos b)^2$$

$$BC^2 = \sin^2 c - 2 \sin b \sin c \cos A + \sin^2 b \cos^2 A + \sin^2 b \sin^2 A$$
$$+ \cos^2 c - 2 \cos b \cos c + \cos^2 b$$
$$= 2 - 2 \cos b \cos c - 2 \sin b \sin c \cos A.$$

But $\frac{1}{2}BC = \sin \frac{1}{2}\angle BOC = \sin \frac{1}{2}a$, and therefore,

$$BC^2 = 4 \sin^2 \frac{a}{2} = 2(1 - \cos a).$$

Hence,

$$2 - 2 \cos a = 2 - 2 \cos b \cos c - 2 \sin b \sin c \cos A,$$

which reduces to the **elliptic law of cosines:**

$$\boxed{\cos a = \cos b \cos c + \sin b \sin c \cos A.} \qquad (80.3)$$

By change of notation one obtains the companion formulas

$$\boxed{\begin{aligned} \cos b &= \cos a \cos c + \sin a \sin c \cos B, \\ \cos c &= \cos a \cos b + \sin a \sin b \cos C. \end{aligned}} \qquad (80.4)$$

A direct consequence of this is the **Pythagorean theorem for elliptic geometry** [obtained by setting $C = \pi/2$ in (80.4)]:

$$\boxed{\cos c = \cos a \cos b.} \qquad (80.5)$$

The use of (80.5) in (80.3) leads to the basic relations in an elliptic right triangle. First write (80.3) in the form

$$\cos A = \frac{\cos a - \cos b \cos c}{\sin b \sin c}$$

Substitute from (80.5) to obtain

$$\cos A = \frac{\cos c/\cos b - \cos b \cos c}{\sin b \sin c} = \frac{\cos c - \cos^2 b \cos c}{\sin b \cos b \sin c}$$

$$= \frac{\cos c \sin^2 b}{\sin b \cos b \sin c} = \frac{\sin b/\cos b}{\sin c/\cos c}.$$

Therefore,

$$\boxed{\cos A = \frac{\tan b}{\tan c}.} \qquad (80.6)$$

(Note the analogy with the Euclidean formula $\cos A = b/c$.)

The formula (80.6) provides a basis for deriving a formula for $\sin A$, still under the assumption $C = \pi/2$:

$$\cos^2 A = \frac{\tan^2 b}{\tan^2 c} = \frac{\sec^2 b - 1}{\tan^2 c} = \frac{\cos^2 a / \cos^2 c - 1}{\tan^2 c} = \frac{\cos^2 a - \cos^2 c}{\sin^2 c}.$$

$$\sin^2 A = 1 - \frac{\cos^2 a - \cos^2 c}{\sin^2 c} = \frac{1 - \cos^2 a}{\sin^2 c} = \frac{\sin^2 a}{\sin^2 c}.$$

That is,

$$\sin A = \frac{\sin a}{\sin c}. \tag{80.7}$$

The formula for $\tan A$ is then an easy application of (80.6) and (80.7):

$$\tan A = \frac{\sin a}{\sin c} \cdot \frac{\tan c}{\tan b} = \frac{\tan a}{\sin b},$$

the details being left for the reader.

One may collect the preceding formulas and, by changing the notation to obtain $\sin B$, $\cos B$, and $\tan B$, exhibit the **formulas relating the parts of a right triangle in elliptic geometry** (Fig. 367):

$$
\boxed{
\begin{array}{ll}
\multicolumn{2}{c}{\cos c = \cos a \cos b,} \\[6pt]
\sin A = \dfrac{\sin a}{\sin c}, & \sin B = \dfrac{\sin b}{\sin c}, \\[10pt]
\cos A = \dfrac{\tan b}{\tan c}, & \cos B = \dfrac{\tan a}{\tan c}, \\[10pt]
\tan A = \dfrac{\tan a}{\sin b}, & \tan B = \dfrac{\tan b}{\sin a}.
\end{array}
}
\tag{80.8}
$$

Figure 367

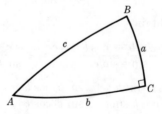

80.9 Example: Show that for sufficiently small triangles each of the relations in (80.8) are approximated to within any predetermined constant by their Euclidean counterparts.

Solution: We illustrate the method by verifying the principle only for the Pythagorean relation and the formula for $\tan A$. Consider the series expansion for $\sin x$ and $\cos x$:

$$\sin x = x - \frac{x^3}{3!} + \frac{x^5}{5!} - \cdots, \qquad \cos x = 1 - \frac{x^2}{2!} + \frac{x^4}{4!} - \cdots.$$

For small x one may ignore all terms of order higher than two. For the Pythagorean relation,

$$\left(1 - \frac{c^2}{2!} + \cdots\right) = \left(1 - \frac{a^2}{2!} + \cdots\right)\left(1 - \frac{b^2}{2!} + \cdots\right)$$

$$1 - \frac{c^2}{2} + \cdots = 1 - \frac{a^2}{2} - \frac{b^2}{2} + \frac{a^2 b^2}{4} + \cdots.$$

$$c^2 = a^2 + b^2 + \{\text{terms of order four or higher}\}.$$

For the other relation, write

$$\sin b \tan A = \tan a$$

or

$$\sin b \cos a \tan A = \sin a$$

$$\left(b - \frac{b^3}{3!} + \cdots\right)\left(1 - \frac{a^2}{2!} + \cdots\right) \tan A = \left(a - \frac{a^3}{3!} + \cdots\right)$$

$$b \tan A = a - \frac{a^3}{3!} + \frac{a^2 b}{2!} \tan A + \frac{b^3}{3!} \tan A + \cdots.$$

Thus,

$$b \tan A = a + \{\text{terms of order three or higher}\}. \;\blacksquare$$

EXERCISES

1. Prove the formula for $\tan A$ in (80.8).
2. Deduce the **elliptic law of sines:**

$$\boxed{\frac{\sin a}{\sin A} = \frac{\sin b}{\sin B} = \frac{\sin c}{\sin C}.} \qquad (80.10)$$

3. Derive the **formula for a cevian of a triangle** in elliptic geometry (Fig. 368):

$$\boxed{\cos d = p \cos b + q \cos c,} \qquad (80.11)$$

where $p = \sin a_1 / \sin a$ and $q = \sin a_2 / \sin a$. *Hint:* You will need the formula for $\sin(a_1 + a_2)$. Apply the elliptic law of cosines to triangles ADB and ADC at D as indicated in Fig. 368.

Figure 368

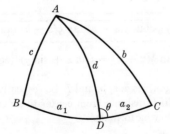

4. Prove the formulas for the medians of an elliptic triangle in terms of its sides:

$$\cos m_a = (\tfrac{1}{2} \cos b + \tfrac{1}{2} \cos c)/\cos \tfrac{1}{2}a,$$
$$\cos m_b = (\tfrac{1}{2} \cos a + \tfrac{1}{2} \cos c)/\cos \tfrac{1}{2}b, \qquad (80.12)$$
$$\cos m_c = (\tfrac{1}{2} \cos a + \tfrac{1}{2} \cos b)/\cos \tfrac{1}{2}c.$$

Hint: Here you will need the identity $\sin a = 2 \sin \tfrac{1}{2}a \cos \tfrac{1}{2}a$.

★5. If s is equal to one half the sum of the elliptic perimeter of triangle ABC prove from the elliptic law of cosines:

$$\sin \frac{A}{2} = \sqrt{\frac{\sin (s - b) \sin (s - c)}{\sin b \sin c}},$$

$$\cos \frac{A}{2} = \sqrt{\frac{\sin s \sin (s - a)}{\sin b \sin c}}. \qquad (80.13)$$

Hint: Use the trigonometric identity $\cos x - \cos y = -2 \sin \tfrac{1}{2}(x + y) \sin \tfrac{1}{2}(x - y)$, and the half-angle formulas for sine and cosine.

6. Using Exercise 5, prove the formula for the inradius of an elliptic triangle in terms of its sides:

$$\tan r = \sqrt{\frac{\sin (s - a) \sin (s - b) \sin (s - c)}{\sin s}}. \qquad (80.14)$$

[Compare with (60.13), Chapter 7.]

7. *Gauss's equations.* From the half-angle formulas (80.13) prove the following formulas (known as **Gauss's equations**):

$$\cos \tfrac{1}{2}c \sin \tfrac{1}{2}(A + B) = \cos \tfrac{1}{2}(a - b) \cos \tfrac{1}{2}C,$$
$$\cos \tfrac{1}{2}c \cos \tfrac{1}{2}(A + B) = \cos \tfrac{1}{2}(a + b) \sin \tfrac{1}{2}C,$$
$$\sin \tfrac{1}{2}c \sin \tfrac{1}{2}(A - B) = \sin \tfrac{1}{2}(a - b) \cos \tfrac{1}{2}C, \qquad (80.15)$$
$$\sin \tfrac{1}{2}c \cos \tfrac{1}{2}(A - B) = \sin \tfrac{1}{2}(a + b) \sin \tfrac{1}{2}C.$$

8. Prove the **elliptic formula for the area of a right triangle in terms of its legs:**

$$\tan \tfrac{1}{2}K = \tan \tfrac{1}{2}a \tan \tfrac{1}{2}b. \qquad (80.16)$$

Hint: Start with the "excess" formula $K = A + B + C - \pi$ and use Gauss's equations (with $C = \pi/2$). The identities involving $\cos x \pm \cos y$ will also be useful here.

★9. *L'Huilier's formula.* Prove the following elliptic analogue of Heron's formula:

$$\tan \tfrac{1}{4}K = \sqrt{\tan \tfrac{1}{2}s \tan \tfrac{1}{2}(s - a) \tan \tfrac{1}{2}(s - b) \tan \tfrac{1}{2}(s - c)}. \qquad (80.17)$$

Hint: Use the identity $\tan \tfrac{1}{2}(x + y) = (\sin x + \sin y)/(\cos x + \cos y)$ and Gauss's equations.

10. In an n-sided regular polygon, let a, r, p, and K be respectively the *apothem* (distance from center to a side), *radius* (distance from center to a vertex), perimeter, and area (Fig. 369). Deduce the following formulas:

$$\sin \frac{p}{2n} = \sin r \sin \frac{\pi}{n},$$

$$\tan \frac{K}{4n} = \tan \frac{a}{2} \tan \frac{p}{4n}.$$

(80.18)

Hint: Use the elliptic trigonometry of right triangle OMB as shown in the figure and (80.16) (Exercise 8).

Figure 369

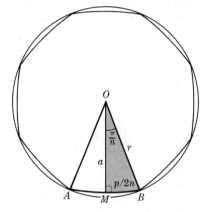

★11. Assuming there is a sequence of regular inscribed polygons of n sides of an elliptic circle of radius r whose perimeters p_n and areas K_n tend to the circumference and area of the circle, respectively, deduce from equations (80.18) the **formulas of the circumference** C **and area** K **of a circle of radius** r **in elliptic geometry:**

$$C = 2\pi \sin r,$$

$$K = 4\pi \sin^2 \frac{r}{2}.$$

(80.19)

Hint: Make use of the limits $\sin x / x \to 1$ and $\tan x / x \to 1$ as $x \to 0$ and the identities $\tan x/2 = (1 - \cos x)/\sin x$ and $1 - \cos x = 2 \sin^2 (x/2)$.

12. Let K be the area of the spherical cap on a sphere of unit radius shown in Fig. 370, with h its height. Since this K is also the area of a spherical circle of radius r, for-

Figure 370

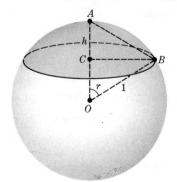

mula (80.19) applies. Show that $h = 2\sin^2 r/2$ and from (80.19) that $K = 2\pi h$, in agreement with the standard formula $K = 2\pi R h$, where R is the radius of the sphere.

81.   The SAS Postulate

The elliptic law of cosines leads to an easy verification of the last remaining axiom of elliptic geometry for our model, Axiom 11. Suppose in "triangles" ABC and $A'B'C'$ (Fig. 371) we have, in standard notation,

$$A = A', \qquad b = b', \qquad \text{and} \qquad c = c'.$$

Then, by the law of cosines,

$$
\begin{aligned}
\cos a &= \cos b \cos c + \sin b \sin c \cos A \\
&= \cos b' \cos c' + \sin b' \sin c' \cos A' \\
&= \cos a'
\end{aligned}
$$

and $a = a'$. It then follows that $B = B'$ and $C = C'$. For example, to prove $B = B'$ write

$$\cos B = \frac{\cos b - \cos a \cos c}{\sin a \sin c} = \frac{\cos b' - \cos a' \cos c'}{\sin a' \sin c'} = \cos B',$$

and hence $B = B'$. ∎

Figure 371

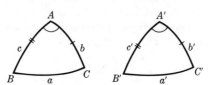

It then follows that all the Axioms 1–12 hold on the sphere S. This warrants the important conclusion that *all theorems previously established from those axioms hold on* S. Not all of those theorems are completely trivial on the sphere (such as the section on "Inequalities in a Triangle," for example). This points up the economy of effort sometimes afforded by a systematic study of an axiomatic system which has several concrete models. Any logical conclusion derived axiomatically is automatically valid in each of the models.

82.   The Angle-Sum Theorem from the Model

Recall the formula for $\tan A$ and $\tan B$ in "right triangle" ABC. It follows that

$$\cot A \cot B = \frac{\sin b \cos a}{\sin a} \cdot \frac{\sin a \cos b}{\sin b},$$

or

$$\cot A \cot B = \cos a \cos b.$$

That is, since $0 < a < \pi$ and $0 < b < \pi$, then

$$\cot A \cot B < 1.$$

Since $\sin A \sin B > 0$ $\quad (0 < A < \pi$ and $0 < B < \pi)$,

$$\cos A \cos B < \sin A \sin B.$$

Thus,

$$\cos (A + B) = \cos A \cos B - \sin A \sin B < 0.$$

Since $0 \leq A + B \leq 2\pi$ and $\cos x$ is negative only for $\pi/2 < x < 3\pi/2$ if x is in the range $0 < x < 2\pi$ (see Fig. 372), then it follows that

$$\frac{\pi}{2} < A + B < \frac{3\pi}{2},$$

or

$$A + B + C > \pi.$$

The result for triangles in general now quickly follows; the details will be left to the reader.

Figure 372

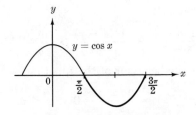

83. The Cosine Inequality for Elliptic Geometry

The chief characteristics of elliptic geometry can be summed up in one revealing principle: *If the sides of a triangle in the elliptic plane have respective lengths equal to those of a triangle in the Euclidean plane, then each of the angles in the elliptic triangle are* greater *than the corresponding angles in the Euclidean triangle.*

Suppose that in Fig. 373 triangle ABC is given in the elliptic plane and triangle $A'B'C'$ is constructed in the Euclidean plane such that, in standard notation,

$$a = a', \qquad b = b', \qquad \text{and} \qquad c = c'.$$

Then it is to be proved that

$$A > A', \qquad B > B', \qquad \text{and} \qquad C > C'.$$

Figure 373

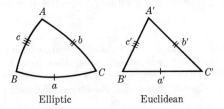

Elliptic   Euclidean

This can be accomplished through the use of two formulas:

$$\sin \frac{A}{2} = \sqrt{\frac{\sin (s - b) \sin (s - c)}{\sin b \sin c}},$$

$$\sin \frac{A'}{2} = \sqrt{\frac{(s' - b')(s' - c')}{b'c'}} = \sqrt{\frac{(s - b)(s - c)}{bc}},$$

$$s' = \tfrac{1}{2}(a' + b' + c').$$

The first was (80.13) (Exercise 5, Section 80) while the second is a standard formula of Euclidean trigonometry. Divide in the above formulas to obtain

$$\frac{\sin \frac{1}{2}A}{\sin \frac{1}{2}A'} = \sqrt{\frac{\sin (s - b)}{s - b} \cdot \frac{\sin (s - c)}{s - c} \cdot \frac{b}{\sin b} \cdot \frac{c}{\sin c}}.$$

Thus it suffices to prove that $\sin \frac{1}{2} A > \sin \frac{1}{2} A'$, or that

$$\frac{\sin (s - b)}{s - b} \cdot \frac{\sin (s - c)}{s - c} \cdot \frac{b}{\sin b} \cdot \frac{c}{\sin c} > 1.$$

That is, it must be proved that

$$\frac{\sin (s - b)}{s - b} \cdot \frac{\sin (s - c)}{s - c} > \frac{\sin b}{b} \cdot \frac{\sin c}{c}. \tag{83.1}$$

But this merely involves the behavior of the function

$$f(x) = \frac{\sin x}{x}, \qquad 0 < x < \pi.$$

It is an elementary calculus exercise to prove that f is strictly decreasing[2] and that therefore, $x_1 < x_2$ implies $\sin x_1/x_1 > \sin x_2/x_2$. Observe that in (83.1) we have

(a) $s - b = \tfrac{1}{2}(a - b + c) < \tfrac{1}{2}(c + c) < c < \pi$
(b) $s - c = \tfrac{1}{2}(a + b - c) < \tfrac{1}{2}(b + b) < b < \pi$.

Hence if $x_1 = s - b$ and $x_2 = c$, then by (a)

$$\frac{\sin (s - b)}{s - b} > \frac{\sin c}{c}.$$

Similarly, (b) implies that

$$\frac{\sin (s - c)}{s - c} > \frac{\sin b}{b}$$

and (83.1) follows, along with the statement in italics.

The particular characteristic of elliptic geometry just described leads to an important *intrinsic* property of the geometry, describable in terms of the objects of S alone.

---

[2] Since $f'(x) = x^{-2} (x \cos x - \sin x)$, one need only examine the inequality $x \cos x < \sin x$. This holds trivially when $\pi/2 \leqq x \leqq \pi$ since $\cos x$ is negative there. For $0 < x < \pi/2$ it becomes $x < \tan x$, which may be verified by observing that x is the area under the curve $y = 1$ from 0 to x while $\tan x$ is that under $y = \sec^2 x$ from 0 to x, the former being obviously less than the latter.

83.2 Cosine Inequality for Elliptic Geometry: In any elliptic triangle ABC with side "lengths" a, b, and c in standard notation, then

$$\boxed{a^2 < b^2 + c^2 - 2bc \cos A.}$$

PROOF: Let triangle $A'B'C'$ be constructed in the Euclidean plane such that $B'C' = a$, $C'A' = b$, and $A'B' = c$. By the property observed earlier, and by the *Euclidean* law of cosines,

$$\cos A < \cos A' = \frac{b^2 + c^2 - a^2}{2bc},$$

which can then be easily put in the desired form.

83.3 Example: Use the cosine inequality to prove that in elliptic geometry the segment joining the midpoints of two sides of a triangle is *greater than* one half the length of the third side.

Solution: First, observe from an application of the cosine inequality in Fig. 374, that if M is the midpoint of $\overline{BC}$,

$$c^2 < \frac{a^2}{4} + d^2 - ad \cos \theta$$

$$b^2 < \frac{a^2}{4} + d^2 + ad \cos \theta$$

from which follows

$$b^2 + c^2 < \frac{a^2}{2} + 2d^2,$$

or

$$d^2 > \tfrac{1}{2}b^2 + \tfrac{1}{2}c^2 - \tfrac{1}{4}a^2. \tag{83.4}$$

Figure 374

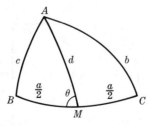

Now apply (83.4) twice in Fig. 375 (where M and N are midpoints):

$$d^2 > \tfrac{1}{2}\left(\frac{b}{2}\right)^2 + \tfrac{1}{2}e^2 - \tfrac{1}{4}c^2$$

$$> \tfrac{1}{8}b^2 + \tfrac{1}{2}(\tfrac{1}{2}a^2 + \tfrac{1}{2}c^2 - \tfrac{1}{4}b^2) - \tfrac{1}{4}c^2$$

$$= \tfrac{1}{4}a^2. \blacksquare$$

Figure 375

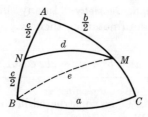

EXERCISES

1. Using elliptic trigonometry prove the following relation between the sides of a Saccheri quadrilateral in elliptic geometry with a, b, and c denoting the base, leg, and summit respectively:

$$\cos b = \frac{\sin \tfrac{1}{2}c}{\sin \tfrac{1}{2}a}. \tag{83.5}$$

   Thus give a trigonometric proof of the theorem: *The summit of a Saccheri quadrilateral in elliptic geometry is less than the base.*

2. Recall that a Lambert quadrilateral (Section 38, Chapter 5) is formed by the common perpendicular bisector of the base and summit of a Saccheri quadrilateral. Hence, with the notation of Fig. 376: (a) Derive the following relations for the parts of a Lambert quadrilateral:

$$\boxed{\begin{aligned} \sin c &= \sin a \cos b, \\[4pt] \cos d &= \frac{\cos a \cos b}{\sqrt{1 - \sin^2 a \cos^2 b}}, \\[4pt] \cos A &= -\frac{\sin a \sin b}{\sqrt{1 - \sin^2 a \cos^2 b}}. \end{aligned}} \tag{83.6}$$

   (b) Interpret the minus sign in the formula for $\cos A$ in (83.6). (c) If $a = \pi/2$ and $b = \pi/4$, use (83.6) to solve for the remaining parts of the quadrilateral, and interpret this case on the sphere.

Figure 376

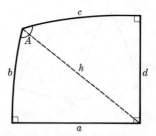

3. Using the cosine inequality prove:

    83.7 Theorem: Each cevian of a triangle in the elliptic plane has length *greater than* that of the corresponding cevian of a triangle in the Euclidean plane whose sides have lengths equal to those of the elliptic triangle and where the cevians cut proportional segments on the sides to which they are joined.

In the following exercises all distances indicated are elliptic.

4. In an elliptic triangle ABC, points D and E are chosen on sides $\overline{AB}$ and $\overline{AC}$ such that

$$AD = \tfrac{2}{3}AB \qquad \text{and} \qquad AE = \tfrac{2}{3}AC.$$

Prove that $DE > \tfrac{2}{3}BC$.

5. Generalize the result of Exercise 4 by the aid of Theorem 83.7 (Exercise 3).

6. In Fig. 377 $\overset{\leftrightarrow}{DE}$ and $\overset{\leftrightarrow}{BC}$ are each perpendicular to $\overset{\leftrightarrow}{AC}$. Prove:

$$\frac{\tan AD}{\tan AB} = \frac{\tan AE}{\tan AC}, \tag{83.8}$$

$$\frac{\sin AD}{\sin AB} = \frac{\sin DE}{\sin BC}. \tag{83.9}$$

Figure 377   **Figure 378**

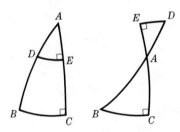   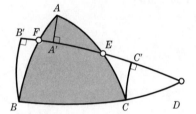

7. *Linearity number and Menelaus' theorem for elliptic geometry.* Let D, E, and F be menelaus points on the sides of triangle ABC and define, in magnitude and in sign,[3]

$$\mathcal{E}\begin{bmatrix} ABC \\ DEF \end{bmatrix} = \frac{\sin AF}{\sin FB} \cdot \frac{\sin BD}{\sin DC} \cdot \frac{\sin CE}{\sin EA}$$

— the **linearity number for elliptic geometry**. Basing your proof on the relation (83.9) (Exercise 6) and the pattern of the argument provided earlier for Menelaus' theorem (Theorem 62.3) prove

83.10 *Menelaus' Theorem for Elliptic Geometry:* The menelaus points D, E, and F of an elliptic triangle ABC are collinear if and only if

$$\mathcal{E}\begin{bmatrix} ABC \\ DEF \end{bmatrix} = -1.$$

(See Fig. 378; recall lemma in Section 7.)

8. *Desargues' theorem for elliptic geometry.* Follow the proof of Desargues' theorem outlined in Section 62 and devise a proof for elliptic geometry, using Theorem 83.10.

9. Prove

83.11 *Ceva's Theorem for Elliptic Geometry:* In an elliptic triangle ABC with menelaus points D, E, and F given, cevians $\overset{\leftrightarrow}{AD}$, $\overset{\leftrightarrow}{BE}$, and $\overset{\leftrightarrow}{CF}$ are concurrent if and only if

---

[3] Directed distance for elliptic geometry may be formulated by simply dualizing the concept of directed angle measure introduced in Section 59. It should be observed that given three collinear points A, B, and C, AB and BC have the same sign if and only if either (ABC) or $(AB'C)$, where B' is the opposite pole of B.

$$\mathcal{E}\begin{bmatrix} ABC \\ DEF \end{bmatrix} = 1.$$

(See lemma in Section 7.)

10. Prove that the medians and altitudes of any elliptic triangle are each concurrent. *Hint:* Use Ceva's theorem (Exercise 9). For the altitudes it will be necessary to use in addition some elliptic trigonometry.

11. Define **elliptic cross ratio** by setting

$$\mathcal{E}[AB, CD] = \frac{\sin AC \sin BD}{\sin AD \sin BC},$$

where A, B, C, and D are any four points on a directed line. Prove that if the harmonic construction be carried out in the elliptic plane for the pairs A, B and C, D (Fig. 379), then

$$\mathcal{E}[AB, CD] = -1.$$

Figure 379

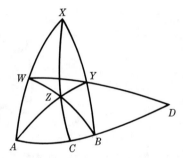

12. Study the properties of the mapping shown in Fig. 380 from the plane to the sphere and show that lines map into "lines." Then obtain a simple proof of Desargues' theorem for elliptic geometry.

Figure 380

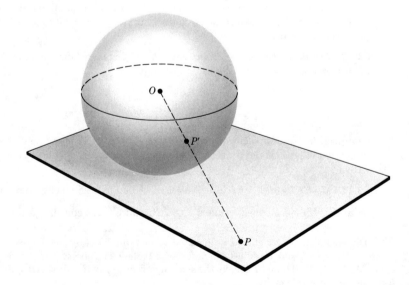

NOTE: It is a significant fact that Desargues' theorem is valid in both the elliptic and hyperbolic geometries, for this indicates that a projective geometry is possible for these two and that it may be developed in much the same manner as it is in Euclidean geometry (by the addition of "points at infinity"). But elliptic geometry has *too many* points, contrary to the situation in Euclidean and hyperbolic geometry where there are not enough. The remedy is to *identify* antipodal points, thus producing the so-called *single* elliptic geometry mentioned earlier.

84. Non-Euclidean Planes and Differential Geometry

Gauss was the first to fully utilize the study of surfaces in three-dimensional Euclidean geometry to shed light on the existence of hyperbolic geometry. The elliptic "plane" is but one example of a plane geometry realizable *on a surface in Euclidean space*. To understand this in some detail let us begin with a surface S (Fig. 381) and take as "points" the ordinary (Euclidean) points of S. One could experimentally determine for a few pairs of "points" the arc (or arcs) having the shortest length which join them. To the proverbial inhabitant on S such a path would appear "straight" in that it would seem to him to be the most direct route on S between the endpoints. This would form a basis for taking certain arcs on S as "segments." These segments could then be extended in both directions to produce the so-called *paths of shortest arc length*, or *geodesics*, on S.[4] One then takes these geodesics as "lines." Now let A and B be any two points and consider all "lines" passing through them (there can be infinitely many). The arc length from A to B along each line is uniquely determined, so define the "distance" from A to B, denoted $\overset{\frown}{AB}$, to be the *greatest lower bound of those arc lengths*. Finally, take "angle measure" to be the measure of the Euclidean angle formed by the tangents to the "sides" of any given "angle."

Figure 381

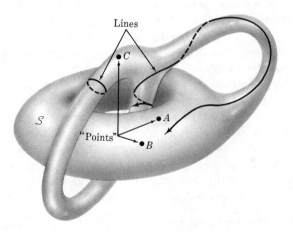

---

[4] Although it would be beyond the scope of this book to do so, these ideas can be supported analytically for a large class of surfaces.

Doing this admittedly leads to many problems which cannot always be resolved satisfactorily, but at least theoretically this procedure describes the **plane geometry associated with** S, or as it is usually called, the **intrinsic geometry** of S. It is conceivable then, that one can produce as many plane geometries as there are surfaces, with the *shape* of the surface controlling the geometric theory. These surfaces would be *models* for various axiomatic systems just as the sphere is our model for elliptic geometry.

Two fundamental problems suggest themselves at this point: (1) Given a surface S, find a characteristic maximal set of independent axioms E which are valid for the intrinsic geometry of S, and (2) given a set of axioms E, find a surface S whose intrinsic geometry is a realization of S. More briefly stated, the problems are: Given a surface as model find the set of axioms which fit it, and, given a set of axioms find a surface which serves as a model. The second problem is known to have no solution in certain cases. For example, the hyperbolic geometry 3C is known *not to have a realization on a surface in Euclidean geometry.*[5]

In considering these two problems it is obviously important to determine how many essentially different surfaces can serve as a model for a particular set of axioms. The term "different" can have various meanings, but for now let it mean simply *noncongruent* (two surfaces are said to be *congruent* if there is a distance-preserving mapping of Euclidean space onto itself which maps one of the surfaces onto the other). One readily finds that there are many "different" surfaces which act as models for elliptic geometry—namely, spheres of *different radii.* Interestingly enough it can be proved by the methods of differential geometry that *any model for elliptic geometry must be a sphere.* It can also be shown that a *plane* provides the *only* model for Euclidean geometry, and as stated before, there is *no* surface which serves as a model for hyperbolic geometry.

Of particular interest in classical differential geometry, however, is the study of *local* properties of surfaces, as opposed to the *global* properties we have been discussing. To illustrate, consider a sphere S and part of another sphere S_1, having the same radius as S (Fig. 382). While the intrinsic geometries of these two surfaces are indeed different from a global point of view (for example, the "lines" of S return upon themselves while those of S_1 do not), certainly all previously deduced formulas for triangles and circles would be valid on both surfaces. That is to say, the intrinsic geometries of S and S_1 coincide *locally.* To be more explicit, we know that each point P of S_1 not on the edge corresponds to some point Q of S such that for some $r > 0$ the intrinsic geometry of S_1 confined to points within a "distance" r of P coincides with that of S confined to points within a "distance" r of Q. Or to put it still another way, elliptic geometry can be realized *locally* on S_1.

If we compromise our previous goal of seeking global interpretations of axiomatic systems in the geometry on surfaces to that of finding only local realizations,

---

[5] This theorem is relatively young, having been proved by G. Lütkemeyer in 1902. It has become known as *Hilbert's theorem,* after the renowned German mathematician David Hilbert (1862–1943). The theorem does not say *there is no model for hyperbolic geometry,* for there are other ways of realizing non-Euclidean geometries in Euclidean geometry.

Figure 382

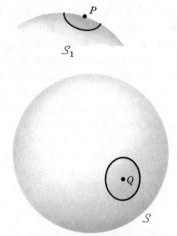

then the tools of classical differential geometry are available, and much is known on the subject. Under this less restrictive requirement then, Euclidean geometry has a local realization not only on the plane but also on the *cone* and *cylinder*, and more generally *on any surface which can be obtained as the envelope of a one-parameter family of planes*, known as a *developable*. And hyperbolic geometry now has a local realization on the horn-like surface of revolution generated by revolving the *tractrix*[6] about its asymptote [Fig. 383(b)], sometimes referred to as the *pseudosphere* (not to be confused with the model for single elliptic geometry which also bore that name). It is beyond the scope of this book to explain even intuitively why the pseudosphere locally depicts hyperbolic geometry, for this requires some fairly refined techniques of differential geometry. We can, however, make the statement about cylinders and cones being local realizations of Euclidean geometry intuitively plausible.

Figure 383

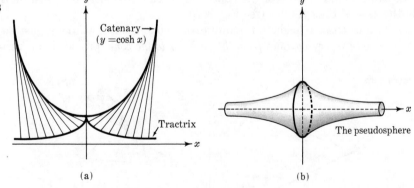

(a)   (b)

---

[6] The tractrix is a plane curve studied in the theory of *involutes*. It can be visualized concretely as the path traced by the end of a piece of string which has been wound about a catenary —so that its initial position is at the vertex of the catenary—and is allowed to unwind tautly. See Fig. 383(a).

Fig. 384 shows a cylinder which is tangent to a plane surface. Consider a small triangle ABC and its exterior angle ACD on the plane. The characteristically Euclidean proposition that $\angle ACD = \angle A + \angle B$ may be "mapped" onto the cylinder by simply "inking" the configuration $ABCD$ and rolling the cylinder over it to make an impression $A'B'C'D'$. Since any curve of length s in the plane will obviously correspond to one of length s on the cylinder under this mapping, *lines* in the plane map into "lines" in the intrinsic geometry of the cylinder. (For if there were a shorter path on the cylinder, say from A' to B', it would have to be the image of a path in the plane from A to B of smaller length than that of the *segment* joining A and B, which is impossible.) Thus the mapping is *locally* "distance" preserving; it is obvious that it is also "angle" preserving. Then on the cylinder we would have $\angle A'C'D' = \angle A' + \angle B'$, and consequently the "angle sum" of every "triangle" on the cylinder would be the measure of a straight angle.

Figure 384

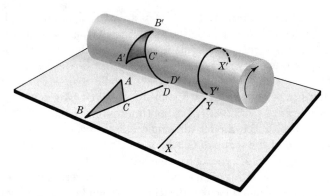

Thus *the intrinsic geometry of the cylinder is locally a realization of Euclidean geometry.* The same would be true of the cone since it can obviously be rolled on a plane as in the case of the cylinder (see Fig. 385).

This discussion leads to the interesting conclusion that there must be a locally "distance"-preserving map *from the cylinder to the cone*, provided the vertex of the

Figure 385

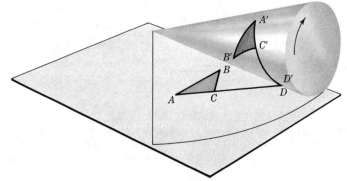

cone is removed. This is true, but again its rigorous proof requires a knowledge of differential geometry.

In this vein we might toy with the possibility of producing a locally "distance"-preserving map *from the plane to the sphere* by the "rolling" method (Fig. 386). Of

Figure 386

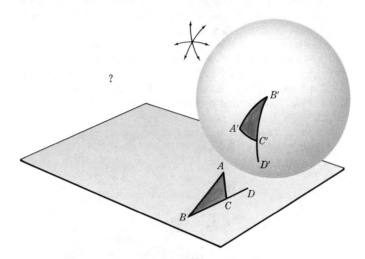

course here one would have to "roll" the sphere in more than one direction. Possibly contrary to our intuitive feelings on the matter however, this mapping process *cannot succeed*, as asserted in the following theorem; it is the formal acknowledgment of the impossibility of solving the ancient problem of producing a planar map of even the smallest portion of the earth's surface without introducing some error:

84.1   Theorem: There exists no locally distance-preserving map from the plane to the sphere.

PROOF: Completely obvious by the cosine inequality (Theorem 83.2). ∎

The interested reader is urged to examine the discussion given by others on this subject. For example, the book by H. Tietze, *Famous Problems of Mathematics* [27] is written for the layman and includes a very readable account of the ideas we have put forth here. Also, for a lucid introduction to differential geometry see H. S. M. Coxeter, *Introduction to Geometry* (Part IV) [4], and for a more detailed analysis, H. Eves, *A Survey of Geometry, Volume Two* [10]. For the serious student of geometry there is no substitute for a thorough knowledge of the classical differential geometry to be found in such books as D. J. Struik, *Lectures on Classical Differential Geometry* [26], and T. J. Willmore, *An Introduction to Differential Geometry* [29] (the latter written for the more advanced student).

EXERCISES

1. Explain why the mapping from a portion of the plane to the cylinder described above is

    not *globally* "distance" preserving. *Hint:* Compare $\overset{\frown}{XY}$ with $\overset{\frown}{X'Y'}$ for the points shown in Fig. 384.

2. In a room 10 ft high with a ceiling 16 ft long and 8 ft wide, there sits a fly, exactly in the center of the narrow wall, 0.8 ft below the ceiling. On the opposite wall, also exactly in the center, sits a spider, 2.4 ft above the ground, that is, 7.6 ft from the ceiling. The fly notices that the spider has just awakened, and addresses it: "Honorable spider, would you care to try to come over here and catch me?" The spider answers: "You would not give me enough time to reach you, but would fly away, just in time." "Well, how fast do you intend crawling over here?" the fly asks, and being told that the spider can't make more than 8 ft a minute, answers sorrowfully: "Too bad I can't wait longer than three minutes; I shall sleep for that length of time and then I must get into the sunlight." Will the spider catch the fly? *Hint:* Solve by flattening out the walls and ceiling of the room into a plane surface. Find the *geodesic* which joins the spider to the fly. (See Fig. 387. Printed by permission from H. Tietze, *Famous Problems of Mathematics* [27, p. 38].)

   Ans: There is a path on the walls and ceiling joining the spider to the fly which is approximately 23.8 ft long.

Figure 387

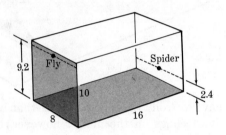

11

A Study
of Hyperbolic Geometry
from the Model
of Poincaré

Because the axioms for absolute geometry were formulated with the sphere and the analytic plane in mind, it is not particularly surprising that those axioms together with the appropriate parallel postulates should be consistent and that they find concrete realizations as geometries on various surfaces. But the existence of a model for the third geometry—hyperbolic geometry—which sprang merely from a *logical case*, seems rather unlikely. However, models for this geometry do indeed exist. We choose to study Poincaré's model since it is easier to develop and to understand from our standpoint. For future reference, let $\mathcal{H}$ denote the "abstract" hyperbolic geometry which consists of all the logical consequences of Axioms 1–11 and 12′ as stated previously.

In order to put the reader in the proper frame of mind, the first section will be devoted to a *method* which can be used to construct a rather general type of non-Euclidean geometry in the Euclidean plane, which will then be applied in constructing the model to be used for hyperbolic geometry.

85. Non-Euclidean Geometries in the Euclidean Plane

There are a variety of ways of creating geometries in the Euclidean plane. One, already mentioned in two previous exercises (Exercise 10, Section 18 and Exercise 6, Section 28), is to take a family of curves in the xy plane for the "lines" and to measure all distances along those "lines." To be more specific, take a "point" as any point $P[x, y]$ in the xy plane and a "line" as any member of the family of curves given by either of the two types of equations

$$x = a = \text{constant},$$
$$y = f(ax + b), \qquad a, b = \text{constant}, \tag{85.1}$$

where f is any fixed, differentiable function with positive derivative at every point, whose range is the set of real numbers, with $ax + b$ restricted to the domain of f. (An example of such a function would be $f(x) = \log x$, $x > 0$.)

It is easy to see that if f is any such function, then f^{-1} exists and $f^{-1}(y)$ is defined for any real y. Hence, if $P_1[x_1, y_1]$ and $P_2[x_2, y_2]$ be any two distinct points, $x_1 \neq x_2$, the system

$$y_1 = f(ax_1 + b),$$
$$y_2 = f(ax_2 + b),$$

has a *unique* solution in a and b, namely,

$$a = \frac{f^{-1}(y_1) - f^{-1}(y_2)^1}{x_1 - x_2} \qquad \text{and} \qquad b = \frac{x_1 f^{-1}(y_2) - x_2 f^{-1}(y_1)}{x_1 - x_2}.$$

Thus, the "line" $y = f(ax + b)$ with the above values for a and b, passes through P_1 and P_2 and is the only one which does. (If $x_1 = x_2$, then the vertical line $x = x_1$ passes through P_1 and P_2.) Therefore, *each pair of "points" lie on a unique "line."*

Then for the concept of "distance," suppose two distinct points P_1 and P_2 are given; let the "distance" from P_1 to P_2 (denoted $\widetilde{P_1 P_2}$) be *the length of the arc between P_1 and P_2 of the unique "line" joining* P_1 and P_2. For "angle measure" the curvilinear angle measure defined previously suffices. It is then almost a routine matter to check which of our axioms would be satisfied in such a geometry. Axioms 1–10, with $\alpha = \infty$, are those which are valid; the SAS axiom (Axiom 11) and the parallel axioms in general are *not* satisfied.

This procedure provides us with many non-Euclidean geometries for the *entire* Euclidean plane. Of course, Euclidean geometry itself may be obtained by simply taking those "lines" generated by the function $f(x) = x$. But now consider the problem of constructing a geometry in a *bounded* region S of the xy-plane.

One might begin by simply restricting the "lines" of any of the non-Euclidean geometries considered above to the region S. That is, let "points" be the points of the plane which lie in S and the "lines" be those arcs of curves (85.1) lying in S. In order to obtain our axioms on S it will be necessary to assume that S is convex *relative to the "segments" being considered*, and that all points on the *boundary* of S are excluded.

It is then easy to verify (if only intuitively) that Axioms 1–4, 6–10 are realized in S. But unless the concept of distance is altered in some way "distances" will be bounded and Axiom 5 will of necessity fail. Since "lines" are certain arcs in S (with their endpoints not in S), suppose one of our "lines" has endpoints M and N, and

---

[1] If $y_1 = y_2$, the relation reduces to $a = 0$. In this case the "line" passing through P_1 and P_2 is represented by $y = f(0 \cdot x + b) = f(b) = \text{constant}$, and is therefore a horizontal Euclidean line.

that A and B are two variable points on that line (Fig. 388). We seek a simple distance function $\widetilde{AB}$ such that:

(a)   As A or B tend to either M or N, then $\widetilde{AB} \to \infty$.

(b)   As A tends to B, then $\widetilde{AB} \to 0$.

Figure 388

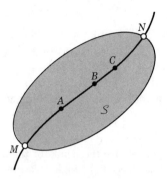

Let $\overline{XY}$ denote the arc length along the "line" from X to Y, the concept of distance as considered previously. Then to obtain a distance function with the above two properties in terms of the function $\overline{XY}$, one needs only observe that $1/\overline{XY} \to \infty$ as $\overline{XY} \to 0$. Many formulas having the desired properties may be constructed. One example is

$$\widetilde{AB} = \left| \frac{1}{\overline{AM}} - \frac{1}{\overline{BM}} - \frac{1}{\overline{AN}} + \frac{1}{\overline{BN}} \right|, \tag{85.2}$$

where absolute values have been inserted to make $\widetilde{AB} \geq 0$. Another is

$$\widetilde{AB} = \left| \log \overline{AM} - \log \overline{BM} - \log \overline{AN} + \log \overline{BN} \right|. \tag{85.3}$$

A desirable feature of these particular formulas is that if A, B, and C occur on the "line" in the order A-B-C (that is, $\overline{AB} + \overline{BC} = \overline{AC}$), then, assuming $\overline{AM} < \overline{BM}$ as indicated in Fig. 388,

$$\widetilde{AB} + \widetilde{BC} = \widetilde{AC}.$$

For, from (85.2),

$$\widetilde{AB} = \frac{1}{\overline{AM}} - \frac{1}{\overline{BM}} - \frac{1}{\overline{AN}} + \frac{1}{\overline{BN}},$$

$$\widetilde{BC} = \frac{1}{\overline{BM}} - \frac{1}{\overline{CM}} - \frac{1}{\overline{BN}} + \frac{1}{\overline{CN}}.$$

Hence,

$$\widetilde{AB} + \widetilde{BC} = \frac{1}{\overline{AM}} - \frac{1}{\overline{CM}} - \frac{1}{\overline{AN}} + \frac{1}{\overline{CN}} = \widetilde{AC}.$$

A similar calculation may be carried out for the formula in (85.3).

It now becomes evident that either of the distance concepts defined in (85.2) and (85.3) would overcome the main difficulty associated with a distance function for S. Since the choice is ours to make, let us take the formula in (85.3) since it may be put in a more compact form. Thus,

$$\widetilde{AB} = \left| \log \frac{\overline{AM} \cdot \overline{BN}}{\overline{AN} \cdot \overline{BM}} \right|. \tag{85.4}$$

The result is a non-Euclidean geometry in S which satisfies Axioms 1–10. If one desires a geometry in S which satisfies, in addition, Axiom 11 (the SAS postulate), then the objects of the geometry cannot be chosen so arbitrarily. It is remarkable that it can be done at all, and even more so that many of the ideas for "making up" non-Euclidean geometries introduced here can be utilized in the effort.

EXERCISES

1. Provide a plausible explanation for the validity of each of the Axioms 1–10 in the geometry for which the "points" are the points of the analytic plane, the "lines" are the curves of (85.1), and "distance" and "angle measure" are defined as above, if

$$f(x) = \log x, \qquad x > 0.$$

   In the course of your discussion work through these examples: (a) What "line" passes through $A[1, 0]$ and $B[-2, \log 7]$? Through $A[1, 0]$ and $C[e, 1]$? (b) Evaluate $\widetilde{\angle BAC}$ (see Section 73). (c) Do the "lines" $y = \log (3x + 4)$ and $y = \log (2x - 5)$ intersect? (d) Under what conditions are the "lines" $y = \log (ax + b)$ and $y = \log (a'x + b')$ parallel? Do any of the parallel axioms 12, 12′, or 12″ hold?

2. (a) Discuss the validity of Axioms 1–11 for the geometry in which

$$\begin{aligned} f(x) &= x && \text{when } x < 0, \\ &= 2x && \text{when } x \geqq 0. \end{aligned}$$

   (b) Which of the parallel axioms holds?

86. The Poincaré Model for Hyperbolic Geometry

One of the simplest ways to construct a geometry which satisfies the hyperbolic hypothesis (Section 42) is by restricting the geometry of the ordinary analytic plane to the interior of a circle ω, taking as "points" the interior points of the circle and as "lines" the subsets of Euclidean lines consisting of points interior to ω. Since the arc length on "lines" in this case would coincide with Euclidean distance ($\overline{AM} = AM$, $\overline{BN} = BN$, $\cdots$ and so on, in Fig. 389), the distance formula (85.4) reduces to

$$\widetilde{AB} = \left| \log \frac{AM \cdot BN}{AN \cdot BM} \right| = |\log [AB, MN]|.$$

As noted before, Axioms 1–10 are satisfied. But it is clear that Axiom 12′ is also satisfied since many "lines" pass through C, for example, which do not meet "line" $\overset{\leftrightarrow}{AB}$ (Fig. 389). However, Axiom 11 would not be satisfied here. A slight alteration of this example will provide the model we are seeking.

Figure 389

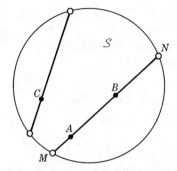

Let ω be a unit circle in the analytic plane and denote by S the set of all interior points of ω. Take as "points" the members of S and as "lines" the intersections of S with the *diameters of ω and with all the circles orthogonal to ω*. (The circle ω is called the **absolute**.) By itself, circle ω *contains no "points"* (see Fig. 390).

Figure 390

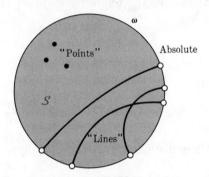

As in previous considerations the concept of distance will depend on the proposition that *each pair of "points" determine a unique line*. An elegant proof is afforded by the analytic method (for a synthetic proof recall Example 73.3).

Let ω be represented by the equation

$$x^2 + y^2 = 1. \tag{86.1}$$

It is easily shown that a circle

$$x^2 + y^2 + ax + by + c = 0 \tag{86.2}$$

Figure 391

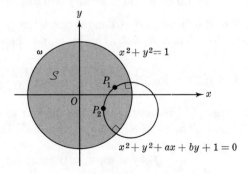

is orthogonal to ω if and only if $c = 1$ (see Exercise 5, Section 75). Hence, if $P_1[x_1, y_2]$ and $P_2[x_2, y_2]$ be any two "points," we have only to impose the condition that the circle represented by (86.2) passes through P_1 and P_2. This leads to the system of equations in the unknowns a and b

$$x_1a + y_1b = -x_1{}^2 - y_1{}^2 - 1,$$
$$x_2a + y_2b = -x_2{}^2 - y_2{}^2 - 1,$$

which has a unique solution if and only if

$$x_1y_2 \neq x_2y_1.$$

But this is equivalent to the requirement that P_1 and P_2 not be collinear with $O[0, 0]$, and in the case when they are collinear the "line" passing through P_1 and P_2 is simply a diameter. Hence, in all cases, each pair of "points" determine a "line" uniquely. ∎

Next, define "distance" in the following manner: If A and B be any two "points" and if $A = B$, set

$$\widetilde{AB} = 0.$$

If $A \neq B$ determine the "line" passing through A and B and let its endpoints on ω be M and N, then put

$$\widetilde{AB} = |\log R[AB, MN]| = \left|\log \frac{|AM||BN|}{|AN||BM|}\right|.$$

The reader should carefully distinguish between $\widetilde{AB}$ (hyperbolic distance) and AB (Euclidean distance).

Since $R[AB, MN] = 1/R[AB, NM]$, then $|\log R[AB, MN]| = |-\log R[AB, NM]| = |\log R[AB, NM]|$, so the order in which the points M and N are specified is immaterial. Thus $\widetilde{AB}$ is solely a function of the "points" A and B. Moreover, similar reasoning would show that $\widetilde{AB} = \widetilde{BA}$, and since $R[AB, MN] \neq 1$ for $A \neq B$, then $\log R[AB, MN] \neq 0$, which implies $\widetilde{AB} > 0$. Thus Axiom 3 is verified, with $\alpha = \infty$. Axioms 1–4 are thereby realized in S.

Consider next Axiom 5. Let A and B be any two distinct "points." Choose the endpoints M and N of the "line" $\overleftrightarrow{AB} = l$ so that $|AM| > |BM|$, and consequently, $|BN| > |AN|$ (Fig. 392). Then $R[AB, MN] > 1$ and therefore, $\log R[AB, MN] > 0$. Define

$$x = f(X) = \log R[AX, MN] = \log \left|\frac{AM \cdot XN}{AN \cdot XM}\right|$$

for all $X \in l$. Note that $f(X)$ is *positive* if X lies interior to arc AM (since in that case $|AM| > |XM|$ and $|XN| > |AN|$) and *negative* if X lies interior to arc AN. Also,

$$f(A) = \log 1 = 0$$

and, as a consequence of the preceding observation,

$$f(B) = \log R[AB, MN] > 0.$$

Figure 392

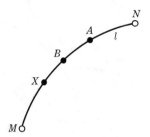

Thus f is a coordinate system for "line" l such that A is the origin and B has positive coordinate. Suppose "points" $X[x]$ and $Y[y]$ be selected at random on l.

Then
$$x = f(X) = -\log R[XA, MN] = -\log \left| \frac{XM \cdot AN}{XN \cdot AM} \right|,$$

$$y = f(Y) = \log R[AY, MN] = \log \left| \frac{AM \cdot YN}{AN \cdot YM} \right|.$$

Therefore,
$$y - x = \log \left| \frac{AM \cdot YN}{AN \cdot YM} \right| + \log \left| \frac{XM \cdot AN}{XN \cdot AM} \right|$$

$$= \log \left| \frac{AM \cdot YN \cdot XM \cdot AN}{AN \cdot YM \cdot XN \cdot AM} \right|$$

$$= \log R[XY, MN].$$

Taking absolute values yields[2]

$$\widetilde{XY} = |x - y|.$$

Thus Axiom 5 holds. It should be noted in this connection that (ABC) holds if and only if B lies on the interior of the arc AC determined by "line" $\overleftrightarrow{AC}$.

Finally, define "angle measure" to be the measure of the *curvilinear angles* determined by any "angle" (see Section 73). Thus, in Fig. 393 the "measure" of "angle" $\widetilde{\angle ABC}$ is defined by

$$\widetilde{\angle ABC} = \angle A'BC'.$$

The validity of Axioms 6, 7, and 8 would therefore follow essentially from the properties of Euclidean angle measure.

Although the direct proof that Axioms 9 and 10 are realized on S is somewhat tedious (the analytic method is probably the most feasible), that verification can be made quite plausible visually.

---

[2] It is clear that the formula $\widetilde{XY} = \log R[XY, MN]$ (without absolute values), produces the concept of *directed* "distance," with $\widetilde{XY} = y - x$.

Figure 393

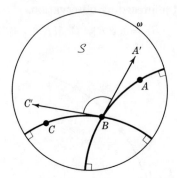

For Axiom 9—the plane separation axiom—let "line" l be given. The arc determined by l obviously determines two disjoint sets H_1 and H_2 of S (H_1 being the
intersection of S with either the *interior* of the *circle* determined by l, or one of the
half planes of the *line* determined by l). If $A \in H_1$ and $B \in H_2$, then the "segment"
$\overline{AB}$ meets l at C such that "(ACB)" [Fig. 394(a)]. Next, if A and B both lie in H_1

Figure 394

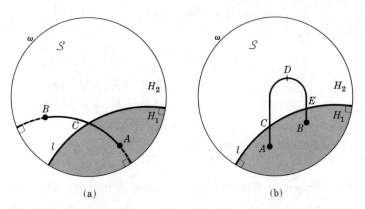

(a)   (b)

(or H_2) and if the "segment" $\overline{AB}$ meets l at C, from the fact that two distinct
circles orthogonal to ω cannot be tangent at a point inside ω, there must be a point
D in H_2 (or H_1). By the first part, "segment" $\overline{BD}$ meets l at an interior point E, and
hence, "lines" $\overleftrightarrow{AB}$ and l have two distinct points C and E in common. Therefore,
$\overline{AB}$ cannot meet l, and again by the first case, it therefore cannot contain any
points of H_2 (or H_1). Thus $\overline{AB} \subset H_1$ (or $\overline{AB} \subset H_2$).

Axiom 10 may be illustrated in Fig. 395, where it is clear that "ray" $\overrightarrow{PB}$ will lie
between "rays" $\overrightarrow{PA}$ and $\overrightarrow{PC}$ if and only if "point" B lies between "points" A and C.

We postpone the verification of Axiom 11, the SAS postulate, until the proper
tools for its proof have been introduced. For Axiom 12', the parallel axiom for hyperbolic geometry, observe "line" l and A any "point" not on it. Let "line" l have
endpoints M and N on ω. Then take $\overleftrightarrow{AC}$ and $\overleftrightarrow{AC^*}$ to be the "lines" through A and

Figure 395

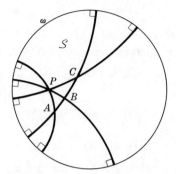

having M and N as endpoints, respectively (Fig. 396). It is then clear that $\overset{\leftrightarrow}{AC}$ and $\overset{\leftrightarrow}{AC}^*$ would be the two *limit parallels* corresponding to the two directions on l. Note that if "ray" $\overset{\rightarrow}{AX}$ is interior to "angle" CAC^*, then "line" $\overset{\leftrightarrow}{AX}$ will meet l; in all other positions $\overset{\leftrightarrow}{AX}$ is a "line" parallel to l, of which there are many. Thus we have Axiom 12′.

Figure 396

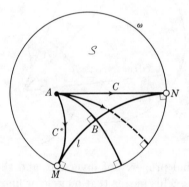

86.3 *Definition:* The model S described in the foregoing analysis is called the **Poincaré model** (after the noted French mathematician H. Poincaré, 1854–1912).

If the results of the next two sections are anticipated, the preceding discussion provides a proof of the following important assertion:

86.4 *Theorem:* The Poincaré model is a model for the hyperbolic plane $\mathcal{H}$.

87. Reflections in S

Let circle ω_1 be orthogonal to the absolute and consider it a circle of inversion in the Euclidean plane. Then ω transforms into itself. Moreover, points inside ω map into points inside ω (that is, S maps into S). To see this, let X be any point on ω and X' its inverse (Fig. 397).

Then $$XO_1 \cdot X'O_1 = r_1^2,$$

where O_1 and r_1 are the center and radius of ω_1. Now if P is an interior point of segment $\overline{XX'}$ and P' its image, then assuming that $XO_1 > 0$,

$$XO_1 < PO_1 < X'O_1,$$

or

$$\frac{r_1^2}{X'O_1} < \frac{r_1^2}{PO_1} < \frac{r_1^2}{XO_1}.$$

Therefore,

$$XO_1 < P'O_1 < X'O_1.$$

That is, P' is also an interior point of $\overline{XX'}$. This also establishes the fact that if H_1 and H_2 are the "half planes" of the "line" l determined by ω_1, then every point P of H_1 maps into a point P' of H_2 and vice versa. Thus the region H_1 is *inverse* to the region H_2, with each "point" of l left fixed under the mapping (since l is on the circle of inversion).

Figure 397

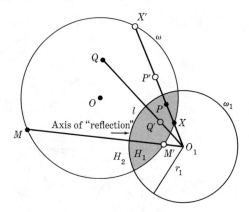

Recall the conformal property of inversion, and the fact that circles or lines map into circles or lines. This means that a circle or line orthogonal to ω maps into another circle or line also orthogonal to ω. Thus "lines" are preserved (Fig. 398).

Figure 398

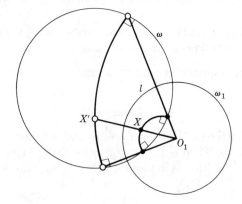

As far as the geometry of S is concerned then, this mapping would be a "linear" ("line"-preserving) transformation of S onto itself.

Now consider any two "points" A and B, with M and N the endpoints of the "line" $\overleftrightarrow{AB}$. Then "line" $\overleftrightarrow{AB}$ maps into a "line" $A'B'$ whose endpoints on ω are M' and N'—the images of M and N under the inversion (see Fig. 399). By a previous theorem,

$$R[AB, MN] = R[A'B', M'N'].$$

Therefore,

$$\widetilde{AB} = |\log R[AB, MN]| = |\log R[A'B', M'N']| = \widetilde{A'B'},$$

and we observe the important fact that the mapping we are considering is *"distance" preserving.* Of course the conformal property of inversion makes the mapping "conformal" in S. That is, the "measure" of each "angle" is invariant under the mapping.

Figure 399

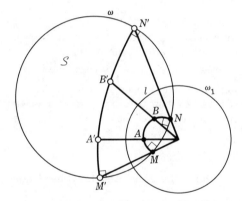

It is therefore concluded that in terms of the geometric objects "point," "line," "distance," and "angle measure" in S, the inversion mapping defined above behaves exactly as a *reflection* in a line in Euclidean geometry. We shall hereafter refer to this mapping as a **"reflection"** in S with "line" l as the **"axis of reflection."**

EXERCISES

1. Show that an "angle" and its "interior" in S may be mapped into part of a Euclidean angle and its interior.

2. If M is any "point" distinct from O (the center of ω), assuming that the radius of ω is unity, find the radius of the "line" l such that O is the "reflection" (inversion) of M in l in terms of a, the Euclidean distance from O to M. Where is the center of inversion located? *Hint:* Make use of the fact that l is orthogonal to ω.   *Ans:* $\sqrt{1 - a^2}/a$;

   on the ray $\overrightarrow{OM}$ *at a distance of* $1/a$ *from* O. *(Thus, it is the inverse of* M *with respect to* ω.)

3. Explain why the following Euclidean construction "bisects" a given "segment" not "collinear" with O: Let $\widetilde{AB}$ be the given "segment" (Fig. 400). Then $\widetilde{AB}$ is an arc of a

circle and the tangents to that circle at A and B meet at O_1, a point not on $\overleftrightarrow{AB}$. Draw $\overleftrightarrow{OO_1}$ and intersect it with $\overleftrightarrow{AB}$ at M. "Point" M is the desired "midpoint." *Hint:* The circle $[O_1, O_1A]$ is *orthogonal* to segment AB. Consider the "reflection" which maps M into O (see Exercise 2). What are the images of AB and circle $[O_1, O_1A]$?

Figure 400

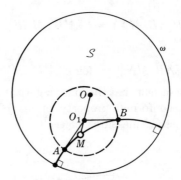

4. Find a Euclidean construction which "bisects" a "segment" AB when A and B are "collinear" with O, making use of "reflections."

★5. "Lines" l_1 and l_2 meet at $M \neq O$, with l_1 passing through O. Prove that the common chord of l_2 and ω meets $\overleftrightarrow{OM}$ at a point A such that $\widetilde{OA} = 2\,\widetilde{OM}$.

6. Use the Poincaré model to show that in hyperbolic geometry there exist "convex" sets which are not the entire "plane" but which nevertheless contain two intersecting "lines."

88. The SAS Postulate; Other Models

The "reflection" principle may now be used to establish Axiom 11. Suppose that "triangles" ABC and XYZ have $\widetilde{AB} = \widetilde{XY}$, $\widetilde{AC} = \widetilde{XZ}$, and $\widetilde{\angle A} = \widetilde{\angle X}$ (Fig. 401). These "triangles" may be "reflected" into "triangles" at O, so that A and X map into O. For, let the circles of "lines" $\overleftrightarrow{AB}$ and $\overleftrightarrow{AC}$ meet again at $\overline{A}$, a point exterior to ω. Put $r_1 = \sqrt{\overline{A}A \cdot \overline{A}O}$ and take $[\overline{A}, r_1] = \omega_1$ as circle of inversion. Since $OA \cdot O\overline{A} = 1$, then $r_1{}^2 = \overline{A}A \cdot \overline{A}O = (\overline{A}O + OA)\overline{A}O = \overline{A}O^2 - 1$ and $\omega_1 \perp \omega$. Further, A and O are inverse points; consequently, "triangle" ABC maps into "triangle" $OB'C'$, with (B, B') and (C, C') inverse pairs. Similarly, with $[\overline{X}, r_2]$ as circle of inversion "triangle" XYZ maps into "triangle" $OY'Z'$, where $r_2 = \sqrt{\overline{X}X \cdot \overline{X}O}$. Since "distance" and "angle measure" are invariant under inversion, then

$$\widetilde{\triangle ABC} \cong \widetilde{\triangle OB'C'}, \qquad \widetilde{\triangle XYZ} \cong \widetilde{\triangle OY'Z'}.$$

Therefore,

$$\widetilde{OB'} = \widetilde{OY'}, \qquad \widetilde{OC'} = \widetilde{OZ'}, \qquad \widetilde{\angle B'OC'} = \widetilde{\angle Y'OZ'}.$$

The desired result then depends on the assertion for "triangles" at O, which is somewhat more obvious. It easily follows that

$$OB' = OY', \qquad OC' = OZ', \qquad \angle B'OC' = \angle Y'OZ'$$

(see Exercise 1 below). Thus Euclidean triangles $B'OC'$ and $Y'OZ'$ are congruent, and there is then a Euclidean rotation or reflection about O which maps $\triangle OB'C'$ into $\triangle OY'Z'$. Under this mapping ω maps into itself and the circle of "line" $\overleftrightarrow{B'C'}$ must map into the circle of "line" $\overleftrightarrow{Y'Z'}$. Since Euclidean rotations or reflections are "distance" and "angle" preserving mappings, we have

$$\widetilde{\triangle OB'C'} \cong \widetilde{\triangle OY'Z'},$$

and it immediately follows that

$$\widetilde{\triangle ABC} \cong \widetilde{\triangle XYZ}. \quad \blacksquare$$

Figure 401

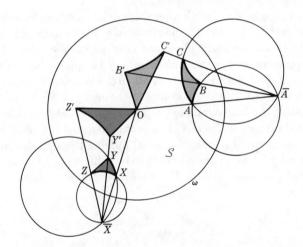

NOTE: We have just shown that any "triangle" is "congruent" to one having the center of the absolute as a vertex, such as "triangle" AOB in Fig. 402. Consider the measures of the "angles" of such a "triangle." What is your observation about the "angle sum" for any "triangle" in S?

Figure 402

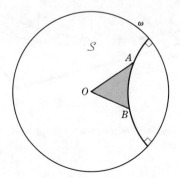

It is not surprising that having found one model for ℋ a variety of other models may be constructed. For example, the *Klein model* (due to Felix Klein, 1849–1929) consists of an *ellipse* as absolute, with all interior points as "points" and all *chords* of that ellipse as "lines." (The construction of this model for the special case in which the ellipse is a circle was actually begun in Section 86.) In that model the concept of "distance" is the same as for the Poincaré model, but the concept of "angle measure" must be altered.

The freedom with which a model for one geometry may be constructed upon another is indeed great, as suggested by our discussion in Section 84 (Chapter 10). In their development of the trigonometric formulas for hyperbolic geometry, both Bolyai and Lobachevski made use of the *limiting surface*, or *horosphere*—the surface of revolution generated by revolving a *limit curve*[3] about a radius in three-dimensional hyperbolic geometry. The intrinsic geometry of such a surface turns out to be, of all things, Euclidean geometry! Surely this must be the ultimate in proving the relative consistency of hyperbolic geometry.[4]

In this vein one might wonder whether there exists a model for spherical geometry which can be realized on the Euclidean plane. There does indeed—provided we agree to adjoin to the ordinary plane *one ideal point*. We take a fixed circle ω as absolute and consider as "points" *all points of the plane* (including the ideal point we added), and as "lines" *the circle ω, all circles passing through the endpoints of diameters of ω, and all lines passing through the center of ω*. To see how this works observe in Fig. 403(a) the stereographic projection indicated from the sphere onto plane π by a central projection from N. If ω is the image of the "equator" ω₁, then all other great circles map into circles (or lines) passing through the endpoints of a diameter of ω. Since such a mapping is known to be conformal, a model of the sphere is obtained on the plane where, like the Poincaré model, angle measure is Euclidean.

Figure 403(a)

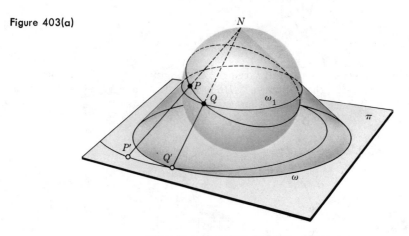

---

[3] Discussed in Section 92.
[4] See H. S. M. Coxeter, *Non-Euclidean Geometry* [5, pp. 197–220].

Figure 403(b)

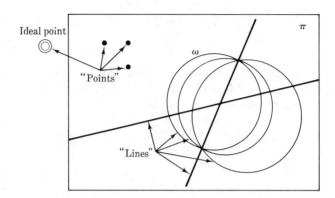

(Thus the angle-sum theorem for spherical geometry may be observed in a direct manner.) Finally, take the "image" of N to be the ideal point, which will then account for all points. For more details on this, the reader is referred to the little paperback by Meschowski, *Non-Euclidean Geometry* [20, Chapter 9].

EXERCISES

1. Prove that if $a = \widetilde{OA}$ for some "point" A, then

$$OA = \frac{e^a - 1}{e^a + 1},$$

where O is the center of ω. Thus for any two "points" A and B show that $\widetilde{OA} = \widetilde{OB}$ if and only if $OA = OB$.

2. Making use of Fig. 396 verify that when A varies on the "perpendicular" at B, as $\widetilde{AB} \to 0$, $\angle BAC \to \pi/2$ and as $\widetilde{AB} \to \infty$, $\angle BAC \to 0$.

3. Using your knowledge of orthogonal systems of circles prove that *any two non-intersectors possess a common perpendicular.*

4. Again referring to Fig. 396, let A be "reflected" into O, the center of ω, under an inversion of S onto itself. Thereby prove that the *two angles of parallelism are congruent.*

89. Hyperbolic Trigonometry of the Right Triangle

Since "triangles" may be studied more easily if two of their sides are Euclidean lines, we use the principle of "reflection" to map a given "right triangle" into one having a vertex at O. Since the given "triangle" is "congruent" to its image, it suffices to derive the desired formulas for the new triangle only. So we assume that "triangle" ABC has right angle at C and that $A = O$. As in elliptic geometry, we adopt standard notation:

$$a = \widetilde{BC}, \qquad b = \widetilde{AC}, \qquad c = \widetilde{AB}$$

$$A = \widetilde{\angle A}, \qquad B = \widetilde{\angle B}, \qquad C = \widetilde{\angle C} = \pi/2.$$

A modified version of a method due to Hajós and Szász[5] provides easy access to the classical formulas. It will be necessary to deal with the **hyperbolic functions,** defined for all real x by the equations

$$\sinh x = \frac{e^x - e^{-x}}{2}, \qquad \cosh x = \frac{e^x + e^{-x}}{2},$$

$$\tanh x = \frac{e^x - e^{-x}}{e^x + e^{-x}}. \tag{89.1}$$

From these may be defined the additional functions $\operatorname{csch} x = 1/\sinh x$, $\operatorname{sech} x = 1/\cosh x$, and $\coth x = 1/\tanh x$, with the obvious qualifying statements regarding the domain in each case. Two important identities will be used repeatedly:

$$\tanh x = \frac{\sinh x}{\cosh x} \tag{89.2}$$

and

$$\cosh^2 x - \sinh^2 x = 1, \tag{89.3}$$

[readily verified from (89.1)]. From (89.3) follow two further useful identities:

$$1 - \tanh^2 x = \operatorname{sech}^2 x, \qquad \text{and} \qquad \coth^2 x - 1 = \operatorname{csch}^2 x.$$

The first step is to obtain the special connection between the Euclidean distance OX and the hyperbolic distance $\widetilde{OX}$ for any point X in S, O the center of the absolute (Fig. 404). For convenience put $r = OX$ and $r' = \widetilde{OX}$. Then

$$r' = |\log R[OX, MN]|$$

$$= \log \frac{|OM||XN|}{|ON||XM|}$$

$$= \log \frac{1(1 + r)}{1(1 - r)}.$$

Hence,

$$e^{r'} = \frac{1 + r}{1 - r} \qquad \text{and} \qquad e^{-r'} = \frac{1 - r}{1 + r}.$$

By simple algebra the following relations hold:

$$\sinh r' = \frac{2r}{1 - r^2}, \qquad \cosh r' = \frac{1 + r^2}{1 - r^2}, \qquad \tanh r' = \frac{2r}{1 + r^2}. \tag{89.4}$$

---

[5] G. Hajós and P. Szász, "On a New Presentation of the Hyperbolic Trigonometry by Aid of the Poincaré Model," *Acta Mathematica Academiae Scientiarum Hungaricae,* **7** (1956), pp. 35–36.

Figure 404

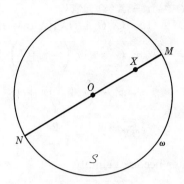

In "right triangle" ABC (Fig. 405) let ω_1 be the circle containing the "line" $\overleftrightarrow{BC}$ and let the Euclidean rays $\overrightarrow{OB}$ and $\overrightarrow{OC}$ meet the Euclidean chord $\overline{MN}$ at B' and C', and circle ω_1 at B'' and C'', respectively. The center O_1 of ω_1 will lie on line $\overleftrightarrow{AC}$.

Figure 405

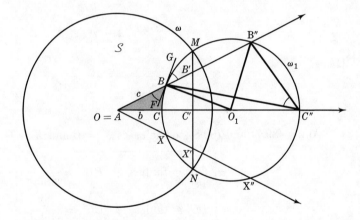

Now suppose X lies on "line" $\overleftrightarrow{BC}$ and that Euclidean ray $\overrightarrow{OX}$ meets chord $\overline{MN}$ at X' and circle ω_1 at X''. As before set $r = OX$, $r' = \widetilde{OX}$.

Observe that since ω_1 is orthogonal to ω, X and X'' are inverse points with respect to ω. Therefore, $OX'' = 1/OX$ or

$$OX'' = \frac{1}{r}. \tag{89.5}$$

To relate OX' to r note that X' lies on the radical axis of ω and ω_1. Hence,

$$\text{Power } X' \text{ (circle } \omega) = \text{Power } X' \text{ (circle } \omega_1).$$

Then in magnitude and in sign, and with $x = OX'$,

$$X'O^2 - 1 = X'X \cdot X'X''$$
$$x^2 - 1 = (X'O + OX)(X'O + OX'')$$
$$= (-x + r)\left(-x + \frac{1}{r}\right)$$
$$= x^2 - \left(r + \frac{1}{r}\right)x + 1.$$

Thus,

$$OX' = x = \frac{2r}{1 + r^2}. \tag{89.6}$$

In view of (89.4) the latter becomes simply

$$\boxed{OX' = \tanh r'.} \tag{89.7}$$

Finally, let XX'' be determined:

$$XX'' = XO + OX''$$
$$= -r + \frac{1}{r}$$
$$= \frac{1 - r^2}{r}.$$

Hence,

$$\boxed{XX'' = \frac{2}{\sinh r'}.} \tag{89.8}$$

Apply (89.7) and (89.8) to the cases $X = B$ and $X = C$:

$$\boxed{\begin{array}{ll} OB' = \tanh c, & BB'' = \dfrac{2}{\sinh c} \\[2ex] OC' = \tanh b, & CC'' = \dfrac{2}{\sinh b}. \end{array}} \tag{89.9}$$

From the Euclidean trigonometry of right triangle $OB'C'$ follows the first relation for hyperbolic trigonometry:

$$\cos A = \frac{OC'}{OB'} = \frac{\tanh b}{\tanh c}. \tag{89.10}$$

To obtain another, consider the fact that in Fig. 405, with $\overleftrightarrow{FG}$ tangent to ω_1 at B,

$$\angle BO_1B'' = 2\angle BC''B'' = 2\angle GBB'' = 2\angle ABF = 2\overset{\frown}{\angle ABC} = 2B.$$

From the Euclidean relation between a central angle and its intercepted chord, and by use of (89.9),

$$\sin B = \sin \frac{\angle BO_1B''}{2} = \frac{BB''}{2\,O_1C} = \frac{BB''}{CC''} = \frac{\sinh b}{\sinh c}. \tag{89.11}$$

At this point it has been proved for any hyperbolic right triangle ABC with right angle at C that

$$\cos A = \frac{\tanh b}{\tanh c} \qquad \text{and} \qquad \sin B = \frac{\sinh b}{\sinh c}.$$

By simply changing notation it also follows that

$$\cos B = \frac{\tanh a}{\tanh c} \qquad \text{and} \qquad \sin A = \frac{\sinh a}{\sinh c}.$$

Using $\cos^2 A + \sin^2 A = 1$ we have

$$1 = \frac{\tanh^2 b}{\tanh^2 c} + \frac{\sinh^2 a}{\sinh^2 c},$$

$$\sinh^2 c = \cosh^2 c \tanh^2 b + \sinh^2 a,$$

$$1 + \sinh^2 c = \cosh^2 c \left(\frac{\sinh^2 b}{\cosh^2 b} \right) + 1 + \sinh^2 a,$$

$$\cosh^2 c \cosh^2 b = \cosh^2 c \sinh^2 b + \cosh^2 a \cosh^2 b,$$
$$\cosh^2 c (\cosh^2 b - \sinh^2 b) = \cosh^2 a \cosh^2 b,$$
$$\cosh^2 c = \cosh^2 a \cosh^2 b.$$

Thus we have obtained a result which may be regarded as **the Pythagorean theorem for hyperbolic geometry:**

$$\boxed{\cosh c = \cosh a \cosh b.} \tag{89.12}$$

The final relation may now be obtained:

$$\tan A = \frac{\sinh a}{\sinh c} \cdot \frac{\sinh c / \cosh c}{\sinh b / \cosh b} = \frac{\sinh a}{\sinh c} \cdot \frac{\sinh c}{\cosh a \cosh b} \cdot \frac{\cosh b}{\sinh b},$$

or,

$$\tan A = \frac{\tanh a}{\sinh b}.$$

We have thus arrived at the **formulas relating the parts of a right triangle in hyperbolic geometry** (Fig. 406):

$$\boxed{\begin{array}{c} \cosh c = \cosh a \cosh b, \\[4pt] \sin A = \dfrac{\sinh a}{\sinh c}, \quad \sin B = \dfrac{\sinh b}{\sinh c}, \\[4pt] \cos A = \dfrac{\tanh b}{\tanh c}, \quad \cos B = \dfrac{\tanh a}{\tanh c}, \\[4pt] \tan A = \dfrac{\tanh a}{\sinh b}, \quad \tan B = \dfrac{\tanh b}{\sinh a}. \end{array}} \tag{89.13}$$

The reader should observe the remarkably close resemblance of the above relations to those of (80.8) for elliptic geometry.

Figure 406

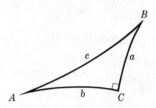

EXERCISES

1. Prove the identity

$$\cosh^2 x - \sinh^2 x = 1,$$

and show that $\sinh x > 0$ and $\cosh x > 1$ for $x > 0$. Sketch the graphs of the curves

$$y = \sinh x, \qquad y = \cosh x, \qquad y = \tanh x.$$

2. Making use of the law for exponents $a^b a^c = a^{b+c}$, prove the following identities for hyperbolic functions:

  (a) $\sinh (x \pm y) = \sinh x \cosh y \pm \cosh x \sinh y$;
  (b) $\cosh (x \pm y) = \cosh x \cosh y \pm \sinh x \sinh y$.

3. Deduce for a general "triangle" ABC the **hyperbolic law of sines**:

$$\boxed{\frac{\sinh a}{\sin A} = \frac{\sinh b}{\sin B} = \frac{\sinh c}{\sin C}.} \qquad\qquad \textbf{(89.14)}$$

4. Prove the **hyperbolic law of cosines**:

$$\boxed{\begin{aligned} \cosh a &= \cosh b \cosh c - \sinh b \sinh c \cos A, \\ \cosh b &= \cosh a \cosh c - \sinh a \sinh c \cos B, \\ \cosh c &= \cosh a \cosh b - \sinh a \sinh b \cos C. \end{aligned}} \qquad \textbf{(89.15)}$$

Hint: First write $\cosh a = \cosh b_2 \cosh h = \cosh (b - b_1) \cosh h$ (Fig. 407) and use one of the identities in Exercise 2; then use $\cos A = \tanh b_1 / \tanh c$, which will yield the first of the identities in (89.15). (Prove also for the case when D falls outside $\overline{AC}$.)

Figure 407

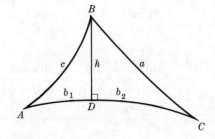

5. Making use of the hyperbolic law of cosines derive the formula for a cevian of a triangle in elliptic geometry (Fig. 408):

$$\cosh d = p \cosh b + q \cosh c.$$ (89.16)

where $p = \sinh a_1/\sinh a$ and $q = \sinh a_2/\sinh a$. *Hint:* The identity for $\sinh (a_1 + a_2)$ is appropriate.

Figure 408

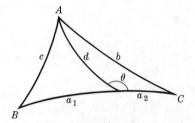

6. Prove the "double angle" formula $\sinh 2x = 2 \sinh x \cosh x$ and use it to derive the formulas for the medians of a hyperbolic triangle in terms of its sides:

$$\cosh m_a = (\tfrac{1}{2} \cosh b + \tfrac{1}{2} \cosh c)/\cosh \tfrac{1}{2}a,$$
$$\cosh m_b = (\tfrac{1}{2} \cosh a + \tfrac{1}{2} \cosh c)/\cosh \tfrac{1}{2}b,$$ (89.17)
$$\cosh m_c = (\tfrac{1}{2} \cosh a + \tfrac{1}{2} \cosh b)/\cosh \tfrac{1}{2}c.$$

7. If s is equal to one half the perimeter of hyperbolic triangle ABC, prove

$$\sin \frac{A}{2} = \sqrt{\frac{\sinh (s - b) \sinh (s - c)}{\sinh b \sinh c}},$$
$$\cos \frac{A}{2} = \sqrt{\frac{\sinh s \sinh (s - a)}{\sinh b \sinh c}}.$$ (89.18)

Hint: First prove the identity $\cosh x - \cosh y = 2 \sinh \tfrac{1}{2}(x + y) \sinh \tfrac{1}{2}(x - y)$, equivalent to $\cosh (x + y) - \cosh (x - y) = 2 \sinh x \sinh y$.

8. Prove the hyperbolic formula for the inradius of a triangle in terms of its sides:

$$\tanh r = \sqrt{\frac{\sinh (s - a) \sinh (s - b) \sinh (s - c)}{\sinh s}}.$$ (89.19)

9. Prove the following sequence of identities for the hyperbolic functions which are analogous to the half-angle formulas in trigonometry:

(a) $\quad \sinh \frac{1}{2}x = \sqrt{\dfrac{\cosh x - 1}{2}} \qquad (x \geq 0),$

(b) $\quad \cosh \frac{1}{2}x = \sqrt{\dfrac{\cosh x + 1}{2}},$

(c) $\quad \tanh \frac{1}{2}x = \sqrt{\dfrac{\cosh x - 1}{\cosh x + 1}} \qquad (x \geq 0),$

(c') $\quad \tanh \frac{1}{2}x = \dfrac{\cosh x - 1}{\sinh x},$

(c'') $\quad \tanh \frac{1}{2}x = \dfrac{\sinh x}{\cosh x + 1}.$

Hint: For (a), (b), and (c) begin with the two identities $\cosh^2 \frac{1}{2}x + \sinh^2 \frac{1}{2}x = \cosh x$ and $\cosh^2 \frac{1}{2}x - \sinh^2 \frac{1}{2}x = 1$. Formulas (c') and (c'') are merely two different rationalizations of (c).

10. Establish the following identities:

   (a) $\quad \sinh x + \sinh y = 2 \sinh \frac{1}{2}(x + y) \cosh \frac{1}{2}(x - y),$
   (b) $\quad \sinh x - \sinh y = 2 \cosh \frac{1}{2}(x + y) \sinh \frac{1}{2}(x - y),$
   (c) $\quad \cosh x + \cosh y = 2 \cosh \frac{1}{2}(x + y) \cosh \frac{1}{2}(x - y),$
   (d) $\quad \cosh x - \cosh y = 2 \sinh \frac{1}{2}(x + y) \sinh \frac{1}{2}(x - y).$

Hint: Prove $\sinh (u + v) + \sinh (u - v) = 2 \sinh u \cosh v.$

11. Prove the **equations of Gauss for hyperbolic geometry:**

$$
\begin{aligned}
\cosh \tfrac{1}{2}c \, \sin \tfrac{1}{2}(A + B) &= \cosh \tfrac{1}{2}(a - b) \cos \tfrac{1}{2}C, \\
\cosh \tfrac{1}{2}c \, \cos \tfrac{1}{2}(A + B) &= \cosh \tfrac{1}{2}(a + b) \sin \tfrac{1}{2}C, \\
\sinh \tfrac{1}{2}c \, \sin \tfrac{1}{2}(A - B) &= \sinh \tfrac{1}{2}(a - b) \cos \tfrac{1}{2}C, \\
\sinh \tfrac{1}{2}c \, \cos \tfrac{1}{2}(A - B) &= \sinh \tfrac{1}{2}(a + b) \sin \tfrac{1}{2}C.
\end{aligned}
\tag{89.20}
$$

★12. Starting with the formula $K = \pi - A - B - C$ for the area of a triangle in hyperbolic triangle, derive the **hyperbolic formula for the area of a right triangle in terms of its legs:**

$$
\tan \tfrac{1}{2}K = \tanh \tfrac{1}{2}a \, \tanh \tfrac{1}{2}b.
\tag{89.21}
$$

Hint: Use the trigonometric identities involving $\cosh x \pm \cosh y$ and Gauss's equations.

13. **Boundedness of area in hyperbolic geometry.** The maximal area of a right triangle in the hyperbolic plane is $\pi/2$. Prove this on the basis of Exercise 12 and the easily derived property of the hyperbolic functions, $\lim\limits_{x \to \infty} \tanh x = 1$. (This may also be seen by appealing directly to the Poincaré model.)

   NOTE: Regarding this phenomenon of hyperbolic geometry Gauss once wrote, "I have arrived at much which most people would consider sufficient for proof, but which proves nothing from my viewpoint. For example, if it could be proved that a rectilinear triangle is possible with an area exceeding any given area, I would be in a position to prove rigorously the whole of [Euclidean] geometry." (See H. E. Wolfe, *Non-Euclidean Geometry* [30, p. 128].)

★14. Prove the following hyperbolic analogue of Heron's formula:

$$\tan \tfrac{1}{4}K = \sqrt{\tanh \tfrac{1}{2}s \, \tanh \tfrac{1}{2}(s - a) \, \tanh \tfrac{1}{2}(s - b) \, \tanh \tfrac{1}{2}(s - c)}.$$ **(89.22)**

Hint: Use the identity $\tan \tfrac{1}{2}(x + y) = (\sin x + \sin y)/(\cos x + \cos y)$ and Gauss's equations (89.20).

15. Using hyperbolic trigonometry derive the following relation between the sides of a Saccheri quadrilateral in hyperbolic geometry:

$$\cosh b = \frac{\sinh \tfrac{1}{2}c}{\sinh \tfrac{1}{2}a},$$ **(89.23)**

where a is the base, b is the length of each leg, and c is the length of the summit. Give a trigonometric proof of the theorem in hyperbolic geometry: *The summit of a Saccheri quadrilateral is greater than the base.*

16. Let $ABCD$ be a Lambert quadrilateral in hyperbolic geometry, right angled at B, C, and D. (a) With the notation of Fig. 409 derive these relations using (89.23):

$$\sinh c = \sinh a \cosh b,$$

$$\cosh d = \frac{\cosh a \cosh b}{\sqrt{1 + \sinh^2 a \, \cosh^2 b}},$$ **(89.24)**

$$\cos A = \frac{\sinh a \sinh b}{\sqrt{1 + \sinh^2 a \, \cosh^2 b}}.$$

(b) Prove the *hypothesis of the acute angle* from the third formula in (89.24). Hence if $b \to \infty$ (with a fixed), $A \to 0$ and $\cosh d \to \coth a$. (Thus d is *bounded!*) If this seems contradictory, interpret on the Poincaré model.

Figure 409

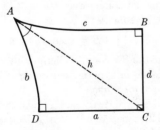

★17. In an n-sided regular polygon, let a, r, p, and K be, respectively, the *apothem* (distance from center to a side), *radius* (distance from center to a vertex), perimeter, and area (Fig. 410). Deduce the following formulas:

$$\sinh \frac{p}{2n} = \sinh r \, \sin \frac{\pi}{n},$$

$$\tan \frac{K}{4n} = \tanh \frac{a}{2} \, \tanh \frac{p}{4n}.$$ **(89.25)**

Hint: Use the hyperbolic trigonometry of right triangle OMB as shown in the figure and (89.21) (Exercise 12).

Figure 410

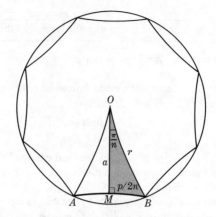

*18.   (a) Assuming $\{P_n\}$ is a sequence of regular inscribed polygons of side n of a hyperbolic circle of radius r such that the perimeter p_n and area K_n of the polygon P_n tend, respectively, to the circumference and area of the circle with increasing n, deduce from (89.25) the formulas for the circumference C and area K of a circle of radius r in hyperbolic geometry:

$$
\begin{aligned}
C &= 2\pi \sinh r, \\
K &= 4\pi \sinh^2 \frac{r}{2}.
\end{aligned}
$$

(89.26)

(b) Is the area of the hyperbolic plane finite or infinite? If your answer seems to contradict the result of Exercise 13 appeal to the Poincaré model to resolve the conflict.

*19.   Derive the formula for the relation between the acute angles of a right triangle and the legs opposite in hyperbolic geometry,

$$\cot A \cot B = \cosh a \cosh b.$$

Making use of the method of Section 82 (Chapter 10) derive a trigonometric proof of the deficiency of all triangles in hyperbolic geometry: *The angle sum of any triangle is less than the measure of a straight angle.*

20.   (a) The least upper bound for the area of all regular polygons having n sides in hyperbolic geometry is $(n-2)\pi$, but is not attained by any of those polygons. Prove. (b) Show that the area of the smallest regular quadrilateral with which one could tile the hyperbolic plane is $\frac{2}{3}\pi$. (c) Show that the area of the smallest regular pentagon with which one could tile the hyperbolic plane, as shown in Fig. 411, is $\frac{1}{2}\pi$. (d) Generalize. (Wolfe [30, p. 130].)

21.   *Euclidean geometry as an approximation to hyperbolic geometry.* Each of the formulas in hyperbolic geometry are, for small distances, approximated by their Euclidean counterparts to within any predetermined constant. Use the series expansions

$$\sinh x = x + \frac{x^3}{3!} + \frac{x^5}{5!} + \cdots, \qquad \cosh x = 1 + \frac{x^2}{2!} + \frac{x^4}{4!} + \cdots$$

to show that if all terms of order higher than two are ignored, then each of the hyperbolic relations for a right triangle in (89.13) reduce to their Euclidean counterparts. (In this connection see Example 80.9.)

Figure 411

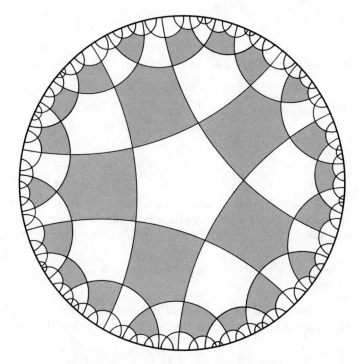

90. The Bolyai-Lobachevski Formula for the Angle of Parallelism

Recall that if $\overset{\leftrightarrow}{AB}$ is the "perpendicular" from A to l and if X varies on l on one side of B such that $BX \rightarrow \infty$, then the least upper bound A of $\theta = \angle BAX$ will be the measure of the angle of parallelism at A corresponding to the distance $a = AB$ (Fig. 412), previously denoted by $\Pi(a)$. One may readily obtain an expression for this function by making use of hyperbolic trigonometry. Put $x = BX$ and $y = AX$. In "right triangle" AXB,

$$\cos \theta = \frac{\tanh a}{\tanh y}. \tag{90.1}$$

Figure 412

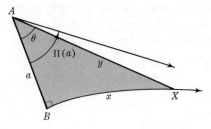

Now since the triangle inequality shows that $y \rightarrow \infty$ as $x \rightarrow \infty$, in taking the limit, (90.1) becomes

$$\cos A = \tanh a, \tag{90.2}$$

by use of the known limit

$$\lim_{y \to \infty} \tanh y = 1.$$

Subtracting the members of (90.2) from 1, yields the relation

$$2 \sin^2 \tfrac{1}{2}A = 1 - \tanh A. \tag{90.3}$$

Similarly, adding 1 to both sides of (90.2) produces

$$2 \cos^2 \tfrac{1}{2}A = 1 + \tanh A. \tag{90.4}$$

Therefore,

$$\tan^2 \tfrac{1}{2}A = \frac{1 - \tanh a}{1 + \tanh a}. \tag{90.5}$$

Using the definition of tanh a, it is now merely a matter of reducing the right-hand member of (90.5) to e^{-2a} by elementary algebra. Since both terms are positive, the desired formula for the angle of parallelism results from taking square roots:

$$\boxed{\tan \tfrac{1}{2}A = e^{-a}.} \tag{90.6}$$

This formula was obtained independently by both Bolyai and Lobachevski. It shows that the function $\Pi(a)$ defined earlier has the form

$$\Pi(a) = 2 \tan^{-1} e^{-a}. \tag{90.7}$$

The decreasing property of $\Pi(a)$ may now be observed from (90.7): $2 \tan^{-1} e^{-a}$ decreases monotonically from $\pi/2$ to 0 as a increases from 0 to ∞.

91.   The Cosine Inequality for Hyperbolic Geometry

The procedure given in Section 83 (Chapter 10) needs only slight modification to prove the characteristic property of hyperbolic geometry:

91.1   Theorem: If the sides of a triangle in the hyperbolic plane have respective lengths equal to those of a triangle in the Euclidean plane, then each of the angles of the hyperbolic triangle are *less than* the corresponding angles of the Euclidean triangle. (See Fig. 413.)

Figure 413

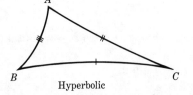
Hyperbolic

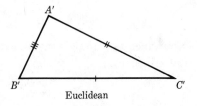
Euclidean

PROOF: The proof is based on the hyperbolic formula (89.18) (Exercise 7, Section 89),

$$\sin \frac{A}{2} = \sqrt{\frac{\sinh (s - b) \sinh (s - c)}{\sinh b \sinh c}},$$

and (as in Section 83) the Euclidean formula

$$\sin \frac{A'}{2} = \sqrt{\frac{(s - b) (s - c)}{bc}}.$$

It can be shown that

$$\frac{\sinh (s - b) \sinh (s - c)}{\sinh b \sinh} < \frac{(s - b) (s - c)}{bc},$$

based this time on the observation that the function

$$f(x) = \frac{\sinh x}{x}, \qquad x > 0$$

is increasing. Thus $A < A'$, which suffices to prove the assertion. ∎

The obvious implication is the

91.2 Cosine Inequality for Hyperbolic Geometry: For any hyperbolic triangle ABC, in standard notation

$$\boxed{a^2 > b^2 + c^2 - 2bc \cos A.}$$

EXERCISES

1. Provide the missing details for the proof of (91.1), then prove (91.2).
2. Prove that the line segment joining the midpoints of two sides of a triangle in hyperbolic geometry is less than one half the length of the third side. [See Example (83.3).]
3. Prove

   91.3 Theorem: Each cevian of a triangle in the hyperbolic plane has length *less than* that of the corresponding cevian of a triangle in the Euclidean plane whose sides have lengths equal to those of the hyperbolic triangle and where the cevians cut proportional segments on the sides to which they are joined.

In the following exercises all distances considered are hyperbolic (it is to be noted that this sequence of exercises is the hyperbolic analogue of Exercises 5–11, Section 83 for elliptic geometry):

4. In a hyperbolic triangle ABC, points D and E are chosen on sides $\overline{AB}$ and $\overline{AC}$ such that

   $$AD = \tfrac{2}{3}AB \qquad \text{and} \qquad AE = \tfrac{2}{3}AC.$$

   Prove that $DE < \tfrac{2}{3}BC$.

5. In Figs. 414(a, b), $\overleftrightarrow{DE}$ and $\overleftrightarrow{BC}$ are each perpendicular to $\overleftrightarrow{AC}$. Prove, in magnitude and in sign

   $$\frac{\tanh AD}{\tanh AB} = \frac{\tanh AE}{\tanh AC}. \tag{91.4}$$

   $$\frac{\sinh AD}{\sinh AB} = \frac{\sinh DE}{\sinh BC}. \tag{91.5}$$

6. *Linearity number and Menelaus' theorem for hyperbolic geometry.* Let D, E, and F be menelaus points on the sides of triangle ABC and define in, magnitude and in sign,

$$\mathcal{K}\begin{bmatrix} ABC \\ DEF \end{bmatrix} = \frac{\sinh AF}{\sinh FB} \cdot \frac{\sinh BD}{\sinh DC} \cdot \frac{\sinh CE}{\sinh EA}$$

as the **linearity number** for hyperbolic geometry. Basing your proof on the relation (91.5) (Exercise 5) and the pattern of the argument presented earlier for menelaus' theorem (Theorem 62.3) prove

91.6 Menelaus' Theorem for Hyperbolic Geometry: The menelaus points D, E, and F of a hyperbolic triangle ABC are collinear if and only if

$$\mathcal{K}\begin{bmatrix} ABC \\ DEF \end{bmatrix} = -1.$$

[See Fig. 414(c); recall lemma in Section 7.]

Figure 414

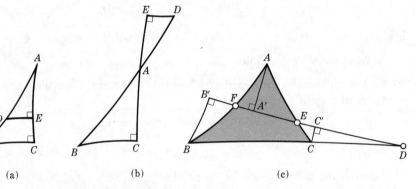

(a)   (b)   (c)

7. *Desargues' theorem for hyperbolic geometry.* Follow the proof of Desargues' theorem outlined in Section 62 and devise a proof for hyperbolic geometry based on Theorem 91.6.

8. Prove

91.7 Ceva's Theorem for Hyperbolic Geometry: In an elliptic triangle ABC with menelaus points D, E, and F given, cevians $\overleftrightarrow{AD}$, $\overleftrightarrow{BE}$, and $\overleftrightarrow{CF}$ are concurrent if and only if any two of them intersect and

$$\mathcal{K}\begin{bmatrix} ABC \\ DEF \end{bmatrix} = 1.$$

9. (a) Prove that the medians of any hyperbolic triangle are concurrent. (b) Prove that the altitudes of a hyperbolic triangle whose angles are each acute are concurrent. *Hint:* Use Ceva's theorem (Exercise 8). For the altitudes it will be necessary to use in addition some hyperbolic trigonometry.

NOTE: It would be interesting to devise strictly synthetic proofs of the theorems stated here.

10. Define **hyperbolic cross ratio** by setting

$$\mathcal{K}[AB, CD] = \frac{\sinh AC \sinh BD}{\sinh AD \sinh BC},$$

where A, B, C, and D are any four distinct points on a directed line. Prove that if the harmonic construction be carried out in the hyperbolic plane for the points A, B, C, and D as indicated in Fig. 415,

$$\mathcal{H}[AB, CD] = -1.$$

Figure 415

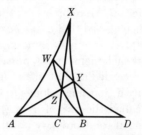

*92. Circles, Limit Curves, and Equidistant Loci

It is the purpose of this section to introduce the reader to a subject which Bolyai used in his development of hyperbolic trigonometry. Since it is about as easy to use the synthetic approach as it is to appeal to the Poincaré model, the former method will be used. In what follows, therefore, all concepts are to be understood as hyperbolic, and the special notation used previously to designate objects in hyperbolic geometry will be omitted.

Let us begin by noticing a property of circles in Euclidean geometry. Consider a circle ω_1 tangent to a fixed line t at a fixed point A (Fig. 416) and let the radius of ω_1 increase without bound. Then ω_1 *will assume the position of line t as a limiting case.* To be more specific, if a perpendicular $\overset{\leftrightarrow}{XX'}$ to t at any fixed point X on t cuts ω_1 at W, and W is nearest X, then W tends to X as limit.

Figure 416

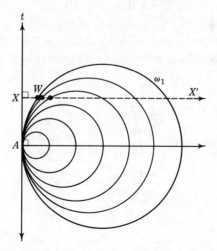

This property is related to the fact that in Euclidean geometry three distinct points are always either *collinear or concyclic*—there are no "intermediate" cases.

Not so in hyperbolic geometry: There is a *third* possibility, which gives rise to a very interesting theory, the object of the present discussion.

To facilitate our development we shall *augment* the hyperbolic plane at the outset by the addition of *ideal* and *ultra-ideal* points. Let 𝔉 be a family of lines which are limit parallel to each other in the same direction.[6] We agree that each member of 𝔉 **passes through a common ideal point** Ω. More precisely, the family 𝔉 can *itself* be taken as the ideal point Ω, and it is then agreed that *a line passes through* Ω *if and only if the line belongs to* 𝔉. Similarly, let 𝒢 be a family of hyperparallels having a fixed perpendicular in common. We say that each member of 𝒢 **passes through a common ultra-ideal point** Ψ. Again 𝒢 itself may be regarded as that ultra-ideal point if it is agreed that a line passes through Ψ if and only if that line belongs to 𝒢. (The common perpendicular of the lines through Ψ is called the **axis** of Ψ or 𝒢.) The original points of the (hyperbolic) plane will be referred to as **ordinary points**.

Thus, the hyperbolic plane is extended in such a way that there is a "point" corresponding to each family of concurrent, limit-parallel, or hyperparallel lines. In the augmented hyperbolic plane *each pair of lines "intersect" in either an ordinary point, an ideal point, or an ultra-ideal point.* For convenience, denote the ordinary points by the capital Latin characters A, B, C, $\cdots$ (as before), the ideal points by the capital Greek omega Ω, Ω′, Ω″, $\cdots$, and ultra-ideal points by the capital Greek psi Ψ, Ψ′, Ψ″, $\cdots$; general "points" will be denoted Π, Π′, Π″, $\cdots$.

92.1   Definition: Three or more lines are said to be **concurrent** if and only if they pass through the same ordinary, ideal, or ultra-ideal point.

Consider now any family 𝒦 of lines concurrent at Π (use *rays* from Π if Π is ordinary) and locate any point A on one of those lines (Fig. 417). Point A' is located on each line l in 𝒦 such that

$$\angle A A' \Pi = \angle A' A \Pi,$$

and in case Π is ultra-ideal, on the same side of the axis of 𝒦 as A. (The proof for the *existence* of a unique A' on each member of 𝒦 is indicated in Exercise 7 below.) The points A and A' are said to **correspond with respect to** Π.

Figure 417

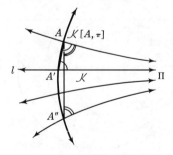

---

[6] Throughout the present discussion the symmetric and transitive properties of limit parallelism of lines will be taken for granted. This fact can, however, be readily observed from the Poincaré model.

As line l varies through Π, A' presumably describes a certain curve, namely one of the orthogonal trajectories of $\mathcal{K}$. The locus of A' will be denoted

$$\mathcal{K}[A, \Pi].$$

The reader should be able to show that $\mathcal{K}[A, \Pi]$ is a *circle* with Π as center if Π is an ordinary point.

92.2   Definition: The locus $\mathcal{K}[A, \Pi]$ just defined is called a **limit curve** (sometimes also called **horocycle**), or an **equidistant locus**[7] according to whether Π is an ideal or ultra-ideal point. The point Π is called the **center.**

The properties of the locus $\mathcal{K}[A, \Pi]$ ultimately depend on the following result.

92.3   Theorem: In the augmented hyperbolic plane the perpendicular bisectors of the sides of a triangle are concurrent.

PROOF: Since the theorem was already established if any two of the perpendicular bisectors meet in an ordinary point (34.5), assume that two of them, say l and m, meet at Ψ, an ultra-ideal point (Fig. 418). Then l and m have a common perpendicular $\overset{\leftrightarrow}{RS}$, the axis of Ψ. Drop perpendiculars $\overset{\leftrightarrow}{AX}$, $\overset{\leftrightarrow}{BY}$, and $\overset{\leftrightarrow}{CZ}$ to $\overset{\leftrightarrow}{RS}$. Since S lies on the perpendicular bisector of $\overline{BC}$, it follows that $\triangle BLS \cong \triangle CLS$ and that, therefore, in right triangles BSY and CSZ, $BS = CS$ and $\angle BSY = \angle CSZ$. Thus $\triangle BSY \cong \triangle CSZ$ and $BY = CZ$. Hence $\Diamond BCZY$ is a Saccheri quadrilateral with base $\overline{YZ}$. In the same manner it may be proved that $\Diamond ACZX$ is a Saccheri quadrilateral with base $\overline{XZ}$. Hence, $AX = BY = CZ$ and $\Diamond ABYX$ is a Saccheri quadrilateral with base $\overline{XY}$. Therefore, the perpendicular bisector n of $\overline{AB}$ is also perpendicular to $\overset{\leftrightarrow}{RS}$ and by definition of ultra-ideal points, passes through Ψ.

To finish the proof, suppose l and m meet at an ideal point Ω. But by the two preceding cases n must also pass through Ω, for if n meets l in an ordinary point O, then $l \cap m \cap n = O = \Omega$ and if n meets l in an ultra-ideal point Ψ, then $l \cap m \cap n = \Psi = \Omega$. Hence, n meets both l and m in an ideal point, which must be Ω. ∎

Figure 418

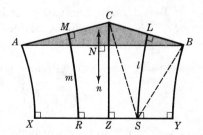

92.4   Corollary: The lines which pass through the vertices of a triangle and are concurrent with the perpendicular bisectors of the sides of that triangle form congruent angles with those sides.

---

[7] See Exercise 6 following.

PROOF: The proposition is obvious if the point of concurrency Π is an ordinary point, and if Π is ultra ideal, then the result follows from the fact that each pair of angles are the summit angles of a Saccheri quadrilateral (Fig. 418). Assume that Π is the ideal point Ω (Fig. 419). But in this case, $\overrightarrow{B\Omega}$ and $\overrightarrow{C\Omega}$ are limit parallel to $\overleftrightarrow{L\Omega}$ and since $\overleftrightarrow{L\Omega} \perp \overleftrightarrow{BC}$, angles $LB\Omega$ and $LC\Omega$ are the angles of parallelism corresponding to the distances LB and CL, respectively. Since $BL = CL$,

$$\angle LB\Omega = \angle LC\Omega.$$

Similarly,

$$\angle MA\Omega = \angle MC\Omega \quad \text{and} \quad \angle NA\Omega = \angle NB\Omega. \quad \blacksquare$$

One more result will be needed.

Figure 419

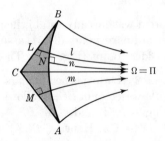

92.5 Lemma: If $\angle AB\Pi = \angle BA\Pi$, the perpendicular bisector l of $\overline{AB}$ passes through Π.

PROOF: If Π is ordinary, then triangle $AB\Pi$ is isosceles with $A\Pi = B\Pi$ and therefore, Π lies on l. If Π is ideal, consider the limit parallel $\overrightarrow{M\Pi}$ to $\overrightarrow{A\Pi}$ and $\overrightarrow{B\Pi}$ at M, the midpoint of $\overline{AB}$ (Fig. 420). Let $\overleftrightarrow{MC}$ and $\overleftrightarrow{MD}$ be the perpendiculars from M.[8] Then $\triangle AMC \cong \triangle BMD$ and therefore, $\angle AMC = \angle BMD$ and $MC = MD$. Since $\overline{\angle CM\Pi}$ and $\overline{\angle DM\Pi}$ are angles of parallelism corresponding to the equal distances MC and MD, $\angle CM\Pi = \angle DM\Pi$. Therefore, $\angle AM\Pi = \angle BM\Pi$ and $\overleftrightarrow{M\Pi}$ is the perpendicular bisector l of $\overline{AB}$. The remaining case follows immediately from the properties of Saccheri quadrilaterals. $\blacksquare$

The fundamental properties of limit curves and equidistant loci may now be derived.

92.6 Theorem: If B is any point of the locus $\mathcal{K}[A, \Pi]$, then $\mathcal{K}[B, \Pi] = \mathcal{K}[A, \Pi]$.

PROOF: Let $C \in \mathcal{K}[A, \Pi]$. Then B and C correspond to A with respect to Π. Hence $\angle BA\Pi = \angle AB\Pi$ and $AC\Pi = \angle CA\Pi$ (refer to Fig. 419). By the lemma,

---

[8] This proof tacitly assumes $\overline{\angle A}$ and $\overline{\angle B}$ are acute; the reader may easily supply the proof in the case when $\overline{\angle A}$ and $\overline{\angle B}$ are obtuse. Actually it can be shown that the latter case never occurs, but this is not obvious from the present development.

the perpendicular bisectors l and n meet at Π. By Theorem 92.3, the remaining perpendicular bisector m passes through Π and by Corollary 92.4, $\angle CB\Pi = \angle BC\Pi$. Hence $C \in \mathcal{K}[B, \Pi]$, and it follows that $\mathcal{K}[A, \Pi]$ is a subset of $\mathcal{K}[B, \Pi]$ and also that

$$\mathcal{K}[A, \Pi] = \mathcal{K}[B, \Pi]. \quad \blacksquare$$

Figure 420

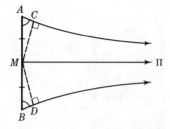

92.7 Theorem: The perpendicular bisector of any chord of a limit curve or equidistant locus $\mathcal{K}[A, \Pi]$ passes through the center Π.

PROOF: Immediate from Lemma 92.5 and the fact that each pair of points of $\mathcal{K}[A, \Pi]$ correspond by Theorem 92.6. $\blacksquare$

92.8 Theorem: Any three noncollinear ordinary points of the hyperbolic plane determine a unique circle, limit curve, or equidistant locus.

PROOF: Left to the reader.

Thus we can see that limit curves and equidistant loci behave formally like ordinary circles, the chief distinction being that their centers are not ordinary points and their radii are not defined.

The situation implicit in Theorem 92.8 may be summed up dynamically: Let $\overleftrightarrow{AB}$ be perpendicular to l at A and take A' any point not on l or $\overleftrightarrow{AB}$ and determine Π on $\overleftrightarrow{AB}$ such that $\angle A'A\Pi = \angle AA'\Pi$ (Fig. 421). Then let $\angle A'AB$ tend to $\pi/2$ as limit. For small values of $\angle A'AB$ the loci $\mathcal{K}[A, \Pi]$ are *circles*. But as $\angle A'AB$ becomes larger Π tends to the ideal point Ω and $\mathcal{K}[A, \Pi]$ approaches a *limit curve* as a first limiting case (thus its name). Then, as $\angle A'AB$ continues to increase, $\mathcal{K}[A, \Pi]$ passes into the *equidistant loci*, which in turn have line l as a limiting case. When $\mathcal{K}[A, \Pi]$ is an equidistant locus, then as $\angle A'AB$ tends to $\pi/2$ the axis of Π also tends to line l as a limiting case.

The following exercises will now guide the reader in visualizing the preceding analysis on the Poincaré model, thereby demonstrating the great perfection with which the model emulates any development in abstract reasoning in hyperbolic geometry.

Figure 421

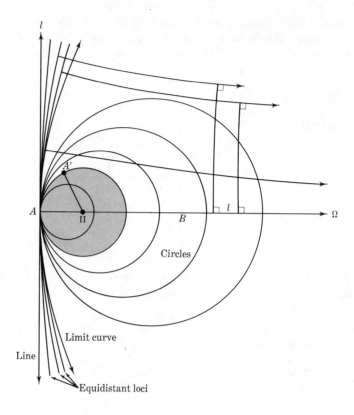

EXERCISES

1. Show by inversion that the Euclidean circles *interior* to ω are the "circles" for S.
2. Consider the three families of concurrent "lines" in S indicated in Fig. 422. Which of the three have an ordinary "point" of concurrency? An ideal point? An ultra-ideal point? Where are the ideal points? The ultra-ideal points?

Figure 422

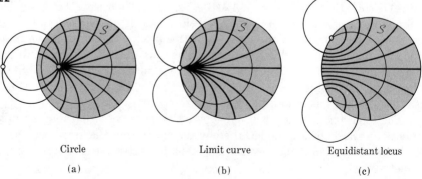

Circle	Limit curve	Equidistant locus
(a)	(b)	(c)

3. Using the fact that the curves $\mathcal{K}[A, \Pi]$ are orthogonal to the lines through Π, find an example of a "circle," a limit curve, and an equidistant locus by observing the three types of concurrent "lines" shown in Fig. 422. How can the limit curves and equidistant loci be identified in S?

4. Starting with a "circle" ω_1 having "center" Π, a fixed "line" as tangent, and a fixed "point" of contact, discuss the crucial positions of ω_1 when Π varies as an ordinary, ideal, and ultra-ideal point. (See Fig. 423.)

Figure 423

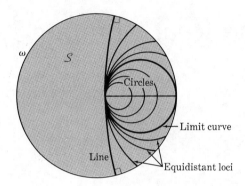

5. Show that Euclidean chords of ω (not diameters) are equidistant loci of S.

6. Using the synthetic method prove that in hyperbolic geometry the locus of points on one side of a line and *equidistant* from it is an *equidistant locus* as defined in (92.2) (thus its name). *Hint:* Recall the argument associated with Fig. 418.

7. (a) Prove that in hyperbolic geometry each point A has a unique point A' corresponding to it on any line through Π, an ordinary or ultra-ideal point. (b) Solve the existence problem when Π is ideal. *Hint:* Let B be *any* point on a line through Π and prove that the bisectors of angles $AB\Pi$ and $BA\Pi$ meet at a point which is equidistant from $\overleftrightarrow{A\Pi}$ and $\overleftrightarrow{B\Pi}$.

8. Prove Theorem 92.8.

appendix 1:
The Real Number System

The elementary properties of the real numbers are used throughout the axiomatic development of geometry in the early chapters, so we outline briefly what constitutes this system. We also mention the basic rules governing inequalities and the absolute value concept.

The most important property of the real numbers system is that it constitutes a **field.** That is, it satisfies the following properties (we denote arbitrary real numbers by a, b, c, $\cdots$ and assume that there are at least *two* such elements):

		FOR ADDITION	FOR MULTIPLICATION
1.	Closure Law:	$a + b$ is a real number for all a and b.	ab is a real number for all a and b.
2.	Commutative Law:	$a + b = b + a$.	$ab = ba$.
3.	Associative Law:	$(a + b) + c = a + (b + c)$.	$(ab)c = a(bc)$.
4.	Existence of Identity:	There is an element 0 such that for any a, $$0 + a = a.$$	There is an element 1 such that for any a, $$1 \cdot a = a.$$
5.	Existence of Inverses:	To each a there corresponds $-a$ such that $$a + (-a) = 0.$$	To each $a \neq 0$ there corresponds a^{-1} such that $$a \cdot a^{-1} = 1.$$
6.	Distributive Law:	For each a, b, and c, $a(b + c) = ab + ac$.	

From these basic laws all the familiar rules of algebra may be developed. For example,

(a) $(-a)b = -ab$ and $a \cdot 0 = 0$.

(b) $a^m \cdot a^n = a^{m+n}$, where m and n are integers.

(c) $(a + b)^2 = a^2 + 2ab + b^2$.

(d) $\dfrac{a}{b} + \dfrac{c}{d} = \dfrac{ad + bc}{bd}$, provided $b \neq 0$ and $d \neq 0$.

(e) For any positive integer n,
$$(a - b)(a^n + a^{n-1}b + a^{n-2}b^2 + \cdots + ab^{n-1} + b^n) = a^{n+1} - b^{n+1}.$$

The real number system is an **ordered** field. That is, there is an order relation $>$ which obeys the following laws:

7. If a is any real number, then either $a > 0$, $a = 0$, or $0 > a$, the cases being mutually exclusive. (One could say less formally that each real number is either *positive, zero,* or *negative.*)

8. $a > b$ if and only if $a - b > 0$.

9. $a > 0$ and $b > 0$ implies that $a + b > 0$.

10. $a > 0$ and $b > 0$ implies that $ab > 0$.

From these can be proved the **transitive law:**

(f) $a > b$ and $b > c$ implies that $a > c$.

For, suppose $a > b$ and $b > c$, or by **8**, $a - b > 0$ and $b - c > 0$. Then by **9**
$$(a - b) + (b - c) > 0.$$
But this simplifies to $a - c > 0$ or $a > c$.

Now the familiar rules for inequalities follow. To illustrate, suppose we want to show that *any counting number*[1] *is positive.*

First, observe that **8** and **9** imply the rule

(g) If $a > b$ and $c \geq d$, then $a + c > b + d$.

For, if $a > b$ and $c \geq d$, then by **8** $a - b > 0$ and either $c - d > 0$ or $c - d = 0$. In either case however, **9** implies
$$(a - b) + (c - d) > 0,$$
or
$$(a + c) - (b + d) > 0,$$
and **g** follows by again using **8**. Note that **g** in turn implies that $a > 0$ holds if and only if $-a < 0$ (add $-a \geq -a$ to $a > 0$ and $a \geq a$ to $0 > -a$).

---

[1] The *counting numbers* may be defined by 1, $1 + 1 = 2$, $1 + 2 = 3$, $\cdots$ where 1 is the element guaranteed to exist by **4**. Assuming that the process continues as indicated by the dots is tantamount to assuming the principle of *mathematical induction.*

Now consider 1. By **7** either $1 > 0$, $1 = 0$, or $1 < 0$. Since $1 \neq 0$ (Why?) then suppose $1 < 0$. Hence $-1 > 0$ and by **9**,

$$(-1)(-1) > 0,$$

or

$$1 > 0,$$

contradicting the assumption $1 < 0$ (by **7** we cannot have *both* cases). Therefore, $1 > 0$. By successive applications of **9** we obtain

$$2 = 1 + 1 > 0, \qquad 3 = 1 + 2 > 0, \qquad \cdots,$$

as was to have been proved.

Further laws for inequalities can be easily proved:

(h) $a > b$ and $c \geq d > 0$ implies $ac > bd$.

(i) $a^2 \geq 0$ for any real a.

EXERCISE

Show that if $a > 0$, then a^{-1} exists and $a^{-1} > 0$.

The final two properties complete the description of the real number system.

11. Archimedian Property: For any two positive real numbers a and b there exists a counting number n such that

$$na > b.$$

12. Completeness Property: Each bounded sequence of real numbers has at least one convergent subsequence.

In dealing with inequalities it is often convenient to introduce absolute value. The inequality $-3 < x < 3$, for instance, can be simplified to $|x| < 3$, where $|x|$ is the absolute value of x. By definition, the absolute value of any real number is its value without regard to its sign. More precisely, define the **absolute value** of a, denoted $|a|$, to be a if $a \geq 0$, and $-a$ if $a < 0$. For example, since $-6 < 0$ we have, with $a = -6$, $|-6| = -(-6) = 6$.

There is an alternate way to define absolute value, welcomed by the novice but avoided by those concerned about fundamentals since this definition cannot be used in an arbitrary field: For each real a,

$$|a| = \sqrt{a^2},$$

where $\sqrt{x}$ denotes the *positive* square root of x.

A few simple properties of absolute value may be cited:

(j) For each a, $|-a| = |a|$ and $-|a| \leq a \leq |a|$.

(k) For $b > 0$, $|a| < b$ if and only if $-b < a < b$.

(l) For $b > 0$, $|a| > b$ if and only if either $a > b$ or $a < -b$.

(m) $|ab| = |a| \, |b|$.

(n) $|a + b| \leq |a| + |b|$.

Proofs to any of these may be readily supplied. Property **j** follows directly by definition, while for **k**, if $|a| < b$, then by **j**,

$$a \leqq |a| < b,$$

or $a < b$. Again **j** implies

$$-a \leqq |a| < b$$

or $a > -b$. Therefore, $-b < a < b$. The converse follows in an analogous manner.

Properties **l** and **m** being left for the reader, we conclude with a proof of **n**: By **j**

$$-|a| \leqq a \leqq |a|$$
$$-|b| \leqq b \leqq |b|.$$

Therefore,

$$-|a| - |b| \leqq a + b \leqq |a| + |b|$$

and by **k**

$$|a + b| \leqq |a| + |b|.$$

appendix 2:
Least Upper Bound

If a set of real numbers is **bounded,** that is, if there is $M > 0$ such that for all x in the set

$$|x| < M,$$

it is apparently a simple observation to assert that there is a *smallest closed interval* $[a, b] = \{u : a \leq u \leq b\}$ *which contains that set as a subset.* Yet this property is equivalent to the completeness property for the reals (**12,** Appendix 1).

Before we proceed to a more formal discussion let us consider a few examples. Suppose

(1) $\quad S = \left\{ \dfrac{1}{n} : n = 1, 2, 3, \cdots \right\}.$

Here, the smallest closed interval containing S is $[0, 1]$. Note that the right endpoint, 1, belongs to S while the left endpoint, 0, does not. Consider next the set

(2) $\quad S = \left\{ \dfrac{1}{2x} - \dfrac{1}{x^2} : x > 1 \right\}.$

In order to find the bounds on S in this case, some calculus is helpful. We conclude that the smallest interval containing the set in (2) is $[-\frac{1}{2}, \frac{1}{16}]$ (let the reader complete the details).

In each of the preceding examples the *right endpoint* b of the smallest interval $[a, b]$ containing the set S enjoys the following two properties:

(a)   For each element $x \in S$, $x \leq b$. That is, b is an **upper bound** of S.
(b)   If M is any other upper bound of S, then $b \leq M$.

To summarize, *the number* b *is an upper bound that is least among all upper bounds.* Thus it is called the **least upper bound** of S. The notation

$$b = \sup S$$

has become traditional, so we use it here ("sup" is short for "supremum").

The *left* endpoint a of the smallest interval containing S is analogously the **greatest lower bound,** with the obvious meaning, and is denoted

$$a = \inf S$$

(where in this case "inf" stands for the term "infimum").

The following theorem is an important property of the real numbers, the proof of which depends very heavily on **12** (Appendix 1).

Theorem: Each bounded set S in the real number system has a least upper bound.

The proof consists of starting with one upper bound of S, say b_0, and constructing a certain sequence of upper bounds. Consider a any fixed element of S and form the interval $I = [a, b_0]$. First subdivide the interval I into successive halves, quarters, eighths, $\cdots$ by setting $c = b_0 - a$ and choosing the points of subdivision $d_{11} = a + c/2$ (for halves), $d_{21} = a + c/4$, $d_{22} = a + c/2$, and $d_{23} = a + 3c/4$ (for quarters), $d_{31} = a + c/8$, $d_{32} = a + c/4$, $d_{33} = a + 3c/8$, $\cdots$, and $d_{37} = a + 7c/8$ (for eighths), and so on. In general define the points

$$d_{m1} = a + c/2^m, \qquad d_{m2} = a + 2c/2^m, \cdots, \qquad d_{mn} = a + nc/2^m, \cdots,$$

for $1 \leq n \leq 2^m - 1$. For each m select the smallest upper bound among the finitely many points d_{mn} and b_0; such a point always exists since the known upper bound b_0 was included among those points and there are only a finite number of decisions to be made in each step. Let our selection in the mth step be b_m. Since $\{b_m\}$ is a bounded sequence, **12** (Appendix 1) implies there is a subsequence of $\{b_m\}$ which converges to some real number b. It is now a simple matter to use the given properties of the sequence $\{b_m\}$ and show that $b = \sup S$. ∎

If the reader feels the above proof belabors a trivial point, he should study the following example.

Example: If the field of rational numbers Q is considered in place of the reals, find an example of a bounded set S in Q which has no least upper bound in Q.

Solution: The set

$$S = \{x: x^2 < 2\}$$

provides the answer. It is obvious that M is an upper bound of S if and only if $M > \sqrt{2}$. But the set of upper bounds of S

$$U = \{M: M > \sqrt{2}\}$$

has no least element in Q, and so S has no least upper bound in Q. ∎

appendix 3:
Euclid's Fifth Postulate

According to T. L. Heath [14] the form of the parallel postulate as found in Euclid's *Elements* was evidently Euclid's own creation and not merely a reformulation of theories of parallelism current in his time. It was his attempt to remove some of the difficulties which had been previously connected with the concept of parallelism. The following translation of that postulate from the original Greek as stated by Heath (p. 202) is obviously an equivalent form of our Theorem 53.1 (printed by permission of Dover Publications, Inc.):

If a straight line falling on two straight lines makes the interior angles on the same side less than two right angles, the two straight lines, if produced indefinitely, meet on that side on which are the angles less than the two right angles.

References

BOOKS:

1. Adler, C. F., *Modern Geometry; an Integrated First Course* (New York: McGraw-Hill Book Company, Inc., 1958).
2. Benson, R. V., *Euclidean Geometry and Convexity* (New York: McGraw-Hill Book Company, Inc., 1966).
3. Brumfiel, C. F., R. E. Eicholz, and M. E. Shanks, *Geometry* (Reading, Mass.: Addison-Wesley Publishing Company, Inc., 1960).
4. Coxeter, H. S. M., *Introduction to Geometry* (New York and London: John Wiley and Sons, Inc., 1961).
5. —————, *Non-Euclidean Geometry* (3rd ed.; Toronto: University of Toronto Press, 1957).
6. Court, N. A., *College Geometry; an Introduction to the Modern Geometry of the Triangle and the Circle* (2nd ed. New York: Barnes and Noble, Inc., 1952).
7. Dubnov, Y., *Mistakes in Geometric Proofs* (Boston: D. C. Heath and Company, 1963).
8. Durell, C. A., *Hints and Solutions to the Exercises in Modern Geometry* (London: Macmillan and Co., Ltd., 1953).
9. Eves, H., *A Survey of Geometry, Volume One* (Boston: Allyn and Bacon, Inc., 1963).
10. —————, *A Survey of Geometry, Volume Two* (Boston: Allyn and Bacon, Inc., 1965).
11. Fetisov, A., *Proof in Geometry* (Boston: D. C. Heath and Company, 1963).
12. Forder, H. G., *The Foundations of Euclidean Geometry* (New York: Dover Publications, Inc., 1927).
13. Gardner, M., *The Scientific American Book of Mathematical Puzzles and Diversions* (New York: Simon and Schuster, Inc., 1959).
14. Heath, T. L., *The Thirteen Books of Euclid's Elements, Volume One* (New York: Dover Publications, Inc., 1956).
15. Hilbert, D., *The Foundations of Geometry* (Chicago, Ill.: The Open Court Publishing Company, 1902).
16. Johnson, R. A., *Advanced Euclidean Geometry: An Elementary Treatise on the Geometry of the Triangle and the Circle* (New York: Dover Publications, Inc., 1960).
17. Lockwood, E. H., *A Book of Curves* (New York and London: Cambridge University Press, 1963).
18. Maxwell, E. A., *Deductive Geometry* (New York: Macmillan-Pergamon, 1963).
19. Meserve, B. E., *Fundamental Concepts of Geometry* (Reading, Mass.: Addison-Wesley Publishing Company, Inc., 1955).
20. Meschkowski, H., *Non-Euclidean Geometry* (New York and London: Academic Press, 1964).
21. Moise, E., *Elementary Geometry from an Advanced Standpoint* (Reading, Mass.: Addison-Wesley Publishing Company, Inc., 1963).

22. Palmer, C. I., and C. W. Leigh, *Plane and Spherical Trigonometry* (New York: McGraw-Hill Book Company, Inc., 1914).
23. Pedoe, D., *Circles* (New York: Pergamon Press, Inc., 1957).
24. Shirkov, P. A., *A Sketch of the Fundamentals of Lobachevskian Geometry* (Groningen, The Netherlands: P. Noordhoff, N.V., 1964).
25. Springer, C. E., *Geometry and Analysis of Projective Spaces* (San Francisco: W. H. Freeman and Company, 1964).
26. Struik, D. J., *Lectures on Classical Differential Geometry* (Reading, Mass.: Addison-Wesley Publishing Company, Inc., 1950).
27. Tietze, H., *Famous Problems of Mathematics* (Baltimore, Md.: Graylock Press, 1965).
28. Verriest, G., *Introduction à la Géométrie non-Euclidienne par la Méthode Élémentaire* (Paris: Gauthier-Villars, 1951).
29. Willmore, T. J., *An Introduction to Differential Geometry* (New York and London: Oxford University Press, 1959).
30. Wolfe, H. E., *Introduction to Non-Euclidean Geometry* (New York: Holt, Rinehart and Winston, Inc., 1945).
31. Wylie, C. R., Jr., *Foundations of Geometry* (New York: McGraw-Hill Book Company, Inc., 1964).
32. Yaglom, I. M., *Geometric Transformations* (New York: Random House, Inc., 1962).
33. ————, and V. G. Boltyanskii, *Convex Figures* (New York: Holt, Rinehart and Winston, Inc., 1961).

RESEARCH ARTICLES:

34. Blundon, W. J., "Generalization of a Relation Involving Right Triangles," *American Mathematical Monthly*, **74** (1967), 566–568.
35. Bouwkamp, C. J., A. J. Duijvestijn, and P. Medema, *Tables Relating to Simple Squared Rectangles of Orders Nine Through Fifteen*. Eindhoven, Netherlands: 1960.
36. Federico, P. J., "A Fibonacci Perfect Squared Square," *American Mathematical Monthly*, **71** (1964), 404–406.
37. Hjós, G., and P. Szász, "On a New Presentation of the Hyperbolic Trigonometry by Aid of the Poincaré Model," *Acta Mathematica Academiae Scientiarum Hungaricae* (Budapest), **7** (1956), 35–36.
38. Joseph, P. C., "A Generalization of Hjelmslev's Theorem," *American Mathematical Monthly*, **74** (1967), 574–575.
39. Lockwood, E. H., "Simson's Line and Its Envelope," *Mathematical Gazette*, **37** (1953), 124–125.
40. Thebault, V., "The Theorem of Lehmus," *Scripta Mathematica*, **15** (1949), 87–88.

Brief Solutions to
Odd-Numbered Exercises

Chapter 1

Section 11 (pp. 25–32): **1.** In Step (4) use: Parallelograms with equal bases and equal altitudes have equal areas. **3.** Use the Pythagorean theorem. **5.** Let ABC be the triangle whose vertices are the centers of the given circles, with $AB = a + b$, $BC = b + c$, and $AC = a + c$. Let D be the foot of the perpendicular from B. Then $AD = a - b$ and $BD = \sqrt{(a+b)^2 - (a-b)^2} = 2\sqrt{ab}$. Assuming the system is stable, the perpendicular CE to AB falls between A and B and that from C to BD cuts BD and BA in F and G (since $a > b$). By Heron's formula, Area $\triangle ABC$ $= \sqrt{abc(a + b + c)}$; then $CE = 2\sqrt{abc(a + b + c)}/(a + b) \cdot BE = \sqrt{BC^2 - CE^2}$ $= ab - ac + bc + b^2$. Since $\triangle GEC \sim \triangle ADB$, $GE = (a - b) \cdot \sqrt{c(a + b + c)}$ $/(a + b)$ and $CG = \sqrt{c(a + b + c)}$. Use $BG = BE - GE$ and $GF/BG = AD/AB$. **7.** $a^2 = m_1^2 + n_1^2 + 2m_1n_1 \cos \theta$, $b^2 = m_2^2 + n_1^2 - 2m_2n_1 \cos \theta$, and similarly for c^2 and d^2; $a^2 - b^2 + c^2 - d^2 = 2(m_1 + m_2)(n_1 \pm n_2) \cos \theta = 2mn \cos \theta$ (minus corresponding to the nonconvex case). **9.** $16K^2 = 4(ac + bd)^2 - (a^2 - b^2 + c^2 - d^2)^2$ $= (2ac + 2bd + a^2 - b^2 + c^2 - d^2)(2ac + 2bd - a^2 + b^2 - c^2 + d^2)$; continue factoring. **11.** $c^2 = a_1^2 + d^2 - 2a_1d \cos \theta$ and $b^2 = a_2^2 + d^2 + 2a_2d \cos \theta$; multiply by a_2 and a_1 respectively, add the resulting equations, and simplify. **13.** 36π in.^{3}; the answer is independent of the radius of the sphere or size of hole. **15.** $\sin 3y = \sin y$ $\cos 2y + \cos y \sin 2y = \sin y (3 \cos^2 y - \sin^2 y) = 4 \sin y \sin (60 + y) \sin (60 - y)$. **17.** Set $UG = a$, $GO = b$; then $HU = a + b = HG - UG = 2GO - UG = 2b$ $- a$. Thus $b = 2a$. **19.** From the given inequalities, $c/a \geqq b/c$ (or a/b) so max $(a/b, b/c, c/a) = c/a$. Likewise, min $(a/b, b/c, c/a)$ is either a/b or b/c. In the first case, $\sigma = (c/a)(a/b) = c/b$ and $a/b \leqq b/c$. By the triangle inequality $1 < c/b$ $< a/b + 1 \leqq b/c + 1$ or $1 < \sigma < 1/\sigma + 1$; $\sigma^2 - \sigma - 1 < 0$ or $1 < \sigma < \mu$. If $1 < r < \mu$ then $a < ar < ar^2$ and $ar^2 < a(r + 1) = ar + a$ so that a, ar, ar^2 are the sides of a triangle, with $a/ar = 1/r = ar/ar^2$ or min $(a/b, b/c, c/a) = 1/r$ and

$\sigma = r$. **21.** Write $f_{n+1}/f_n = [\mu^{n+1} + (-1)^n\mu^{-n-1}]/[\mu^n - (-1)^n\mu^{-n}] = [\mu + (-1)^n /\mu^{2n+1}]/[1 - (-1)^n/\mu^{2n}]$; use $\mu > 1$ in taking the limit. **23.** For an inductive argument write $f_nf_{n+1} = f_n(f_n + f_{n-1}) = f_n^2 + f_nf_{n-1} = f_n^2 + (f_{n-1}^2 + \cdots + f_2^2 + f_1^2)$. **25.** Set $x = f_{n-1}$, $y = f_n$, and $z = f_{n+1}$ for any $n = 2, 3, \cdots$; Ex. 24 is the case $n = 6$. **27.** Use L-shaped regions. **29.** Take B, D, G as points of pivot to obtain a rectangle $HKK'H'$; since $BE^2 = AB \cdot BF = \text{Area } (HKK'H')$, it is a square. **31.** The polar coordinates of the points mentioned are $(1, 0)$, $(1/\mu, \pi/2)$, $(1/\mu^2, \pi)$, $(1/\mu^3, 3\pi/2)$, $(1/\mu^4, 2\pi)$, $(1/\mu^5, 5\pi/2)$, and $(1/\mu^6, 3\pi)$, each pair satisfying $r = \mu^{-2\theta/\pi}$.

Chapter 2

Section 13 (p. 38): **1. (a)** The equation of MO is $y - \frac{1}{2}(a + \sqrt{a^2 - u^2}) = (u + b)(a - \sqrt{a^2 - u^2})^{-1}[x - \frac{1}{2}(u - b)]; y = \frac{1}{2}bu/(\sqrt{a^2 - u^2} - a) = -\frac{1}{2}b(\sqrt{a^2 - u^2} + a)/u \to -\infty$. **(b)** $y' = -\frac{1}{2}b\sqrt{a^2 - u^2}/u$ implies $y < y'$ for each $u > 0$. Thus line OA *never* intersects segment BC. **3.** Not without restriction.

Section 17 (p. 44): **1. (a)** 3. **(b)** $5!/3! = 10$. **(c)** All axioms except Ax. 4 hold. **3.** Similar to Ex. 1.

Section 18 (pp. 49–50): **1. (a)** 19. **(b)** 19. **(c)** 11. **(d)** 1. **3.** If either $a < b < c$ or $c < b < a$; No. **5.** $AB = 10$, $BC = 4$, $AC = 6$. Therefore $AC + CB = 6 + 4 = 10 = AB$ and (ACB). As α increases AB and BC are unchanged but AC increases to 14 (for $\alpha \geq 14$) and (ABC) holds. **7. (a)** Either $|x - 11| = 20$ or $40 - |x - 11| = 20$; $|x - 11| = 20$ gives $x - 11 = \pm 20$ or $x = 31, -9$. Since $x < \alpha = 20$, $x \neq 31$. **(b)** $|x - 11| = 19$ yields $x = 30, -8$ or, $x = -8$. The other case yields $x = -10$. **(c)** $1, -19$. **(d)** 8, 14. **(e)** 10, 12. **9.** It is commutative, associative, 0 is identity, and the inverse of a is $-a$. In addition there is the *closure* law: $a \oplus b$ lies on the range $-\alpha < x \leq \alpha$ if a and b do. Thus $x' = x \oplus a$ defines a "translation" on $-\alpha < x \leq \alpha$ such that for any two numbers x, y, $|x - y| = |x' - y'|$ and thus Axiom 5 is satisfied in the new coordinate system. Therefore, one coordinate system (x') may be obtained from another (x) by an equation of the form $x' = \pm x \oplus a$. When $\alpha = \infty$ $a \oplus b = a + b$. **11.** Four points A, B, C, D are necessary with, say, $AB = 1$, $BC = 2$, and $AC = 3$. If these points were "Euclidean" then A, B, and C would be collinear and regardless of the position of D there would have to be at least one additional positive distance $d \neq 1, 2, 3, 5$.

Section 19 (pp. 52–53): **1.** If $a \geq 0$ then $b \geq 0$ and we add $0 \leq a \leq \alpha$ to $-\alpha \leq -b \leq 0$ to obtain $-\alpha \leq a - b \leq \alpha$ or $|a - b| \leq \alpha$. Similarly if $a < 0$. **3.** If $AB + BC \leq \alpha$ choose a coordinate system such that B is origin, and let

a, c be the coordinates of A, C. By the result of Ex. 1 $AB = |0 - a| = |a|$ and $BC = |0 - c| = |c|$. Then $|a - c| \leq |a| + |c| = AB + BC \leq \alpha$ and Thm. **19.1** applies, regardless of the ordering of a, b, c. **5.** Let A be the origin, $b > 0$ and c the coordinates of B and C. $|c - 0| = \alpha$ implies $c = \pm \alpha$. Since $c \neq -\alpha$ we must have $c = \alpha$. Apply Thm. **19.1**. **7.** Not reasonable. **9.** Since $c > b \geq \alpha/2 > 0$, $CB = |c - b| = c - b$. $a \leq -\alpha/2$ implies $-a \geq \alpha/2$. Hence $|b - a| = b - a \geq \alpha/2 + \alpha/2 = \alpha$; similarly $|c - a| \geq \alpha$. By Ax. 5 $AB = 2\alpha - b + a$ and $AC = 2\alpha - c + a$. Hence $AC + CB = AB$.

Section 20 (pp. 57–58): **1.** $\alpha = \pi r$; the set would be the *entire sphere* minus the points A and B. **3.** Take $C[b/3]$ and $D[2b/3]$ where $A[0]$ and $B[b]$, $b > 0$, are the given endpoints. **5.** See definitions. **7.** If $X \in \overset{\leftrightarrow}{AB}$, $X \neq A$, $X \neq B$, then by Thm. 19.3 either (XAB), (AXB), or (ABX). Hence, either $X \in \overset{\rightarrow}{BA}$ or $X \in \overset{\rightarrow}{AB}$. **9.** If $X \in l$, then by Thm. 19.6 either $X = A$, $X = B$, $X = C$, (XAB), (AXB), (BXC), or (BCX); thus $X \in \overset{\rightarrow}{BA} \cup \overset{\rightarrow}{BC}$.

Chapter 3

Section 22 (pp. 64–65): **1.** $h[0]$, $k[120]$, $u[-120]$. **3.** $hk = 188$, $hu = 176$, $hv = 61$, $ku = 66$, $kv = 179$, $uv = 115$. **5.** *Five concurrent rays.* **7.** $hk + ku > hu + uv$. **9.** Dualize Ex. 3, Sec. 20.

Section 24 (p. 72): **3.** In pseudospherical geometry, if A is the "north pole," then every point on the "equator"—of which there are infinitely many—is at a distance α from A. **5.** Since $0 < AB < \alpha$, $\overset{\leftrightarrow}{AB}$ is a line through A; by Thm. 24.5 $C \in \overset{\leftrightarrow}{AB}$ and hence A, B, C are distinct collinear points. The rest follows as in Ex. 5, Sec. 19. **7.** The difficulty is that without Ax. 9 there is no way to define hemispheres (or half planes) abstractly. **9.** Locate B on one of the lines through A at a distance α from A (Ax. 5) and by Thm. 24.5 the other line passes through B; if there were two points A' and A'' at a distance α from A, then by the result of Ex. 5, both $(AA'A'')$ and $(AA''A')$. **11.** If $AB < \alpha$ the assertion follows directly from parts (b) and (c) of Ax. 9; if $AB = \alpha$ then for $X \in l$ (AXB) follows by the result of Ex. 5. **13.** If l is any line, define a relation $\sim$ between pairs of points not on l by: $A \sim B$ *if and only if there exists no point* $X \in l$ *such that* (AXB). Show that $\sim$ is an *equivalence relation* (for any A, B, C, $A \sim A$, $A \sim B$ implies $B \sim A$, and $A \sim B$, $B \sim C$ imply $A \sim C$), define $H_1 = \{X : X \sim P\}$ and $H_2 = \{X : X \sim Q\}$ where P

and Q are any two points not on l such that $P \nsim Q$ and prove that H_1, H_2 have the desired properties.

Section 25 (pp. 77–78):   1. It was defined as the intersection of two convex sets. 3. (a) Same as the interior of that angle in Euclidean geometry; (b) The interior of $\angle AOB$. 5. Let x' be the coordinate of k'. Then $kk' = |x - x'|$ or $kk' = 360 - |x - x'|$. Therefore $x - x' = \pm 180$, and $x' = x - 180$ easily follows. 7. Apply the crossbar principle. 9. Let $\epsilon > 0$ with $0 \leq x_0 \leq BC$ and $X_0[x_0]$. Let ϵ' be the lesser of $\frac{1}{2}\angle X_0 AC$ and ϵ. Let $(\overrightarrow{AX_0} \ \overrightarrow{AY_1} \ \overrightarrow{AC})$ with $\angle X_0 A Y_1 = \epsilon'$ and suppose $\overrightarrow{AY_1}$ meets $\overline{BC}$ at X_1. Choose $\delta = X_0 X_1$ and show that $\theta(x) < \theta(x_0) + \epsilon$ thus proving that the right-hand limit $\lim_{x \to x_0^+} \theta(x)$ is $\theta(x_0)$. The left-hand limit is dealt with in a similar manner. 11. Apply the angle-construction theorem.

Chapter 4

Section 28 (pp. 89–90):   1. Six ways, one of them always occurring, two of them when the triangle is equilateral, and three when the triangle is isosceles. 3. If $\overline{AB} = \overline{XY}$ then either $A = X$ and $B = Y$ or $A = Y$ and $B = X$, so that $AB = XY$.

5. If $A[x_1, y_1]$, $B[x_2, y_2]$, $C[x_3, y_3]$ be the given points, then $\widetilde{AC} = |x_1 - x_3| + |y_1 - y_3| = |(x_1 - x_2) + (x_2 - x_3)| + |(y_1 - y_2) + (y_2 - y_3)| \leq |x_1 - x_2| + |x_2 - x_3| + |y_1 - y_2| + |y_2 - y_3| = \widetilde{AB} + \widetilde{BC}$. 7. (b) By the crossbar principle $\overrightarrow{AP}$ meets $\overline{BC}$ at Q, and either (APQ) or (AQP). But (AQP) implies $P \in H_2(A, \overleftrightarrow{BC})$, a contradiction. Now apply the postulate of Pasch. 9. $AB = AC - BC = AE - DE = AD$ so by SAS $\triangle ADC \cong \triangle ABE$ and $CD = BE$. 11. By Ax. 5 there is F on the opposite ray of $\overrightarrow{BC}$ such that $CB = BF$. Since $\angle ABC = 90$ $\triangle ABC \cong \triangle ABF$ so that $\angle BAF = \angle CAB = \angle BAE$. But $\overrightarrow{AF}$ and $\overrightarrow{AE}$ lie on the same side of $\overleftrightarrow{AB}$ and there can be only one such ray (Ex. 11, Sec. 25). Therefore, $\overrightarrow{AE} = \overrightarrow{AF}$. 13. The assumption (AEC) was made which might not hold even if the triangles were Euclidean. 15. The concept of perpendicular bisectors would be useful; the *existence* of equilateral triangles can be established (see Ex. 3, Sec. 29), but in order to let the base $\overline{AB}$ be prescribed the continuity of the function AX as X varies on the perpendicular bisector of $\overline{AB}$ is needed.

Section 29 (pp. 94–96):   1. Using (CEA) and (BED) we have $\angle ACB = \angle ECB = \angle EBC = \angle DBC$ (by Thm. 28.2). Also $\angle ABC = \angle BCD$ and $BC = BC$, so by

ASA $\triangle ABC \cong \triangle DCB$ and $AB = CD$. **3.** Apply the result of Ex. 10, Sec. 29 for isosceles triangles; using coordinates for the rays at O locate $h[0]$, $k[120]$, $u[-120]$ and take points $A \in h$, $B \in k$, $C \in u$ such that $OA = OB = OC$. **5.** Either $(\overrightarrow{BA}\ \overrightarrow{BD}\ \overrightarrow{BE})$, $\overrightarrow{BD} = \overrightarrow{BE}$, or $(\overrightarrow{BA}\ \overrightarrow{BE}\ \overrightarrow{BD})$ (use coordinates and apply Ex. 7, Sec. 25). Fig. 114 illustrates the first of these; $(\overrightarrow{BA}\ \overrightarrow{BD}\ \overrightarrow{BE})$ and $(\overrightarrow{BA}\ \overrightarrow{BD}\ \overrightarrow{BC})$ (Lemma 22.9) imply $(\overrightarrow{BA}\ \overrightarrow{BD}\ \overrightarrow{BE}\ \overrightarrow{BC})$ so $\angle ABE = \angle ABD + \angle DBE = \angle EBC + \angle DBE$ $= \angle DBC$. By SAS $\triangle ABE \cong \triangle DBC$. **7.** (a) With A on l locate B on l such that $AB = \alpha/2$; at B construct a right angle ABC, $BC < AB$. Then show that $\angle B = \angle C = 90$ (Lemma 30.2), but $\angle A \neq 90$ since $AB \neq BC$. (b) No; Thm. 29.9, since no segment can have length $\alpha/2$ here. **9.** By SSS $\triangle OBA \cong \triangle OBC$ so that $\angle OBA = \angle OBC$. Since $O \in$ Interior $\overline{\angle ABC}$, $(\overrightarrow{BA}\ \overrightarrow{BO}\ \overrightarrow{BC})$ and it follows by simple algebra that $\angle OBA = \angle OBC = \frac{1}{2}\angle ABC$. Since $(\overrightarrow{CB}\ \overrightarrow{CO}\ \overrightarrow{CD})$ and $OC = OB$ then $\angle OCD = \angle BCD - \angle OCB = \angle ABC - \angle OBC = \frac{1}{2}\angle ABC = \angle OBA$. Then $AB = CD$ implies $\triangle OCD \cong \triangle OBA$ and therefore $OD = OA$. **11.** If $X \in$ Interior $\overline{hk} \cap$ Interior $\overline{ku}$ and $v = \overrightarrow{OX}$ (where O is the origin of the given rays) then both (hvk) and (kvu). But (hku) would then imply, with (kvu), the relation $(hkvu)$ or (hkv), a contradiction. Therefore by definition $\overline{hk}$ and $\overline{ku}$ are adjacent. The converse is not true. **13.** Since (hkh') by Lemma 22.9 we have by Thm. 22.4 either (uhk), (huk), (kuh'), or $(kh'u)$. The result of Ex. 12 denies (uhk) or (huk), so either (kuh') or $(kh'u)$ and $h'u = \pm(h'k - ku) = \pm(180 - hk - ku) = \pm(180 - 180) = 0$. Hence $u = h'$. **15.** If u and v are the bisectors of $\overline{hk}$ and $\overline{kh'}$ (h' opposite h) then (hkh') and (huk) imply (ukh'). (kvh') then yields (ukv). Hence $uv = uk + kv = \frac{1}{2}hk + \frac{1}{2}kh' = \frac{1}{2}\cdot 180 = 90$.

Section 31 (pp. 107–109): **1.** Equality must hold in each step of the proof of Thm. 31.4, so we get $(P_1P_2P_3)$, $(P_1P_3P_4)$, $(P_1P_4P_5)$, $\cdots$, $(P_1P_{n-1}P_n)$. Hence the points P_k must be collinear and must occur in the order P_1, P_2, P_3, P_4, $\cdots$, P_n. **3.** Let $\triangle ABC$ have right angle at C, with D such that (BCD). Then $\angle A < \angle ACD = 90$, and similarly $\angle B < 90$. For Cor. 30.5' let M be the midpoint of $\overline{BC}$ and use Fig. 127. **5.** Either (DBM), (BDM), (MDC), or (MCD). If (DBM) then $\angle MBA > \angle ADB = 90$, a contradiction, or if (MCD) then $\angle MCA > \angle ADC = 90$, a contradiction. **7.** (a) $2AX \leq (AB + BX) + (AC + CX) \leq BX + XC + 2r = BC + 2r \leq BA + AC + 2r \leq 4r$. (b) The inequality cannot be improved. **9.** Choose A and B on the sides of angle C such that $AC = BC < a/2$; $AB \leq AC + CB < a/2 + a/2 = a$. **11.** Locate W on $\overleftrightarrow{XZ}$ such that (XWZ) and $XW = AC$. Then $\triangle XYW \cong \triangle ABC$ and $\angle C = \angle XWY > \angle WZY = \angle Z$. **13.** (a) 1/10; 1/100. (b) $PX_0 - PX_n \leq X_0X_n$ and $PX_n - PX_0 \leq X_0X_n$ yield $|PX_0 - PX_n| \leq X_0X_n = 1/n$, so therefore $\lim_{n\to\infty} |PX_0 - PX_n| = 0$. **15.** (a) $d_{\frac{1}{4}} < \frac{1}{2}(b + d_{\frac{1}{2}}) < \frac{1}{2}[b + \frac{1}{2}(a + b)] = \frac{1}{4}a + \frac{3}{4}b$; $d_{\frac{3}{4}} < \frac{1}{2}(a + d_{\frac{1}{2}}) < \frac{1}{2}[a + \frac{1}{2}(a + b)] = \frac{3}{4}a + \frac{1}{4}b$.

Section 34 (pp. 117–119): **1.** If $AC = XZ$ then $\triangle ABC \cong \triangle XYZ$. Otherwise, it may be assumed that $AC > XZ$. On $\overline{AC}$ lay off $AD = XZ$; $\triangle ABD \cong \triangle XYZ$ so that $BC = YZ = BD$. Since (ADC), $\angle CDB + \angle BDA = 180$. Therefore $\angle C + \angle Z$ $= 180$. **3. (a)** Use the result of Ex. 2 to show that $\overrightarrow{AD}$ and $\overrightarrow{AE}$ cannot be trisectors. **(b)** Yes, if $AB = \alpha/2$. **5.** Consider a trirectangular triangle on the sphere. **7.** If $\alpha < \infty$ take $AA' = \alpha$ and locate B and C any two points such that $AB = AC$ $= 2\alpha/3$. If L and M are the midpoints of $\overline{AB}$ and $\overline{AC}$, show that $\triangle ALM \cong \triangle A'BC$ and therefore $LM = BC$. **9.** Let $AB = AC$ and suppose $\angle B = \angle C < 90$. If A' is the extremal of A, show that $AB \geqq \alpha/2$ implies that $\angle A'BC \leqq 90$. **11.** Let AB $= AC$ and suppose altitudes $\overline{BE}$ and $\overline{CF}$ meet at H (by the result of Ex. 5, Sec. 31, the crossbar principle, and Ax. 10). Show that H lies on the perpendicular bisector of $\overline{BC}$. **13.** As in Ex. 11, suppose $\overline{BM}$ and $\overline{CN}$ meet at G; show that $\triangle NBC \cong \triangle MCB$. **15.** Since $\angle DBC > \angle DCB$, $DC > DB$ and by Ex. 12, Sec. 31 $CZ > CF > BY$. **17.** If $IA = IB = \alpha/2$, then $\angle IAB = \angle IBA = 90$. But $\angle IAB = \frac{1}{2}\angle BAC < 90$, so $IB \neq \alpha/2$. Hence, the foot F of I on $\overleftrightarrow{AB}$ is unique and $IF < \alpha/2$. Apply Ex. 5, Sec. 31.

Chapter 5

Section 37 (pp. 126–127): **3.** If $\Diamond ABCD$ is not convex, then $\overline{AC} \cap \overline{BD} = \phi$ and $[ABDCA]$, $[ACBDA]$ are quadrilaterals. **5. (a)** Valid: Let $\Diamond ABCD$ and $\Diamond XYZW$ have $AB = XY$, $\angle B = \angle Y$, $BC = YZ$, $\angle C = \angle Z$, and $\angle D = \angle W$. By SAS $\triangle ABC \cong \triangle XYZ$ so $AC = XZ$ and $\angle BCA = \angle YZX$. By Cor. 36.5, $\angle ACD$ $= \angle BCD - \angle BCA = \angle YZW - \angle YZX = \angle XZW$. Then by SAA, $\triangle ACD$ $\cong \triangle XZW$. **(b)** Not valid: A square and a rectangle with one dimension the same provides a counterexample. **(c)** Valid. **(d)** Not valid. **7.** If $\Diamond ABCD \cong \Diamond XYZW$ then $\triangle ABC \cong \triangle XYZ$ by SAS. **9. (a)** If in $\Diamond ABCD$, $AB = CD$ and $\angle B = \angle C$, then by SASAS $\Diamond ABCD \cong \Diamond DCBA$ and therefore $\angle A = \angle D$. By Ex. 7 also $AC = DB$. **(b)** Assume $\angle A = \angle B$, $\angle C = \angle D$, $BC > AD$. Locate E on $\overline{BC}$ such that $CE = AD$, making $\Diamond AECD$ isosceles. Then $\angle A = \angle BAD > \angle EAD =$ $\angle AEC > \angle ABC = \angle B$, in contradiction.

Section 39 (pp. 133–134): **1. (a)** Prove $\triangle AA'M \cong \triangle BB'M$ and $\triangle AA'N$ $\cong \triangle CC'N$. Then $BB' = AA' = CC'$ and $\angle B'BC + \angle BCC' = \angle MAA' + \angle ABC$ $+ \angle A'AN + \angle ACB$. **3.** If $\overleftrightarrow{AA'}$ meets $\overline{MM'}$ at a point C then (MCM') implies $\overleftrightarrow{AA'}$ meets $\overline{BM}$ at D or $\overline{BM'}$ at E, by the postulate of Pasch. Since $\overleftrightarrow{AA'}$ cannot meet $\overline{BM}$ it must meet $\overline{BM'}$. In like manner, it follows that $\overleftrightarrow{AA'}$ meets $\overline{BB'}$, impossible. Hence $\overleftrightarrow{AA'}$ does not meet $\overline{MM'}$; by the convexity of $\Diamond A'ABB'$, $\overleftrightarrow{AM}$ does not meet $\overline{A'M'}$ and $\overleftrightarrow{A'M'}$ does not meet $\overline{AM}$. Apply Thm. 36.3. **5.** $\frac{1}{2}(n^2 - 3n)$.

7. (a) Diagonal $\overline{P_1P_{k+1}}$ forms two further convex polygons $\bigcirc P_1P_2 \cdots P_{k+1}$ and $\bigcirc P_1P_{k+1}P_{k+2} \cdots P_{2k}$. Since $\angle P_2 = \angle P_{2k}$, $\angle P_3 = \angle P_{2k-1}, \cdots, \angle P_k = \angle P_{k+2}$ and $P_1P_2 = P_1P_{2k}$, $P_2P_3 = P_{2k}P_{2k-1}, \cdots, P_kP_{k+1} = P_{k+2}P_{k+1}$ the SASAS $\cdots$ proposition for convex polygons (Ex. 6) implies $\bigcirc P_1P_2P_3 \cdots P_{k+1} \cong \bigcirc P_1P_{2k}P_{2k-1} \cdots P_{k+1}$ and hence $\angle P_{2k}P_1P_{k+1} = \angle P_{k+1}P_1P_2$, $\angle P_{k+2}P_{k+1}P_1 = \angle P_1P_{k+1}P_k$. It follows that $\overrightarrow{P_1P_{k+1}}$ bisects $\angle P_{2k}P_1P_2$ and $\overrightarrow{P_{k+1}P_1}$ bisects $\angle P_kP_{k+1}P_{k+2}$. **(b)** Since $(\overrightarrow{P_1P_2}\ \overrightarrow{P_1P_{k+1}}\ \overrightarrow{P_1P_{k+2}})$ the crossbar principle implies that $\overrightarrow{P_1P_{k+1}}$ meets $\overline{P_2P_{k+2}}$ at some point O. In the same manner $\overrightarrow{P_2P_{k+2}}$ meets $\overline{P_1P_{k+1}}$ at O', and since it would follow that $OO' < \alpha$, we must have $O = O'$. Thus, (P_1OP_{k+1}) and (P_2OP_{k+2}) and the rest follows by Ax. 10 and (25.4). **9.** It may be defined as the intersection of the interiors of the angles of the polygon, or alternatively, as the intersection of the half planes determined by the sides and containing the remaining vertices.

Section 42 (pp. 140–142): **1.** Thm. 40.2: Let segment $\overline{AB}$ pass through the center O of a circle ω, with A and B on ω. Then (AOB) and by definition $AO = OB$. Thm. 40.3: The perpendicular bisector l of chord $\overline{AB}$ is the locus of points equidistant from A and B; the center O is such a point so $O \in l$. Thm. 40.4: If O is the center, A and B are the ends of the chord, and M is the midpoint of $\overline{AB}$, $\triangle AOM \cong \triangle BOM$ by SSS and therefore $\angle AMO = \angle OMB$ so that $\overleftrightarrow{OM} \perp \overleftrightarrow{AB}$. **3.** Let $\overleftrightarrow{PA}$ and $\overleftrightarrow{PB}$ be tangents with A and B the points of contact, O the center. Use Ex. 1, Sec. 34 to show that $\triangle APO \cong \triangle BPO$. **5. (a)** By Thm. 34.2 $IX = IY = IZ$ and hence X, Y, Z lie on a circle, and the sides of the triangle are tangent to that circle. **(b)** $a + b + c = (y + z) + (z + x) + (x + y) = 2x + 2y + 2z$; $s = x + y + z = x + a$. **7.** Apply the intermediate-value theorem in the manner of the proof of Thm. 41.4. **9. (a)** All except Ax. 5, 9, 10 are valid. **(b)** The line $y = \sqrt{2}/2$ and the circle $x^2 + y^2 = 1$. **11.** If O is the center of the circle, locate the rays $h_1, h_2, \cdots, h_n$ from O whose respective coordinates are 0, $360/n = \theta$, 2θ, $3\theta, \cdots, m\theta$, $-m\theta$ (omitted if $m\theta = 180$), $-(m - 1)\theta$, $-(m - 2)\theta, \cdots, -2\theta$, and $-\theta$, where m is the last integer such that $m\theta \le 180$. If P_i is the intersection of h_i and circle O for each i, it remains to show that $[P_1P_2P_3 \cdots P_nP_1]$ is a regular polygon. **13.** By Ex. 1, 2, Sec. 39 $c' > \frac{1}{2}c$; $a'^2 + b'^2 = \frac{1}{4}a^2 + \frac{1}{4}b^2 = \frac{1}{4}c^2 < c'^2$.

Chapter 6

Section 44 (pp. 147–148): **1.** Apply Thm. 43.3. **3.** If $\triangle ABC$ is the given triangle, and (BCD), use the fact that angles ACD and ACB are supplementary. **5.** If $(A_0A_1A_2)$ and D_0, D_1, D_2 are the feet of the perpendiculars from A_0, A_1, A_2 on $\overleftrightarrow{BC}$ it follows that $A_0D_0 = A_1D_1 = A_2D_2$ and $D_0D_1 = D_1D_2$. By SASAS the Saccheri quadrilaterals $A_0D_0D_1A_1$ and $A_1D_1D_2A_2$ are congruent, so therefore, $\angle A_0A_1D_1 = \angle D_1A_1A_2 = \frac{1}{2}(180) = 90$.

Section 48 (pp. 157–158): 1. Take A as the pole of $\overleftrightarrow{BC}$ with C chosen so that $BC = \pi/2$. Then $\overleftrightarrow{AB} \perp \overleftrightarrow{AC}$ and $AB = \pi/2$ implies that B is a pole of $\overleftrightarrow{AC}$. Similarly, C is a pole of $\overleftrightarrow{AB}$ which proves that $\triangle ABC$ is self-polar. Self-polar triangles are equilateral and trirectangular. **3.** 120. **5.** Show that the plane is the union of the sides and interiors of four right angles, each of which is the union of two trirectangular triangles and their interiors. Thus $Total\ Area = 8(3\cdot90 - 180) = 720$. **7.** No, it is a circle. **9. (a)** If S denotes the area of the sphere, $L/S = \angle A/360$ and hence, $L/4\pi = \angle A/360$ or $L = \pi\angle A/90$. **(b)** (BAB') and $(AB'A')$ imply $AB + AB' = \pi = AB' + A'B'$ and hence $AB = A'B'$. Similarly $AC = AC'$, so $\angle A = \angle A'$ implies $\triangle ABC \cong \triangle A'B'C'$. Then $2\pi = \text{Area}(K \cup I \cup II \cup III) = \text{Area}(K \cup I) + \text{Area}(K' \cup II) + \text{Area}(K \cup III) - 2\,\text{Area}(K) = \pi\angle C/90 + \pi\angle A/90 + \pi\angle B/90 - 2K$. Now solve for K. Thus, the concept of area introduced in Def. 48.2 agrees with the familiar one, but for the constant $\pi/180$ which reflects merely the choice of the unit of measure.

Section 52 (p. 166): 1. Construct $\overline{BC}$ such that $BC = AD$ and $(\overrightarrow{CD}\ \overrightarrow{CA}\ \overrightarrow{CB})$; then $\angle ACB = 90 - \angle ACD = \angle DAC$, $BC = AD$, $AC = AC$ so $\triangle ABC \cong \triangle CDA$. **3.** Not valid, for otherwise two adjacent, congruent Saccheri quadrilaterals could be constructed, which must then be rectangles. **5.** Define the defect of $\bigcirc P_1P_2 \cdots P_n$ to be the number $(n - 2)180 - \angle P_1 - \angle P_2 - \cdots - \angle P_n$; the rest is routine. **7.** In Fig. 176 let C be the midpoint of $\overline{A'B}$; by the result of Ex. 2, Sec. 39 and the triangle inequality, $MM' \leq MC + CM' < \frac{1}{2}AA' + \frac{1}{2}BB' = AA'$. **9.** Suppose that $\overline{\angle CAB}$ and $\overline{\angle C'A'B}$ are congruent angles of parallelism with respect to A, A' and $\overleftrightarrow{BD}$, with C, C', D on the same side of $\overleftrightarrow{AB}$ and $(AA'B)$. If M is the midpoint of $\overline{AA'}$ and E and F are the respective feet of M on $\overleftrightarrow{AC}$ and $\overleftrightarrow{A'C'}$, then show that (EMF) and that $E \in \overrightarrow{AC}$, $A' \in \overrightarrow{FC'}$. Without loss of generality it may be assumed that (AEC) and $(FA'C')$. Now prove that any ray between $\overrightarrow{EF}$ and $\overrightarrow{EC}$ meets $\overrightarrow{FC'}$ and hence $\overline{\angle FEC}$ is an angle of parallelism, a contradiction.

Section 55 (pp. 173–175): 1. Suppose $l_1 \not\parallel l_3$. Then l_1 intersects l_3 at some point P and by our convention $l_1 \neq l_3$. Hence l_1 and l_3 are two distinct lines through P parallel to l_2, a denial of Ax. 12''. **3.** In Fig. 240, let $\overleftrightarrow{DE'} \parallel \overleftrightarrow{BC}$ $(E' \in \overline{AC})$ and $AD/AB = AE/AC$. Then by Thm. 55.1 $AD/AB = AE'/AC$ which implies $AE = AE'$. Hence $E = E'$ and $\overleftrightarrow{DE} \parallel \overleftrightarrow{BC}$. **5.** Let $AB/XY = BC/YZ = AC/XZ$, and suppose that $XY < AB$, $XZ < AC$; locate $D \in \overline{AB}$, $E \in \overline{AC}$ with $AD = XY$, $AE = XZ$. Now prove that $\triangle ADE \cong \triangle XYZ$ by use of Thm. 55.7 and Cor. 55.6.

7. If $\overline{AD}$ and $\overline{BE}$ are altitudes to $\overleftrightarrow{BC}$ and $\overleftrightarrow{AC}$ then by Thm. 54.1 the midpoint M is equidistant from A, B, D, and E. Thus $\overline{DE}$ is a chord of circle M whose perpendicular bisector must pass through M. **9. (a)** Let $\Diamond ABCD$ be a parallelogram. Since B and D lie on opposite sides of $\overleftrightarrow{AC}$, $\angle BAC$ and $\angle ACD$ are alternate interior angles (convexity used here). Hence $\angle BAC = \angle ACD$. Similarly $\angle BCA = \angle DAC$, and by ASA $\triangle ABC \cong \triangle CDA$. Therefore, $BC = AD$ and $AB = CD$. **(b)** Reverse the proof in (a); use SSS. **(c)** Use SAS. **(d)** Since $\Diamond ABCD$ is convex $\overline{AC}$ meets $\overline{BD}$ at E, an interior point on each diagonal. By ASA $\triangle AED \cong \triangle BEC$ and $AE = EC$. **(e)** Reverse proof in (d); use vertical angles. **11.** If in $\triangle ABC$ $a^2 + b^2 = c^2$ in standard notation, $\angle C = 90$. To prove this, let D be the foot of A on $\overleftrightarrow{BC}$; by Thm. 55.9 $AD^2 = c^2 - BD^2 = b^2 - CD^2$. Use algebra to show that in all cases $CD = 0$. **13.** Construct $\overline{BD}$ which meets $\overline{EF}$ at some point G; $m = EG + GF = pa + CD$ $(BG/BD) = pa + qb$. **17.** As in Ex. 9(e), E is the midpoint of $\overline{AC}$; since $AB = BC$ triangle ABC is isosceles and hence $\overleftrightarrow{BE} \perp \overleftrightarrow{AC}$. **19.** Let the points of trisection of $\overline{AB}$, $\overline{BC}$, $\overline{CD}$, and $\overline{DA}$ be, respectively, J and K, U and V, R and S, and X and Y. Then by Thm. 55.7 $\overleftrightarrow{JY} \parallel \overleftrightarrow{BD} \parallel \overleftrightarrow{VR}$ and $\overleftrightarrow{KU} \parallel \overleftrightarrow{AC} \parallel \overleftrightarrow{XS}$.

Chapter 7

Section 57 (pp. 188–189): **1.** Let the vertices be $A[0, 0]$, $B[a, 0]$, $C[b, c]$, $D[b - a, c]$. The midpoint of $\overline{AC}$ is $M[\frac{1}{2}b, \frac{1}{2}c]$, which is also the midpoint of $\overline{BD}$. **3.** Let the line be $y = a = $ constant, the circle, $x^2 + y^2 = r^2$, and the point inside the circle, $P[b, c]$. Then $b^2 + c^2 < r^2$ and since $c = a$, $a^2 + b^2 < r^2$. Now solve simultaneously $y = a$, $x^2 + y^2 = r^2$ and interpret the result geometrically. **5.** Translate by $x' = x - r$, $y' = y - s$ so that A and B have new coordinates $(0, 0)$ and $(u - r, v - s) = (p, q)$. The perpendicular bisector of $\overline{AB}$ has the equation $y - q/2 = -(p/q)$ $(x - p/2)$. Intersect with the circle $(x - p/2)^2 + (y - q/2)^2 = (\sqrt{3}/2)^2(p^2 + q^2)$ to obtain the points $(\frac{1}{2}p \pm \frac{1}{2}\sqrt{3}q, \mp \frac{1}{2}\sqrt{3}p + \frac{1}{2}q)$ or $(\frac{1}{2}[u - r] \pm \frac{1}{2}\sqrt{3}[v - s]$, $\mp \frac{1}{2}\sqrt{3}[u - r] + \frac{1}{2}[v - s])$. Translate back by $x = x' + r$, $y = y' + s$ to obtain final answer. **7.** By SAS $\triangle A'CA \cong \triangle BCB'$; therefore, $AA' = BB'$. Similarly, $BB' = CC'$. **9.** Under reflection through O, since O is the midpoint of $\overline{AC}$ and $\overline{BD}$ and since distances and angle-measure are preserved, square $ABEF$ maps into square $DCKJ$ and hence X maps into Z. Similarly, Y maps into W, and O becomes the midpoint of both $\overline{XZ}$ and $\overline{YW}$. Apply the result of Ex. 8. **11.** Let T be the point of tangency of either tangent from P. Since $O[h, k]$ is the center of the circle then

$F(x_0, y_0) = OP^2 - r^2 = TP^2$. **13.** $H[0, -bc/a]$, $G[(b + c)/3, a/3]$, and $O[(b + c)/2,$ $(a^2 + bc)/2a]$; $(a^2 + 3bc)x - (b + c)ay = (b + c)bc$. **15.** Let the vertices be $A[0, \sqrt{3}a]$, $B[-a, 0]$, $C[a, 0]$ $(a > 0)$, and let $P[x_0, y_0]$ be the given interior point. Using Example 56.7 the distances from P to the three sides are $\pm \frac{1}{2}\sqrt{3}x_0 - \frac{1}{2}y_0$, $+ \frac{1}{2}\sqrt{3}a$, and y_0. The sum is the constant $\sqrt{3}a$.

Section 58 (pp. 193–194): **1.** $CD + DA = CA$ so a substitution gives $AB + AB$ $+ BC + CA$, which is zero by Thm. 58.2(b). **3.** $[AB, DC] = (AD \cdot BC)/(AC \cdot BD)$ $= [(AC \cdot BD)/(AD \cdot BC)]^{-1}$. By Euler's theorem $(AB \cdot CD)/(AD \cdot CB) + (AC \cdot DB)$ $/(AD \cdot CB) - 1 = 0$ or $[AC, BD] + [AB, CD] = 1$. The rest follows by the observations $[AC, DB] = [AC, BD]^{-1}$, $[AD, BC] + [AB, DC] = 1$, and $[AD, CB]$ $= [AD, BC]^{-1}$. **5.** If $\lambda = [AC, BD]$ then $\lambda/AB = CD/(AD \cdot CB) = (CA + AD)$ $/(AD \cdot CB) = CA/(AD \cdot CB) + 1/CB = 1/AD + BA/(AD \cdot CB) + 1/CB = 1/AD$ $- \lambda/CD + 1/CB$. Thus $\lambda/AB + 1/BC + \lambda/CD + 1/DA = 0$. **7.** $\angle OPO' + \angle O'PF$ $+ \angle FPE + \angle EPO = 2\pi$ and thus $\angle OPO' = \pi - 2\theta$. Continuing in this manner prove that $\angle CPB = \theta$, then use (58.10) to show that $[AC, BD] = \csc^2\theta$.

Section 59 (p. 197): Let $\overrightarrow{OA} = a$, $\overrightarrow{OB} = b$, $\overrightarrow{OC} = c$, $\overrightarrow{OD} = d$, and similarly for the primed symbols, $\overrightarrow{OA'} = a'$, and so on. Then by (59.3) and the discussion following, $[AB, CD] = (\sin ac \sin bd)/(\sin ad \sin bc) = (\sin a'c' \sin b'd')/(\sin a'd' \sin b'c')$ $= [A'B', C'D']$.

Section 60 (pp. 202–204): **1.** $d_1^2 = \frac{1}{3}a^2 + \frac{2}{3}b^2 - \frac{2}{9}c^2$ and $d_2^2 = \frac{2}{3}a^2 + \frac{1}{3}b^2 - \frac{2}{9}c^2$; $d_1^2 + d_2^2 = a^2 + b^2 - \frac{4}{9}c^2 = \frac{5}{9}c^2$. **3,5.** Use (60.5). **7.** By (60.12) and (60.15), $r r_a r_b r_c = r^4 s^4/s(s - a)(s - b)(s - c) = K^4/K^2 = K^2$. **9.** Since $\triangle ARI \sim \triangle AR'I_a$ and $AR = s - a$, $AR' = s$ then $AI/AI_a = (s - a)/s$. Let the feet of I and I_a on $\overleftrightarrow{BC}$ be X and Y. Then $ID/DI_a = r/r_a = (s - a)/s$. **11.** $s = 156/\sqrt{455} = 12 \cdot 13/\sqrt{5 \cdot 7 \cdot 13}$, $s - a = 12 \cdot 7/\sqrt{5 \cdot 7 \cdot 13}$, $s - b = 12 \cdot 5/\sqrt{5 \cdot 7 \cdot 13}$, and $s - c = 12/\sqrt{5 \cdot 7 \cdot 13}$. Then $K = \sqrt{12^4 \cdot 13 \cdot 7 \cdot 5/5^2 \cdot 7^2 \cdot 13^2} = 144/\sqrt{455}$ and $h_a = 2 \cdot 144/72$, and so on. **13. (a)** Let the sides opposite be a, b, c. Then $a^2 + b^2 = 4K^2/h_a^2 + 4K^2/h_b^2 = 4K^2(3^2 + 4^2)$ $= 4K^2 \cdot 5^2 = 4K^2/h_c^2 = c^2$. **(b)** $\frac{1}{4}$, $\frac{1}{3}$, $\frac{5}{12}$. **15.** $1/h_a + 1/h_b + 1/h_c = a/2K + b/2K$ $+ c/2K = 2s/2K = 2s/2rs = 1/r$. **17.** Set $x = a^2$, $y = b^2$, and $z = c^2$. Then solve the linear system resulting from equations (60.5). A more clever approach makes use of Ex. 6: The medians of $\triangle XYZ$ are $3/4$ the lengths of the sides of $\triangle ABC$. Thus, for example, $m_a' = \frac{3}{4} a$. But $m_a' = \sqrt{\frac{1}{2}m_b^2 + \frac{1}{2}m_c^2 - \frac{1}{4}m_a^2}$, which proves the first formula in (60.21). **19.** Since the sides of $\triangle XYZ$ are m_a, m_b, m_c (60.22) gives the result directly. (There is also a very simple synthetic solution to this problem.)

Section 61 (p. 208): **3.** $K = \triangle + \triangle_1 + \triangle_2 + \triangle_3 = \triangle + p_3q_2K + p_1q_3K + p_2q_1K$ so that $\triangle = K(1 - p_3q_2 - p_1q_3 - p_2q_1) = K[1 - p_3(1 - p_2) - (1 - q_1)q_3 - p_2q_1]$, and so on.

Section 62 (pp. 212–213): **1.** $|BD|/|DP| = |AB'|/|AP|$ and $|DP|/|DC| = |AP|$ $/|AC'|$. Multiply to obtain $|BD|/|DC| = |AB'|/|AC'|$. $\triangle BEC \sim \triangle B'EA$ implies

$|CE|/|EA| = |BC|/|AB'|$ and $\triangle BFC \sim \triangle AFC'$ implies $|AF|/|FB| = |AC'|/|BC|$. Now multiply the latter three equations. $\begin{bmatrix} ABC \\ DEF \end{bmatrix} = \pm 1$ then follows, and the argument may then be easily completed. **3.** From Ex. 5, Sec. 42 $ZB = BX = s - b$, $XC = CY = s - c$, and $YA = AZ = s - a$. Hence $\begin{bmatrix} ABC \\ XYZ \end{bmatrix} = 1$. **5.** Label the measures of the angles at A, B, C in counterclockwise order by (α, β), (λ, μ), and (ρ, σ). As in Ex. 7, Sec. 57 $\triangle ABA' \cong \triangle C'BC$, $\triangle ACA' \cong \triangle B'CB$, and so on, and hence $AA' = BB' = CC'$ and $\angle BC'F = \beta$, $\angle EB'C = \alpha$, $\angle AB'E = \sigma$, $\cdots$. Now use (59.2). **7.** By Menelaus' theorem each linearity number equals -1. Upon multiplying and simplifying it is found that $\begin{bmatrix} PRQ \\ MLN \end{bmatrix} = -1$. **9.** $|BD|/d = |EA|/e = \cot \alpha$; $|DC|/d = |AF|/f = \cot \beta$; $|CE|/e = |FB|/f = \cot \gamma$. Then $\begin{bmatrix} ABC \\ DEF \end{bmatrix} = \pm 1$. Since an odd number of points are exterior to their corresponding sides, the negative occurs.

Section 63 (pp. 216–217): **1. (a)** Since a, h, b form a harmonic sequence, $1/a = a'$, $1/h = a' + d$, $1/b = a' + 2d$. Hence $\frac{1}{2}(1/a + 1/b) = \frac{1}{2}(a' + a' + 2d) = 1/h$. **(b)** $\sqrt{ab} = \sqrt{\frac{1}{2}(a + b)(1/2a + 1/2b)^{-1}}$. **3.** $AC \cdot BD = AD \cdot CB$; $(AD + DC)BD = AD(CD + DB)$; $2AD \cdot BD = CD(BD + AD) = CD(BO + OD + AO + OD) = CD(2\,OD)$. **5.** Use Thm. 63.5: $AC/AD = (AO + OC)/(AO + OD) = (\sqrt{OC \cdot OD} + OC)/(\sqrt{OC \cdot OD} + OD) = \sqrt{OC}(\sqrt{OD} + \sqrt{OC})/\sqrt{OD}(\sqrt{OC} + \sqrt{OD}) = \sqrt{OC}/\sqrt{OD}$. **7.** Let $A' = \overleftrightarrow{BC} \cap \overleftrightarrow{Y'W'}$. By the harmonic construction $[AB, CD] = [CD, BA'] = -1$. But $[AB, CD] = [CD, AB] = 1/[CD, BA]$ and $[CD, BA] = [CD, BA']$.

Section 64 (pp. 223–225): **1.** Consider $X[x, 0]$ and $\overline{X}[x', 0]$; by Thm. 64.4 $\overleftrightarrow{PX} \parallel \overleftrightarrow{P'X'}$. But $X' \in \overleftrightarrow{OX}$, so $\overline{X} = X'$. Then $OX = x$ and $OX' = x'$ and the first equation in (64.5) follows from Def. 64.3. **3. (a)** If $P_1 \neq P_2$ let $\triangle P_1 P_2 P_3$ be any nondegenerate triangle. By definition $\triangle P_1 P_2 P_3 \sim \triangle P_1' P_2' P_3'$, so $P_1' \neq P_2'$. Suppose (ABC) and consider $D \not\in \overleftrightarrow{AB}$. Let a, b, c be the proportionality factors between the similarities: $\triangle ABD \sim \triangle A'B'D'$, $\triangle ACD \sim \triangle A'C'D'$, $\triangle BCD \sim \triangle B'C'D'$. Hence $B'D' = a\,BD = c\,BD$, $A'D' = a\,AD = b\,AD$, or $a = b = c$. Thus $A'C' = a\,AC = a(AB + BC) = A'B' + A'B'$, so by the triangle inequality (Thm. 31.3), $(A'B'C')$. **(b)** Let A and B be any two fixed points and define $a = A'B'/AB$. $\triangle ABX \sim \triangle A'B'X'$ implies $B'X' = a\,BX$ and $\triangle BXY \sim \triangle B'X'Y'$ implies $X'Y' = a\,XY$. **(c)** No; any rotation provides a counterexample. **5.** An isometry is a special kind of similitude since $\triangle XYZ \cong \triangle X'Y'Z'$ always follows by SSS. Since betweenness is preserved, $\overline{ZXYZ}$ maps into $\overline{ZX'Y'Z'}$ with $\angle XYZ = \angle X'Y'Z'$. **7. (a)** Let $y = mx + d_k$ $(k = 1, 2)$ be two parallel lines. Then the parametric form of their images would be [substituting into (64.7)] $x' = (a_1 + b_1 m)x + b_1 d_k + c_1$, $y' = (a_2 + b_2 m)x + b_2 d_k + c_2$, each of which are lines having the slope

$(a_2 + b_2m)/(a_1 + b_1m)$. **(b)** Use the special case $x' = ax$, $y' = by$ to show this. The x-axis maps into itself while the line $y = x$ maps into $y' = (b/a)x'$. **(c)** Since parallelograms map into parallelograms it suffices to prove this for three points on $y = mx$: $P_k[x_k, mx_k]$, $k = 1, 2, 3$. Use the distance formula. **(d)** Use the theory of matrices. **(e)** In view of (d) it suffices to prove this for simple affine transformations only: It is known that the area K of the triangle whose vertices are (x_k, y_k), $k = 1, 2, 3$ is given by the determinant of the matrix whose first and second columns are the $x_k's$ and $y_k's$ respectively, and whose third column consists of ones. Then the area K' of the image triangle would be

$$K' = \begin{vmatrix} ax_1 & by_1 & 1 \\ ax_2 & by_2 & 1 \\ ax_3 & by_3 & 1 \end{vmatrix} = ab \begin{vmatrix} x_1 & y_1 & 1 \\ x_2 & y_2 & 1 \\ x_3 & y_3 & 1 \end{vmatrix} = \begin{vmatrix} a & 0 \\ 0 & b \end{vmatrix} K$$

Section 65 (pp. 227–228):   **1.** If $A[0, a]$, $B[b, c]$ and $C[x, 0]$ the total length of pipe is given by $d(x) = \sqrt{x^2 + a^2} + \sqrt{(x - b)^2 + c^2}$. The only permissible root of $d'(x) = 0$ is $ab/(a + c)$, and the minimum does not occur at the endpoints. Hence $C[ab/(a + c), 0]$ is the desired point. If B' is the reflection of B in l (the river bank) then $AC + CB = AC + CB' \leqq AB'$, which is minimal when $C \in \overleftrightarrow{AB'}$. **3.** Let l and m intersect at B ($B \neq A$) and let x_0 be an arbitrarily chosen initial position of the line x, meeting l, m at X_0, Y_0, and take Z_0 such that $[AX_0, Y_0Z_0] = k$. Using a perspectivity from line x_0 to line x with center B prove that Z varies on $\overleftrightarrow{BZ_0}$. **5.** Recall two previous formulas (60.25), (60.26) and use $HG = 2\,GO$. **7.** The centroid of $\triangle HPP'$ is G since $\overleftrightarrow{HO}$ bisects $\overline{PP'}$ and $HG = \frac{2}{3}\,HO$.

Chapter 8

Section 66 (pp. 232–233):   **1.** By the secant theorem, Thm. 41.4. **3.** To find the center of the desired circle, simply intersect the perpendicular to the tangent at the given point of contact with the bisector of the angle of the tangents. **5.** $x = 4$ by Cor. 66.3; *Radius* $= \sqrt{65}/2$. **7.** $6\sqrt{2}$. **9.** $\sqrt{221}/2$; $\sqrt{317}/2$. **11.** Let M be the midpoint of $\overline{BC}$ and P the remaining intersection of $\overleftrightarrow{AG}$ with the circle. $GM \cdot MP = BM^2$ and hence $MP = BM^2/\frac{1}{3}AM$. $AP = AM + MP = AM + 3\,BM^2/AM$; Power A $= AG \cdot AP$; the rest is simple algebra. **13.** With X on $\overleftrightarrow{PQ}$, the common chord of the circles $[A, a]$ and $[B, b]$, and $C = \overleftrightarrow{AB} \cap \overleftrightarrow{PQ}$, then $XA^2 = XC^2 + AC^2 = XC^2 + a^2 - PC^2$, and similarly for XB^2.

Section 68 (pp. 238–240): **1.** If both $PA/PB = PC/PD$ and $PA \cdot PB = PC \cdot PD$ then $PA^2 = PC^2$ or $PA = PC$. **3.** Let E fall on B and F fall on C (Fig. 315). Then $\overleftrightarrow{AB} \parallel \overleftrightarrow{CD}$ if and only if $\angle E = \angle ABC = \angle BCD = \angle F$. In Fig. 314(b) as $AP \to \infty$, $\angle P \to 0$ and therefore $\angle E - \angle F \to 0$. **5. (a)** Well-defined by Thm. 68.1. Let the endpoints of ω_1, ω_2 be A and B, B and C. If $\omega_1 \cup \omega_2$ is not the entire circle choose $P \in \omega$, $P \notin \omega_1 \cup \omega_2$. Then $m(\omega_1 \cup \omega_2) = 2\angle APC = 2\angle APB + 2\angle BPC = m(\omega_1) + m(\omega_2)$. The remaining case follows by Example 68.5. **7.** If the endpoints of ω_1, ω_2 are A and B, C and D then $m(\omega_1) = m(\omega_2)$ implies $\angle AOB = \angle COD$. Thus, $\triangle AOB \cong \triangle COD$ and $AB = CD$. **9.** Use (68.7), (60.10), and factoring. **11.** Since $r = s - c$, then $s = r + c = r + 2R$. **13.** $BD = BS$, $ST = TD$, $BT = BT$ so $\triangle BST \cong \triangle BDT$. Then $\angle SBT = \angle TBD$. $\overleftrightarrow{BS}$ is the perpendicular bisector of $\overline{RT}$.

Section 69 (p. 241): **1.** It is possible to circumscribe a nonequilateral quadrilateral about a square whose vertices bisect the sides of that quadrilateral. **3.** If Y is the midpoint of chord $\overline{AX}$ of circle O, $\angle OYA$ is a right angle. Therefore, Y varies on the circle having $\overline{OA}$ as diameter. **5.** Ex. 10, Sec. 63 applies since $\angle CPD = \pi/2$; conversely, $PA/PB = AC/CB$ implies that $\overleftrightarrow{PC}$ bisects $\angle APB$ and hence the perpendicular $\overleftrightarrow{PE}$ to $\overleftrightarrow{PC}$, $E \in \overleftrightarrow{AB}$, bisects $\angle QPB$. Then by Ex. 9, Sec. 63 $[AB, CE] = -1 = [AB, CD]$ and $D = E$.

Section 70 (p. 244): Compare Figs. 321 and 322.

Section 71 (pp. 246–247): **1.** See Ex. 17, Sec. 11. **3.** $YZ = \frac{1}{2}BC = MN$; similarly, $XY = LM$ and $XZ = LN$. **5.** Use the notation of Fig. 323, with L the fixed midpoint. The perpendicular to $\overleftrightarrow{BC}$ at L remains fixed as the triangle varies so O varies continuously on that perpendicular. Now use the fact that U is the midpoint of $\overline{HO}$. **7.** $\triangle ABC$ is the orthic triangle of $\triangle I_a I_b I_c$ and circle O is therefore its nine-point circle. **9.** With a figure analogous to Fig. 324, replace I by I_a, r by r_a, and define $\angle 3 = \angle I_a BT$, $\angle 4 = \angle I_a BC$. Then $\angle 3 = \angle 1 + \angle 2 = \angle 4$ and $\angle 2 = \angle 5$ as before. $I_a V = BV = VC$ and therefore $2r_a R = VC \cdot I_a A = I_a V \cdot I_a A = $ Power I_a.

Section 72 (pp. 252–253): **1.** Use the formula $\dfrac{dy}{dx} = \dfrac{dy}{dt} \bigg/ \dfrac{dx}{dt}$; the rest is routine calculation. **3.** First, compute the (directed) angle λ between $\overleftrightarrow{BC}$ and $\overleftrightarrow{RQ}$ of the Morley triangle (Fig. 17) by use of (7.10), obtaining $\lambda = \frac{1}{3}(A + 2B - \pi) = \frac{1}{6}(2\alpha - \beta - \gamma + 2m\pi)$ for some integer m $(A < 90, -\pi < \beta \le 0 < \alpha < \pi)$. The line through O perpendicular to $\overleftrightarrow{RQ}$ then has inclination $\frac{1}{2}(\beta + \gamma) + n\pi + \lambda = \frac{1}{3}(\alpha + \beta + \gamma) + k\pi/3$, where n and k are integers.

Chapter 9

Section 75 (pp. 262–263): 1. It follows directly from Def. 66.5 and Thm. 75.2. 3. Provided their radical axes are concurrent (not parallel); see the following exercise. 5. (a) Apply condition (c) in Sec. 73, with centers $A[0, 0]$, $B[-\frac{1}{2}a, -\frac{1}{2}b]$ and radii r, $s = \frac{1}{2}\sqrt{a^2 + b^2 - 4c}$. Hence, $r^2 + s^2 - AB^2 = r^2 - c$. (b) Set $b = 0$; the line $y = mx$ cuts $x^2 + y^2 + ax + c = 0$ at $U[x_1, y_1]$ and $V[x_2, y_2]$ where $x_k = (-a \pm \sqrt{a^2 - 4r^2 - 4m^2r^2})/(2 + 2m^2)$. $OU \cdot OV = x_1x_2(1 + m^2) = r^2 = OY^2$. (c) A family of nonintersecting circles with centers on the x-axis each orthogonal to circles through $(\pm 1, 0)$; $(1, 0)$. 7. The system corresponding to $t = t_1$ and $t = t_2$ ($t_1 \neq t_2$) has no real roots. Intersect the line of centers $ay = bx$ with $ax + by + c = 0$ to obtain $O[-ac/(a^2 + b^2), -bc/(a^2 + b^2)]$; find the equation of the circle centered at O and orthogonal to $t = 1$: $(a^2 + b^2)(x^2 + y^2) + 2acx + 2bcy = 0$, and intersect with $ay = bx$. 9. (a) The center of each member is given by $0[-\frac{1}{2}a(t), -\frac{1}{2}b(t)]$, which varies on a line if and only if $a(t)$ and $b(t)$ are linear, say $a(t) = at + a'$, $b(t) = bt + b'$. The radical axis between the member $t = 0$ and an arbitrary member is then $atx + bty + [c(t) - c'] = 0$ where $c' = c(0)$, which is fixed if and only if $[c(t) - c']/t = c = $ constant. 11. Let $P = \overleftrightarrow{A_1B_1} \cap \overleftrightarrow{A_2B_2}$. Then P lies on the radical axis of ω_1, ω_3 and also that of ω_2, ω_3. 13. Let the circles be $\omega_k: x^2 + y^2 + a_kx + b_k = 0$ ($k = 1, 2$) and the point $P[x_0, y_0]$. Then $Power\ P$ (circle ω_1) $- Power\ P$ (circle ω_2) $= (a_1 - a_2)x_0 + (b_1 - b_2) = k$. If $(x_0', 0)$ is on the radical axis of ω_1, ω_2 then $x_0' = -(b_1 - b_2)/(a_1 - a_2)$; show that $2(x_0 - x_0')(\frac{1}{2}a_1 - \frac{1}{2}a_2) = k$.

Section 78 (pp. 274–276): 1. Locate P any point on l and let ω_3 be any circle orthogonal to ω_2 whose center is P. Then $l \perp \omega_3$ and hence $\omega_2 \perp \omega_3$ since ω_2 is the image of l; the assertion then follows easily. 3. Invert $(1, 0)$, $(2, 0)$, and $(3, 0)$ with respect to the unit circle. 5. Let $x^2 + y^2 + ax + by + r^2 = 0$ be orthogonal to the circle of inversion $x^2 + y^2 = r^2$. Show that equations (78.4) are self-inverting, that is, $x = r^2x'/(x'^2 + y'^2)$ and $y = r^2y'/(x'^2 + y'^2)$, and substitute into $x^2 + y^2 + ax + by + r^2 = 0$. 7. $OP \cdot OP' = (OC + CP)(OC + CP') = OC^2 - CP^2 = (OA^2 - AC^2) - (AP^2 - AC^2) = OA^2 - AP^2$. 9. (a) Since $\overleftrightarrow{AY}$ and $\overleftrightarrow{AZ}$ are the tangents to circle I from A, $\overleftrightarrow{AI}$ is the perpendicular bisector of $\overline{YZ}$, and similarly for $\overleftrightarrow{BI}$ and $\overleftrightarrow{CI}$. $A'Y^2 = AA' \cdot A'I = AI \cdot A'I - A'I^2$. Since $A'Y^2 + A'I^2 = IY^2 = r^2$, A and A' are inverse points; similarly with B, B' and C, C'. (b) Apply (78.5), with $A = O$, $B = N$, $C = I$, $a = R$, $b = NA' = r/2$, and $c = r$. (c) Let excircle I_a be the circle of inversion, let X, Y, Z be the points of contact with the sides of $\triangle ABC$, and apply an argument similar to (b).

Chapter 10

Section 80 (pp. 283–286): 1. Use $\cos c = \cos a \cos b$ in the formula $\sin A/\cos A = \sin a \tan c/\sin c \tan b$. 3. Add the two equations $\cos b/\sin a_2 = \cos d \cos a_2/\sin a_2$

$+ \sin d \cos \theta$, $\cos c / \sin a_1 = \cos d \cos a_1 / \sin a_1 - \sin d \cos \theta$, then clear fractions. **5.** $\sin A/2 = \sqrt{\tfrac{1}{2}(1 - \cos A)} = \sqrt{\tfrac{1}{2}[1 - (\cos a - \cos b \cos c)/\sin b \sin c]} = \sqrt{(\sin b \sin c - \cos a + \cos b \cos c)/2 \sin b \sin c} = \sqrt{[\sin \tfrac{1}{2}(a + b - c) \sin \tfrac{1}{2}(a - b + c)]/\sin b \sin c}$. **7.** Use the addition formulas for $\sin \tfrac{1}{2}(A \pm B)$, (80.13), the identities involving $\sin x \pm \sin y$, and simple factoring. **9.** $\tan \tfrac{1}{4}K = [\sin \tfrac{1}{2}(A + B) + \sin \tfrac{1}{2}(C - \pi)]/[\cos \tfrac{1}{2}(A + B) + \cos \tfrac{1}{2}(C - \pi)]$. Make the obvious substitutions for $\sin \tfrac{1}{2}(C - \pi)$ and $\cos \tfrac{1}{2}(C - \pi)$ and use Gauss's equations for remainder. **11.** From (80.18) derive the equation $p_n(\sin \lambda_n)/\lambda_n = 2\pi \sin r(\sin \mu_n)/\mu_n$ where $\lambda_n = p_n/2n$ and $\mu_n = \pi/n$, and take the limit as $n \to \infty$. Similarly, write $K_n(\tan \lambda_n)/\lambda_n = p_n \tan \tfrac{1}{2}a_n(\tan \mu_n)/\mu_n$ with $\lambda_n = K_n/4n$ and $\mu_n = p_n/4n$. [Use $\cos a_n = \cos r/\cos(p_n/2n)$ to show that $a_n \to r$ as $n \to \infty$.]

Section 83 (pp. 290–293): **1.** In Fig. 182 set $\theta = \angle 1$, $h = AB'$. Then $\cos c = \cos b \cos h + \sin b \sin h \cos \theta$ and $\cos \theta = \sin \angle 2 = \sin b/\sin h$; thus $\cos c = \cos b \cos h + \sin^2 b$. But $\cos h = \cos a \cos b$, so $\cos h$ may be eliminated; use the identity $1 - \cos x = 2 \sin^2 \tfrac{1}{2} x$. $\cos b < 1$ implies $c < a$. **3.** (Refer to Fig. 368.) The cevian in the corresponding Euclidean triangle has length $d' = \sqrt{pb^2 + qc^2 - pqa^2}$ where $p = a_1/a$, $q = a_2/a$. But from the cosine inequality, $pb^2 < pd^2 + p(qa)^2 - 2p(qa) \cos \theta$ and $qc^2 < qd^2 + q(pa)^2 + 2q(pa) \cos \theta$. Now sum. **5.** Let $AD = p\, AB$ and $AE = p\, AC$; set $d = DE$, $d_1 = BE$, $q = 1 - p$; $d^2 > pd_1^2 + q(pb)^2 - pqc^2 > p(pa^2 + qc^2 - pqb^2) + p^2 q b^2 - pqc^2 = p^2 a^2$. **7.** $\sin |AF|/\sin |FB| = \sin |AA'|/\sin |BB'|$.

Obtain two analogous equations and multiply. **9.** In Fig. 282 (without $\overset{\leftrightarrow}{B'C'}$), label angles at P starting with $\angle BPD$ by $\alpha, \beta, \lambda, \mu, \rho, \sigma$ in counterclockwise order, and either angle at D, E, and F by θ, ϕ, ψ, with $x = AP$, $y = BP$, $z = CP$. Then in all cases $\sin \alpha = \sin \mu$, $\sin \beta = \sin \rho$, and $\sin \lambda = \sin \sigma$; apply the elliptic law of sines. **11.** Prefix each linearity number by "$\mathcal{E}$" in the argument accompanying Fig. 287 (Sec. 63).

Section 84 (pp. 297–298): **1.** If XY is larger than the circumference of the cylinder and if X', Y' lie on a circular arc on the cylinder, then $X'Y' < XY$.

Chapter 11

Section 85 (p. 302): **1.** (a) $y = \log(-2x + 3)$; $y = \log x$. (b) $\tan^{-1} 3$. (c) No. (d) $a \ne a'$ and $(ab' - a'b)/(a - a') \geqq 0$. (e) No; Ax. 12' holds in all cases except horizontal lines.

Section 87 (pp. 309–310): **1.** Let the "sides" of $\angle ABC$ meet at B and D (D outside ω). Then $\overset{\leftrightarrow}{BD}$ passes through the center of ω, since that center must lie on the radical axis of circles $[B, A, D]$ and $[B, C, D]$. Let $\omega_1 \perp \omega$, with D the center of ω_1, and invert with respect to ω_1. **3.** Since the "reflection" mapping M into O preserves orthogonality, and circle O_1 maps into some line or circle ω_1', $\omega_1' \perp \overline{A'B'}$, where $\overline{A'B'}$ is the image of "segment" $\overline{AB}$. O, the image of M, must then be the center of ω_1' and hence $\widetilde{A'M'} = \widetilde{M'B'}$ which proves $\widetilde{AM} = \widetilde{MB}$. **5.** Let W be the center of l_2 and locate U and V, the points of intersection of l_2 and ω. Since $l_2 \perp \omega$ the circle ω_1 having $\overline{OW}$ as diameter passes through U. Now show that ω_1 is the inverse of chord $\overline{UV}$ with respect to circle l_2.

Section 88 (p. 313): **1.** Let U, V, be the diameter of ω containing $\overline{OA}$, with the order $(VOAU)$. Then $a = \overline{OA} = \log \dfrac{OU \cdot AV}{OV \cdot AU} = \log \dfrac{1 + OA}{1 - OA}$. Solve for OA. **3.** Let l_1 and l_2 be the nonintersectors, also denoting nonintersecting circles orthogonal to ω. The system $\mathfrak{F}$ of circles orthogonal to l_1 and l_2, which are themselves members of a family $\mathfrak{G}$, has a common chord $\overline{AB}$ of ω. The desired "perpendicular" is the circle through A and B orthogonal to ω.

Section 89 (pp. 318–322): **3.** (Fig. 407.) $\angle BAD = A$ or $\pi - A$ so that $\sin \angle BAD = \sin A$; $\sin \angle BCD = \sin C$. Apply (89.13) to right triangles BAD and BDC. **5.** Add the two equations $\cosh b/\sinh a_2 = \cosh d \cosh a_2/\sinh a_2 - \sinh d \cos \theta$ and $\cosh c/\sinh a_1 = \cosh d \cosh a_1/\sinh a_1 + \sinh d \cos \theta$ to obtain $\cosh b/\sinh a_2 + \cosh c/\sinh a_1 = \cosh d (\cosh a_2/\sinh a_2 + \cosh a_1/\sinh a_1)$, then clear fractions. **7.** $\sin A/2 = \sqrt{\tfrac{1}{2}(1 - \cos A)} = \sqrt{\tfrac{1}{2}[1 - (\cosh b \cosh c - \cosh a)/\sinh b \sinh c}$ $= \sqrt{\tfrac{1}{2}[\cosh a - \cosh(b - c)]/\sinh b \sinh c}$, and so on. **9.** In Ex. 2(b) set $x = y$ to give $\cosh 2x = \cosh^2 x + \sinh^2 x$. Then add and subtract the equation $1 = \cosh^2 x - \sinh^2 x$. **11.** Use the formulas for $\sin \tfrac{1}{2}(A \pm B)$, $\cos \tfrac{1}{2}(A \pm B)$, the identities involving $\sinh x \pm \sinh y$, (89.18), and simple factoring. **13.** If $a \to \infty$, $b \to \infty$, then $\tanh a/2 \tanh b/2 \to 1$ from below so that $\tan K/2 \to 1$ and therefore $K \to \pi/2$ from below. **15.** In Fig. 182, set $\theta = \angle 1$, $h = AB'$. Then $\cosh c = \cosh b \cosh h - \sinh b \sinh h \cos \theta$ and $\cos \theta = \sin \angle 2 = \sinh b/\sinh h$; thus $\cosh c = \cosh b \cosh h - \sinh^2 b$. But $\cosh h = \cosh a \cosh b$, so $\cosh h$ may be eliminated; use the identity $\cosh x - 1 = 2 \sinh^2 \tfrac{1}{2}x$. $\cosh b > 1$ implies $c > a$. **17.** The right triangle relation $\sin \pi/n = \sinh p/2n/\sinh r$ yields the first formula. **19.** Apply (89.13). **21.** From the first relation (89.13), $1 + c^2/2! + c^4/4! + \cdots = (1 + a^2/2! + a^4/4! + \cdots)$ $(1 + b^2/2! + b^4/4! + \cdots)$; $1 + \tfrac{1}{2}c^2 = 1 + \tfrac{1}{2}a^2 + \tfrac{1}{2}b^2 + \{$terms of higher order$\}$.

Section 91 (pp. 325–327): **1.** In Fig. 413, Thm. 91.1 implies $\cos A > \cos A'$ and hence the Euclidean law of cosines applied to $\triangle A'B'C'$ yields $a^2 = b^2 + c^2 - 2bc \cos A' > b^2 + c^2 - 2bc \cos A$. **3.** See the answer to Ex. 3, Sec. 83. **5.** $\cos \angle A = \tanh AE/\tanh AD = \tanh AC/\tanh AB$, and $\sin \angle A = \sinh DE/\sinh AD = \sinh BC/\sinh AB$. **7.** Prefix each linearity number in Example 62.4 by "$\mathfrak{C}$." **9. (b)** For the altitudes (in the usual notation), $\sinh |AF|/\tanh h_c = \cot A$, $\sinh |FB|/\tanh h_c = \cot B$, and hence $\sinh |AF|/\sinh |EA| = \cot C/\cot A$. Continue in this manner and prove that $\mathfrak{C} \begin{bmatrix} ABC \\ DEF \end{bmatrix} = 1$. (Why is the linearity number positive?)

Section 92 (pp. 332–333): **1.** Let ω_1 be any circle interior to ω and take ω_2 any circle orthogonal to both, whose center lies on l, the line of centers of ω and ω_1. Invert with

respect to a circle orthogonal to ω and centered at the point of intersection of l and ω_2 exterior to ω. ω_1 maps into a circle centered at O which is therefore a "circle."
3. The "circles" are ordinary Euclidean circles interior to ω (Ex. 1), the limit curves are circles tangent to and in the interior of ω, and the equidistant loci are portions of circles in ω which intersect ω. **5.** Let the endpoints A and B determine a family $\mathcal{G}$ of nonintersecting circles orthogonal to ω, where the family $\mathcal{F}$ orthogonal to $\mathcal{G}$ has $\overline{AB}$ as common chord. Then any orthogonal trajectory of $\mathcal{G}$, such as chord $\overline{AB}$, will be an equidistant locus. **7. (a)** For the case of an ultra-ideal point, obtain a suitable Saccheri quadrilateral. **(b)** Show that the bisectors mentioned meet at some point I. Then I is equidistant from the two lines. If D and E are the feet of I on $\overset{\leftrightarrow}{A\Pi}$ and $\overset{\leftrightarrow}{B\Pi}$, locate A' on $\overset{\leftrightarrow}{B\Pi}$ so that $A'E = AD$ and $\overrightarrow{A'\Pi} = \overrightarrow{B\Pi}$. Then show that $\angle A'A\Pi = \angle AA'\Pi$.

List of Symbols

Index